God—or Gor

MEDICINE, SCIENCE, AND RELIGION IN HISTORICAL CONTEXT

Ronald L. Numbers, *Consulting Editor*

GOD—OR GORILLA

Images of Evolution in the Jazz Age

CONSTANCE ARESON CLARK

The Johns Hopkins University Press

Baltimore

Johns Hopkins Paperback edition, 2012
2 4 6 8 9 7 5 3 1

The Johns Hopkins University Press
2715 North Charles Street
Baltimore, Maryland 21218-4363
www.press.jhu.edu

The Library of Congress has cataloged the hardcover edition of this book as follows:
Clark, Constance Areson, 1949–
God—or gorilla : images of evolution in the jazz age /
Constance Areson Clark.
p. cm. — (Medicine, science, and religion
in historical context)
Includes bibliographical references and index.
ISBN-13: 978-0-8018-8825-0 (hardcover : alk. paper)
ISBN-10: 0-8018-8825-5 (hardcover : alk. paper)
1. Human beings—Origin. 2. Human evolution—Public opinion.
3. Human evolution—Caricatures and cartoons. 4. Art and science.
5. Public opinion—United States—History—20th century.
6. United States—History—20th century. I. Title.
GN281.C513 2008
569.9—dc22 2007041258

A catalog record for this book is available from the British Library.

ISBN-13: 978-1-4214-0776-0
ISBN-10: 1-4214-0776-0

*Special discounts are available for bulk purchases of this book. For more information,
please contact Special Sales at 410-516-6936 or specialsales@press.jhu.edu.*

The Johns Hopkins University Press uses environmentally friendly
book materials, including recycled text paper that is composed of at least
30 percent post-consumer waste, whenever possible.

In memory of Richard Areson Clark

CONTENTS

Science, like art, suggests radically new ways of seeing. This is a book about scientists' and artists' attempts to offer the public new ways to imagine the human evolutionary past; about how scientists responded to the evolution debates of the 1920s in the United States; and about the central importance in those debates of visual images. It focuses on the changing appearance of evolutionary theory as it passed through a series of different lenses into popular culture.

The evolution debates of the 1920s neither began nor ended with the 1925 Scopes trial, though that event did a great deal to shape historical—and historians'—memories of the controversy. By the time John Thomas Scopes came to trial in Dayton, Tennessee, in 1925, charged with violating the state's law against teaching evolution, evolution had been a topic of intense newspaper and magazine conversation for several years. This conversation drew on a rich symbolic vocabulary. Much of that vocabulary was visual, and visual images held a prominent and problematic place in the controversy.

Science had been one of the great shibboleths in American culture in the years before World War I. By the 1920s, however, anxieties about modern life fostered new uneasiness about science. In this decade when the authority of science was at issue, when the details of evolutionary theory remained in dispute among scientists, and when the very boundaries of science often seemed remarkably permeable, scientists found themselves faced with passionate public controversies over evolution. Which scientists responded, and how they responded, to these controversies are among the subjects of this book. Its themes include the issues of scientific authority, the role of publicity and media, and the changing place of science in American culture in the 1920s. But pictures form the heart of the story.

Scientists understood how difficult it could be for lay readers to follow the complex arguments of scientific theory couched in an esoteric lexicon, and they often relied on scientific illustration to bridge the gap. Communication through scientific illustration, translated and mistranslated through the lenses of popular culture, turned out to be a complicated matter, however. Interpretation of images raised difficult questions about the very definition of science and of its boundaries.

Scientists offered the public new ways to visualize the human past, in the conviction that these visual images would not only clarify—and serve as evidence in support of—new discoveries about human evolution but would also provide solace and comfort. Diagrams demonstrating the optimistic view that the evolutionary story was a tale of ineluctable progress, and portrayals of early humans as dignified and worthy ancestors, would, evolutionists hoped, cast evolution in a hopeful light. They would suggest new and positive ways to visualize an ennobled human past, human history as a grand pageant.

Anti-evolutionists were not mollified. Indeed, they focused many of their objections to evolution precisely on those visual images. When scientists invoked "life's splendid drama," anti-evolutionists mocked them as "dramatists of evolution."[1]

In a sermon published in the year of the Scopes trial, the Reverend John Roach Straton, ascribing modern immorality to "the sinister shadow of Darwinism," insisted that "if man is a descendant of the beast instead of a child of God, then we need not be surprised if we find him inclined to live like a beast. Monkey men make monkey morals."[2] If one accepted the teachings of evolutionists, Straton insisted in another sermon, it would mean that sin was no longer sin but merely "a survival of the brute in man."[3] Straton emphasized the visual over and over again in his attacks on evolution—not only because so much of the most compelling evidence scientists marshaled was visual but also because, for Straton, the visual could be suspect: under a subheading, "The Lust of the Eyes," he warned that "the things we see with the eyes, that we are led by the Devil to admire and long for, become destructive forms of worldliness."[4] While he certainly understood how to use challenges to visual images effectively, Straton also attacked images of evolution because he understood them to be powerful. They were.

A liberal minister, the Reverend Harry Emerson Fosdick, an evolutionist, disagreed with Straton about many things, but he, too, confronted the significance of visual images for religion, and the significance of the visual implications of modern science, writing of "the collapse of the old imaginative frameworks in which our fathers commonly thought of God. What a cozy stage was furnished by the old cosmology with its flat earth and its close, convenient heaven, on which the religious imagination could picture its gods, their entrances and exits!"[5] Like Straton, Fosdick noted that these "imaginative frameworks" were profoundly visual: "Our conceptions of God have been shaped by picture-thinking set in the

framework of the old world-view." Spinning out this line of reasoning, Fosdick opined, "The marvel is . . . that now, when there is no longer any up or down, or heaven beyond the clouds, men on this whirling planet in the sky should still be preserving in religious imagination what they have discarded everywhere else."[6]

Scientists tried to offer reassuring pictures to people who would not, in many cases, have been persuaded by any kind of visual images but who put the pictures to their own rhetorical uses. Anti-evolutionists were able to do that effectively because such pictures were profoundly ambiguous, depending upon the perspective of the viewer. Visual images of human evolution drew on complex iconographic traditions, including some popular-culture traditions that made them evocative in a multitude of unpredictable ways. They could be made humorous, which gave them enormous potential for satire and social commentary. They were also evocative because they made many people shudder. For many people— and not only anti-evolutionists—images of simians suggested grotesque caricatures of humanity, and images of "primitive" people seemed to confirm deeply felt anxieties about the modern.

Critics targeted visual images of evolution because these images were so evocative and so protean and because the evolution debate was about so much more than the substance of science. Images of evolutionary ideas suggest key themes among the cultural concerns of the time, including issues of racial and ethnic hierarchy, assumptions about gender, and the human role in nature. And illustrations of scientific ideas often revealed far more than their authors intended. Comparison of scientific illustrations with their use in popular and mass culture highlights the substantial symbolic freight carried by evolution. Visual images of scientific ideas held a central place in the debate because the debate was about so much more than evolutionary theory. Images mattered because the evolution debates were, ultimately, about symbols.

Scientific illustrations served a variety of purposes, and they grew out of long visual traditions, both within science and without. They expressed multiplicities of ideas, in visual languages that could be understood differently by different audiences. This became evident during the 1920s debates over evolution as visual images of evolution drew constant fire from anti-evolutionists. There was a lot at stake in how one pictured the human past. This book is about the complexities for scientists of this debate, about why those pictures mattered so much, and how it was that every picture could tell so many different, and even conflicting stories.

ACKNOWLEDGMENTS

This book had a lucky start in life: from the beginning it has benefited from the generosity of an extraordinary group of friends and colleagues. Mark Pittenger, Fred Anderson, Phil Deloria, John Enyeart, Susan Jones, and Erica Doss all read complete drafts (in some cases several complete drafts!) at various stages in the book's progress, always perceptively and in detail. John may have read every draft, and there have been many. In addition, they have been willing to talk with me endlessly about this project over a long period of time. I have learned a great deal from all of them, not only about scholarship and writing but also about the gift of belonging to a generous community of scholars.

It is difficult to find words adequate to express my gratitude to—and admiration for—my series editor, Ron Numbers, who has generously read several iterations of this book, offering wise counsel, detailed and astute suggestions, and encouragement and support throughout a long process. I am very grateful for his interest, enthusiasm, and support. Michael Lienesch thoughtfully read the entire penultimate draft and offered perceptive and extremely helpful suggestions for revision; his comments have made this a better book. An anonymous reviewer of the book proposal contributed detailed and perceptive suggestions which proved extremely helpful. This book is much better for the attention and generosity of these good people, and I thank all of them.

It is a real pleasure as well to thank friends and colleagues who have given me helpful comments—in some cases so long ago that they may not even remember—on parts of the manuscript and on related projects that found their way into this book. Warm thanks go to Julie Brown, Carol Byerly, Katherine Clark, David Gross, John S. Haller, Jr., Peter Hansen, Sharon Kingsland, Tom Krainz, Franklin Reeve, Gerry Ronning, Ed Ruestow, Mike Sokal, Rickie Solinger, Laura Stevenson, and Kathleen Whalen.

I am also glad at last to be able to thank good friends and generous colleagues with whom I have had enlightening conversations about the topics that went into the book, and who have offered suggestions, references, and support in many forms: Naomi Amos, Virginia Anderson, Susan Bednarczyk, Steve Bullock, Carol Clark, Ted Davis, John d'Entremont, Kristin Donnan, Christine Drew, Abby Dyer, Jeffery Forgeng, Julie Greene, Jim Hanlan, Martha Hanna, Laura Katzman, Susan Kent, Laura Knotts, Tom Krainz, Heidi Kunz, Patty Limerick, Kent Ljungquist, Rhonda Miller, Scott Miller, Richard Milner, Marc Ordower, Kathleen Placidi, Jon Roberts, Tom Robertson, David Schwartz, Adam Shapiro, Deb Smith, David Stiegerwald, Dan Stiffler, Karin Warren, and Tom Zeiler.

Many of the ideas in the book were tested in papers and in conversations at conferences. Co-panelists and commentators at these conferences offered comments, discussions, and suggestions that have been helpful. I thank Paul Conkin, Ted Davis, Ann Fabian, Martin Fichman, Peter Holleran, Ed Larson, Bernie Lightman, Jeff Moran, Edward O'Donnell, Ed Rafferty, Jon Roberts, and David Wagner for their participation in and contributions to these conference panels and conversations. I would also like to thank the many colleagues with whom I've had helpful and enlightening conversations at conferences and by email, who have offered suggestions and references, or who have shared their work with me.

Remaining errors, inconsistencies, and flaws are, of course, no one's fault but my own.

I am grateful to friends, colleagues, and students at the University of Colorado at Boulder, at Randolph-Macon Woman's College (now Randolph College), and at Worcester Polytechnic Institute. Warm thanks especially to the generous and welcoming community among the faculty and staff in the Humanities and Arts Department at WPI.

Long ago a brilliant lecture in an undergraduate class on seventeenth-century Europe by Richard Rapp made such a profound impression on me that I decided to study the history of science instead of marine biology. I turned for advice to Ruth Schwartz Cowan, and I could not have found a better or more gracious guide. For a long time I have looked forward to the opportunity to thank both of these individuals in print, and I am very happy to do so now.

No one could ever do history without the help of librarians and archivists. I wish to thank all librarians, especially the extraordinary Christine Drew at Gordon Library at WPI, and the ace interlibrary loan librarians Laura Hanlan at WPI and Marty Covey at the University of Colorado at Boulder. Heartfelt thanks also to the librarians and archivists at the New-York Historical Society, Princeton University Library, the American Philosophical Society Library, the Columbia University Rare Books and Manuscripts Collection, the archives of the Science Service at the Smithsonian Institution, Harvard University, and the New York Public Library. It has been a pleasure as well as a privilege to work in these great archives.

Thanks particularly to the librarians, archivists, and others at the American Museum of Natural History in New York, who always make me feel as if I belong in that extraordinary place, a very great gift indeed. Thank you especially to the altogether remarkable Barbara Mathé of the American Museum Library for her interest, generosity, and friendship, for making the priceless archives of the museum available to me, and for helping me to negotiate the intricacies of an immense collection. And thanks to her and to Chip Marks for a wonderful excursion to the Osborn house at Castle Rock at Garrison, New York. Thanks to Eleanor Schwartz for her excellent organization of the Osborn Papers at the American Museum. I also want to thank the American Museum of Natural History for permission to reproduce illustrations from its collection.

And no one could do this kind of work without money and time. I am extremely grateful indeed to the National Endowment for the Humanities for a fellowship that allowed me the luxury of time for extended research and revision. I am grateful also to the Gilder Lehrman Institute of American History for awarding me a research fellowship to help with the cost of doing research in that magical but expensive city, New York, and to the Virginia Council of the Arts and Humanities for a Mednick Fellowship, which made it possible for me to spend valuable time in the archives of the American Philosophical Society Library in Philadelphia. The University of Colorado provided a number of fellowships that gave me time in archives and at my computer. I am very grateful for the Davaney Fellowship, the Emerson Humanities Fellowship, the George F. Reynolds Fellowship, the Lefforge Fellowship, the Pile Research Fellowship, and the Beverley Sears Grant for research and travel, and to the History Department at Colorado, especially Scott Miller, Mark Pittenger, Bob Ferry, Martha Hanna, Tom Zeiler, and Susan Kent, for making these things happen.

I am very grateful to the Pelzer Prize Committee for 2000, especially David Paul Nord and George Roeder, for their insightful suggestions for revision of the paper that represented the initial foray into this territory, and to Susan Armeny, from whom I learned a good deal about the art of revision.

Thanks to the Organization of American Historians for permission to use parts of my article "Evolution for John Doe: Pictures, the Public, and the Scopes Trial Debate," *Journal of American History* (March 2001), and to Left Coast Press for permission to use sections of my essay "Ignoring the Elephants: Visual Images and Jazz Age Critics," *Museums and Social Issues* (Spring 2006).

I thank the *New Yorker* for permission to reproduce a cartoon from one of its earliest issues, and for the efficient organization of its permissions protocols.

It has been a very great pleasure to work with Jacqueline Wehmueller, my editor at the Johns Hopkins University Press. Many thanks to Jackie for her warmth, wit, encouragement, and enthusiastic support. Debby Bors has guided the book—and me—through the production process with great kindness, good humor, and patience. Thanks also to the rest of the very capable, enthusiastic, and supportive people at the Johns Hopkins University Press who have worked on this book: Juliana McCarthy, Carol Zimmerman, Jennifer Gray, Ashleigh McKown, Brendan Coyne, and Robin Rennison.

I have also benefited from cheerful and expert technical help with the reproduction of images. I thank Bob Bakker, Kelli Anderson, Emily Fincher, Sunni Joshi, Dewardrick Mack, Eric Remy, Kushboo Shrestha, and the excellent and friendly team at WPI's Academic Technology Center, especially Jim Cormier, James Monaco, and Ellen Lincourt, for their able and good-natured assistance.

I have more things to thank Bob Bakker for than I can list here. Bob has been supportive and generous for many years, spending hours helping me with illustrations, talking with me about science, scientific illustration, and museums—and about Henry Fairfield Osborn and the American Museum. I thank him for all of this, and for his ability to ask original and provocative questions. I will always treasure our many long walks in Colorado and Wyoming talking about all these things.

Finally, and most important: In my family we are not much given to public expressions of private emotions, so I will just say that my deepest gratitude is to my family. My warmest thanks to Betty Clark, Kathy Clark, Bill Bissell, Priscilla Lyons Robertson, and Homer Clark. I hope and trust that they know how I feel. And thanks most of all to my father, Richard Areson Clark. I hope he knew. This book is dedicated to his memory.

God—or Gorilla

CHAPTER ONE

The Caveman and the Strenuous Life

Monkeys were everywhere in the 1920s. In a July 1925 cartoon in the magazine *Judge*, a congregation of friendly looking animals watched an appealing little monkey set off to go somewhere alone. The caption read, "The upstart." Anyone reading *Judge* that summer would immediately have understood that the little monkey was heading off to evolve and that the cartoon referred to the coming trial of John Thomas Scopes in Dayton, Tennessee, for teaching evolution. There were good reasons for calling the trial the "monkey trial." And almost everyone did, it sometimes seemed. Monkeys, apes, ape-men, cavemen, and all manner of missing links would populate newspapers, newsreels, magazines, and metaphors that summer.

Images of ideas about evolution galvanized the media, kindled debate, and shaped the thinking of both scientists and the public in the decade of the Scopes trial. Pundits and commentators of the time remarked frequently on the number of books and articles about evolution. Scientists and science popularizers in the 1920s generated an enormous number of words intended to enlighten the public about evolution; newspaper commentators sometimes referred to the evolution debate as a "battle of the books." But it was more than a battle of books; it was also a contest among images.

Images of the participants in the debate proliferated, of course. Perhaps no one was more often paired with monkeys in the 1920s than William Jennings Bryan, the most prominent advocate for the anti-evolutionist cause. Bryan was a one-time congressman from Nebraska, a three-time candidate for U.S. president, and secretary of state under Woodrow Wilson. Called the "Great Commoner," he was a populist advocate, during the early twentieth century, for woman suffrage, a graduated income tax, and direct election of senators. Bryan also lent his considerable fame to the cause of Prohibition. And early in the 1920s, he took up the mantle for anti-evolutionism. In 1925 he offered his services to the Scopes trial prosecution. Cartoon monkeys offered him grape juice, haunted his cartoon dreams, and thanked him for disavowing any relationship to them.[1]

Primate jokes multiplied with the heating up of the evolution debate: monkey jokes, ape jokes, and caveman jokes. There was something serious in all these

"The Upstart." The weekly magazine of humor *Judge* ran this cartoon on the first page of the "Evolution Issue," in honor of the Scopes trial, which journalists dubbed "the monkey trial." *Judge*, "Evolution Issue," July 18, 1925, 1.

simian references, too, however. Evolutionists sought "missing links," specimens that would fill the gaps in the fossil record, intermediaries between known species of organisms. In the case of human evolution, the obvious links, given the anatomical similarities, would seem to have connected humans with the families of monkeys and the great apes, a thought that many anti-evolutionists found repellent. In popular literature, the search for missing links among fossil lineages became the search for "*the* missing link," as if there were only one. Monkeys, apes, and "missing links" served a multitude of symbolic purposes. Monkeys could be funny, of course, but they could also serve as vehicles for commentary about weighty, even solemn issues. Cartoons often featured monkeys laughing at the antics of humans and, in some cases, lamenting the bigotry and intolerance of humans. In one of the most stunning evolution cartoons, published in the African American newspaper the *Chicago Defender* in 1925, a pair of horrified monkeys—clinging to each other in a tree as they watched a lynching below—refused to believe that such fiends as humans could be related to them.[2]

The little monkey in *Judge* was a rather benign version of the monkey theme, but many people found references to monkeys, apes, and ape-men horrifying or repulsive, evocative of things nasty and brutish—or of brutish potential within humans. There were good historical reasons why some people perceived non-

Cartoon William Jennings Bryan and Cartoon Monkey. Probably no one was more often paired with monkeys in the 1920s than William Jennings Bryan, the leader of the anti-evolution crusade. *Judge*, July 25, 1925, 14.

human primates and images of human ancestors as caricatures or parodies of humanity at its worst. From the very beginnings of European awareness of gorillas in the mid-nineteenth century, for example, these gentle animals often carried sensational, even salacious, metaphoric freight—and racial connotations—in European popular culture.[3] Missing-link jokes and references alarmed many people because they reflected old visual and metaphoric traditions that held complex connotations. Anti-evolutionists were not the only people to respond viscerally to missing-link images; even some vigorous defenders of evolution understood or shared such concerns.

Many people found the overwhelming of the Scopes trial by missing-link irreverence frustrating. The trial should have showcased significant issues; instead, it seemed to become a parody, a circus or a carnival. Still, a number of editorialists, in lamenting the national embarrassment they perceived the Scopes trial to have been, tempered their criticism with the idea that there had been one bright spot. All the publicity about evolution must have been educational. For example, the trial had brought "to thousands of thinking people knowledge of primitive man for the first time," according to one Chicago paper. "Many who have used the phrase 'cave man' glibly the last twenty years have had no inkling of what it meant. Now comes curiosity and wonderment."[4] The term *caveman* certainly had been used glibly, and often. Images of cavemen, references to cavemen, and debates about the meanings of cavemen held a prominent place in the evolution debates that unfolded in the decade of the Scopes trial. Cavemen had proliferated in American popular culture in the late nineteenth century, and by the 1920s they had acquired an evocative iconographic history.

Monkeys, monkey jokes, and monkey motifs had followed the publication of Darwin's *On the Origin of Species* in 1859, along with apes and ape-men. Darwin himself turned up in a cartoon image as a hybridized Darwin-ape.[5] Cavemen were not so prominent in the popular frenzy over Darwinism in the 1860s, though: the first Neanderthal specimen had been discovered only in 1857, and scientists did not at first agree on its meaning. It might, they thought, be an aberration, an anomalous individual. Even so, cavemen quickly entered the popular visual lexicon of cartoons and caricatures. By the time Eugène Dubois discovered the specimen he named *Pithecanthropus erectus* and the newspapers called "Java Man," in 1891, the notion of cavemen had become unavoidable and compelling. Following post-Darwinian discoveries of fossil hominids, cavemen joined monkeys, apes, and other missing links in a crowded symbolic vocabulary. With discoveries in the early twentieth century of the Piltdown specimens—apparently large-brained hominids with primitive jaws found in England in 1912—and of the spectacular cave art of the Cro-Magnon people, Neanderthals, Cro-Magnon people, and "cavemen" in general became symbols as well as objects of curiosity, but the process had started even earlier.

A new genre, Stone Age fiction, flourished in the late nineteenth and early twentieth centuries. Cavemen, ape-men, and other variations on the theme of the "missing link" appeared in novels, movies, and cartoons as well as in museums. According to one count, the twentieth century saw the production of some 150 caveman films, including movies, animated cartoons, and television programs, and the process was well under way by the 1920s.[6] The invention of Stone Age and Lost World fiction came at a time of increasing sensationalism in newspapers, of expanding outlets for publication of visual images, and of growth for museums. The beginnings of the twentieth-century caveman coincided with the opening up of the visual culture that the poet and critic Vachel Lindsay characterized as a "hieroglyphic civilization."[7] The possibilities for reproducing visual images multiplied, and the caveman literature of the early twentieth century took advantage of them. It also drew on and enlarged pictorial traditions that made it possible—even necessary—to visualize the human past in a new variety of ways. Anti-evolutionists would increasingly challenge the claims to transparency of visual arguments about evolution, but writers of evolution-inspired fiction, many of them enamored of a romantic view of science, emphasized the idea that "seeing is believing" in constructing narratives about the human past.

Modern Tales of a Primitive Past

"Pictures! Pictures! Pictures! Often, before I learned, did I wonder whence came the multitudes of pictures that thronged my dreams; for they were pictures the like of which I had never seen in real wake-a-day life." So began Jack London's 1906 tale of human prehistory, *Before Adam*. The novel's narrator insisted on the visual essence of his dreamtime memory: in his life in the distant past "when the world was young," he and his protohuman companions had limited language, and so his memories were profoundly and disturbingly visual, more visceral and more elemental than dreams mediated by language could be. The narrator emphasized that he had been plagued by these dreams of "the Younger World" all through his childhood. There were no humans in the dreams; not even the dreamer himself was, in his dreams, yet human. His mother, he recalls, "was like a large orang-utan . . . or like a chimpanzee, and yet, in sharp and definite ways, quite different."[8] All of this must have seemed incredible, he conceded, but it was true, nonetheless: "For know that I remember only things I saw myself, with my own eyes, in those prehistoric days."[9] To have seen the thing oneself, London suggested, was the most convincing form of proof.

The night terrors that plagued London's narrator as a child mystified him until he attended college. There he learned, finally, of evolution. His dreams, he discovered, "dated back to our remote ancestors who lived in trees." According to his professor, shocking experiences caused "molecular changes in the cerebral

The protagonist in Jack London's Stone Age novel claimed to be an atavism, a person whose evolutionary memory of his evolutionary past remained intact and returned to him in highly visual form in his dreams. His dream self, not yet human, sported opposable big toes and found himself surrounded by brutal enemies. Jack London, *Before Adam* (New York: Macmillan, 1906).

cells. These molecular changes were transmitted to the cerebral cells of progeny, became, in short, racial memories." Combining several versions of evolutionary theory, mostly implying the Lamarckian assumption that traits acquired during an individual's lifetime could be inherited by its descendants, the professor convinced him that "some strains of germplasm carry an excessive freightage of memories— are, to be scientific, more atavistic than other strains." The narrator had been cursed with such a strain: "I am a freak of heredity, an atavistic nightmare."[10]

Jack London was far from alone in stressing the notion of atavism: atavisms preoccupied fiction writers of all kinds. H. G. Wells imagined a mad scientist producing atavistic missing links—cavemen as fabricated and tragically defective— and long after it was published, *The Island of Dr. Moreau* continued to inspire movie versions of its story. Science fiction writers were not the only ones who used the concept: it appears even in Edith Wharton. In *The Glimpses of the Moon*, Wharton, who took a strong interest in Darwinism, portrayed a young arriviste— one of the nouveaux riches so familiar in her work—as something like an evolutionary throwback: "There was something harsh and bracing in her blunt primitive build, in the projection of her black eyebrows that nearly met over her thick straight nose, and the faint barely visible black down on her upper lip."[11] Eugene O'Neill also employed the idea of the primitive as a metaphor for class. The stage directions for his 1922 play *The Hairy Ape* specify that the workers stoking the fires in "the bowels" of an ocean liner "should resemble those pictures in which the appearance of Neanderthal Man is guessed at. All are hairy-chested, with long arms of tremendous power, and low, receding brows above their small, fierce, resentful eyes."[12]

Images of the primitive in the late nineteenth and early twentieth centuries everywhere implied evolutionary and racial themes—and, as the cases of Edith Wharton and Eugene O'Neill suggest, class as well. This was, after all, an age of imperialism and of sweeping generalizations about the differences between the "civilized" and the "primitive." When Jack London related tales of atavisms, he drew on an evolutionary metaphor ubiquitous among American and British Victorians.

The development metaphor—the analogy of various forms of change with the growth and development of the individual—had a long history, but it acquired a special significance among the Victorians. American Victorians described all manner of growth as "just like" the development of individuals through childhood to maturity. In its general form, the metaphor radiated outward not from biology but from the French Enlightenment, especially Condorcet's description of the stages of civilization. Nineteenth-century anthropologists such as Lewis

Henry Morgan took up the theme, portraying peoples they called "primitive" as representative of "the childhood of the race," or stages in a development leading inexorably toward social adulthood, as manifested in American and European civilization.[13] The ways in which the development metaphor legitimated many forms of biological determinism have been well documented. Theories of hereditary criminality, for example, deployed the analogy to argue that habitual criminals stagnated at "primitive" stages of development. They had, as a result of genetic aberrations, failed to achieve moral adulthood. Criminality thus originated from inherited atavism. Similarly, like certain salamanders, women illustrated the principle of neoteny: their development slowed down so that they reached maturity without attaining adulthood. They were, in a word, childlike.

The same word could be handily applied to people cast as primitive, as well. The Harvard anthropologist Frederic Ward Putnam designed a series of ethnological exhibits for the 1893 World's Columbian Exposition which he intended as a spatial representation of the trajectory of human societies from primitive to civilized as he understood it. Putnam's plan was to map the stages-of-civilization model dominant in the anthropology of the time onto the fairgrounds, so that exhibits of people from around the world—ethnological exhibits—would follow a trajectory starting with the primitive people of Dahomey (now known as Benin) through increasingly highly evolved peoples and culminating in the White City, which represented the most advanced—the modern Euro-American— stage of human evolution. Putnam's didactic intent may have been blunted by the chaotic reality of the "Midway" section of the fair, where visitors were probably overwhelmed by exotic sounds, tastes, and smells and especially by novelties such as ice cream cones and the first ever Ferris wheel. Even so, ethnological exhibits modeled on those at the World's Columbian returned in later expositions; even after anthropologists, beginning with Franz Boas, challenged the stages-of-civilization model, fairgoers became accustomed to seeing other human beings on exhibit. One of the people brought to the Louisiana Purchase Exposition at St. Louis in 1904, a Pygmy by the name of Ota Benga, was subsequently brought to New York and put on display at the Bronx Zoo. Zoo visitors could watch Ota Benga play with his friends the chimpanzees.[14]

Cavemen also played a part in the romanticizing of the primitive; they became emblems of a sturdier time. If Stone Age times had been dangerous, at least the dangers cavemen faced had been authentic. Stone Age people had been strong, brave, and competent; otherwise, they could not have survived. They must have been alert, and they would not have suffered from modern neuroses. For many people, the newly enhanced reputation of the "primitive" grew out of a sense that modern life had swollen to cumbersome dimensions—the complexity of an urban civilization run amok. Cave people, or at least cavemen, seemed to represent the romantic possibilities of an ancient past in contrast to the compromised present.[15]

Conservationists often combined the nineteenth-century notion of recapitulation with the popular psychological metaphor of an "inner caveman" and with Progressive Era themes such as scouting, nature study, and Theodore Roosevelt's notion of the Strenuous Life to suggest that modern civilization threatened American virility. This was one of the underlying tenets of the "back to nature" movement in the early twentieth century. Boys, by the logic of the development metaphor, remained closer to their inner cavemen than did adult men, and in order to avoid the feminizing influences of modern urban life, they needed to avail themselves of opportunities to make themselves into naturalists, pioneers, explorers, and playground and campground "Indians," learning the skills that had so perfectly equipped Stone Age boys for the difficulties of primitive life.[16]

Tarzan made a good model, as did many of the heroes of Stone Age fiction. Edgar Rice Burroughs's creation, Tarzan exemplified the possibilities of good breeding—his parents, who had perished, were British aristocrats—combined with the kind of masculine qualities necessary to survival in the jungle.[17] Burroughs also created a tale in the Lost World genre of primitive people who reenacted—quite literally—the development metaphor in their individual lives. The Stone Age people living on a remote island in *The Land That Time Forgot* recapitulated, as individuals, human evolutionary history; each individual struggled to progress from lower to higher stages of human development.[18]

The psychologist G. Stanley Hall, a professor at Clark University who worried about the ostensibly debilitating and feminizing influences of modern life, introduced the idea that human beings did indeed recapitulate the evolutionary history of the species in their lifetimes, so that young people—his concern was with boys—were actually, not metaphorically, primitive. Hall suggested that encouraging boys to relive our evolutionary history by acting out the strenuous outdoor life, by engaging in "primitive" adventures, would protect them from excessive civilizing influences later in their lives. It would, in a sense, inoculate them with a form of primitive masculinity.[19]

G. Stanley Hall also introduced Freud to the American public, bringing him to Clark University for a series of lectures in 1909. Freud enunciated his own theories of the psychological stages of human development, including the idea that religion was a product of the primitive stage of human history. Freud, and popularizations of his theories, became immensely popular in American literary and middle-class culture by the 1920s.[20] A popular psychology self-help book by William J. Fielding, *The Caveman within Us*, elaborating on the idea of recapitulation, suggested that modern people could return to the child within by finding their inner cavemen.[21] The "sinister and subtle" disorders labeled "neurosis," "neurasthenia," and "nervous exhaustion," Fielding maintained, "indicate a lack of coordination between the primitive and cultural components of our nature. They represent a struggle between the Caveman and the Socialized Being."[22] The book urged readers to understand the caveman origins, and therefore the evolutionary utility, of such phenomena as fear, humor, daydreams, and even "the universal mother-in-law jokes."[23] Witch hunts, intolerance, and "Puritanical obsessions" all derived from imbalances between the primitive and the social. So did the profound pathologies characterizing, it seemed, most of the men of genius who had, according to Fielding, shaped history. All humans suffered from the dual nature—primitive and rational—that remained as the residuum of their evolutionary history. The only way "to acquire a healthy organism" was "to bring these two factors of the personality into harmonious relations."[24]

The newspaper frenzy over the heavyweight championship boxing match between Gene Tunney and Jack Dempsey similarly featured the caveman motif: newspapers predicted "the greatest battle since the Silurian Age." Journalists cast Tunney as "brainy" and refined, referring to him as "the student" and even the "scientific boxer," in contrast with Dempsey, portrayed as a "caveman" and as a "Neanderthal." Intriguingly, Dempsey's identification as a caveman contributed to his popular allure. Many writers, characterizing the fight as pitting "student" against "caveman," favored Dempsey—the caveman. Paul Gallico, writing for the *New York Daily News*, called him "unspoiled, natural, himself, one of us, the generation's best-loved athlete" and caricatured Tunney as "a no account clerk in an arrow collar," an "affected prig," claiming that most people "wanted to see the book reading snob socked back to Shakespeare."[25] The caveman grew in stature as a contrast to the perceived inadequacies of the pampered modern, his "real" knowledge a superior counterpoint to the vain intellectual knowledge of the educated challenger.

Cavemen could also be used to highlight human progress, however. An essay in the *Montana Socialist*, "From Cave-Man to Edison," pointed out: "It is a long way from the hairy cave man, who, with blood dripping jaw, sucked the marrow from the bones of a slain man, to Thomas A. Edison. A long and painful way, leading over a thousand generations into slavery, feudalism and capitalism." The magnitude of the distance implied that the next step in the progression was imminent: "Tomorrow comes Socialism."[26]

Rather than claiming that moderns represented unalloyed progress, however, Stone Age romances often aspired to rehabilitate early humans in the eyes of the public. The cave people in Stanley Waterloo's 1897 novel *The Story of Ab*—both male and female characters—were strong and intelligent. They were inventive and enjoyed steadfast family attachments. True, their ears moved much more than those of modern humans, they had opposable big toes for climbing, and they got about best by swinging like gibbons from branch to branch of trees. But they formed deep family attachments, they were loyal and brave, and Ab himself was an especially clever lad. It was he who domesticated wolves and invented the bow and arrow.[27] Similarly, a 1916 children's book, *The Cave Twins*, emphasized the humor and bravery of the cave children and the warm affection with which they teased their old grandmother.[28]

Jack London's 1906 caveman tale, *Before Adam,* based on Waterloo's novel, in turn inspired D. W. Griffith's 1912 film *Man's Genesis* and its 1913 sequel, *Brute Force,* films in the tradition of the caveman as "noble savage." Moviegoers could watch as the protagonist in *Man's Genesis* invented the stone ax; Griffith intended the film, his biographer Richard Schickel tells us, as a parable about reason and unreason, and though advisers warned Griffith that "cave men dressed in animal skins and brandishing clubs could be used only for comic purposes," the finished movie was considered "a formidable 'art' film," as well as a success at the box office.[29] Charlie Chaplin's 1914 Stone Age spoof, *His Prehistoric Past,* in contrast, played broadly for laughs. In Chaplin's dream sequence all the characters wore the one-shouldered garments that had become a caveman cliché, and Chaplin himself sported a tail. The film was set in the Solomon Islands, where, according to the movie caption, every man had a thousand wives. Chaplin joked, "A thousand wives for every man? I wish I had brought a bigger club!"[30]

The term *caveman* was clearly not intended simply as a male universal.[31] The gender implications of the caveman expanded with increasing uneasiness about gender roles in the decade of the 1920s. Playacting at being an "Indian," an explorer, or a cave boy was supposed to help young boys grow into masculine adulthood, but the caveman also continued to convey the gender-role meaning suggested in Chaplin's prehistoric dream movie and implied in a 1911 cartoon depicting a henpecked modern man in an art gallery gazing enviously at a painting of a caveman and woman. The caveman in the painting pulled the woman's hair while brandishing a club over her kneeling figure.[32]

Complaining of the ubiquity of cavemen in popular culture, the critic Simeon Strunsky wrote in the *New York Times Book Review* that readers had been inun-

"WHAT MAN HAS DONE, MAN CAN DO."

The popular image of the caveman as a brute often reflected cultural concerns about changing gender roles, as in this 1911 cartoon. Cartoons with similar themes were commonplace in the 1920s. This cartoon is from *Caricature: Wit and Humor of a Nation in Picture, Song and Story*, 8th ed. (New York: Leslie-Judge Co., 1911); the book was not paginated, and no author's credit was given to either the book or the cartoon.

dated with images of the caveman as a club-wielding, wife-beating brute.[33] Newspapers referred often to "caveman" types, assuming the meaning to be understood. For example, a 1921 story about a man who had kidnapped a woman in order to force her to marry him carried the headline: "He No Cave Man, Wife Willing Bride, He Says." The man in the story denied having "used any cave man tactics."[34] A similar story appeared in 1925, this time about a man who staged a "playful" kidnapping of his paramour, reported in the newspapers as the "caveman bookkeeper."[35] Stories in women's magazines regularly used the term *caveman* to refer to pushy, oafish, boorish, and generally unenlightened husbands—louts.

The kidnapping of women by cavemen echoed an older popular motif of abductions of human women by apes. Gorillas began abducting women in European art almost as soon as gorillas became familiar to Europeans, in the mid-nineteenth century. The gorilla abduction theme would have been familiar in the 1920s through a famous World War I propaganda poster of a German ape carrying a club—labeled "Kultur"—and making off with a ravished-looking woman.

It was also familiar to readers of the Tarzan books. Tarzan first won Jane by rescuing her from an ape with alarming intentions: "Jane Porter—her lithe, young form flattened against the trunk of a great tree, her hands tight pressed against her rising and falling bosom, and her eyes wide with mingled horror, fascination, fear and admiration—watched the primordial ape battle with the primeval man for possession of a woman—for her."[36] Combining the themes of primitive masculinity, race, and concern about a shaky modern civilization, the Tarzan books and movies attracted an immense audience.[37] Tarzan movies, based on Burroughs's successful Tarzan novels, became staples of American popular culture. Whether they seemed scandalous may have been a matter of perspective. Many viewers of *King Kong*, in which a giant gorilla tried to make off with a blond woman, would see it as a Beauty and the Beast fable.[38]

Cavemen turned out to be rather protean symbols, adapting easily to Progressive Era preoccupations, and by the time of the revitalized evolution debates of the 1920s, they had been fixed in a common cultural vocabulary. Caveman dramas appeared regularly in the 1920s. The ape-men in the 1925 movie version of Arthur Conan Doyle's 1921 novel *The Lost World*, sinister and obviously malevolent brutes—played by white actors in blackface—traveled in the company of monkey and chimpanzee comrades—or cousins.

Cavemen and evolution entered the theatrical repertoire—a play called *Survival of the Fittest* appeared in the Greenwich Village Theater in New York in 1921. The cover of the libretto for a 1929 "prehistoric operetta" called *Cave-Man Stuff* featured wrestling men clad in one-shouldered spotted furs and waving clubs.[39] A reviewer for the *New York Times* complained in 1922 that "it has suddenly become the 'swanky' thing to be enormously amused by the primitive stuff," noting that "primitive" humor depended for its bite upon connotations of social class.[40] Notions of the primitive carried a lot of freight. And cavemen contained all kinds of possibility for dissent, especially through irreverent humor. The insistently irreverent *New Yorker* made fun of New York high society and its concern for ancestry with a spoof on an Ivy Leaguer reminiscing about his grandfather, a Cro-Magnon, who was president of the Fifth Avenue Fossil Club and was known as "Old Cro."[41] Even that most sophisticated of moderns, Cole Porter, contributed a caveman song, "Find Me a Primitive Man," in which the singer yearned for a "primitive man" who was not "the kind that belongs to a club / But the kind that has a club that belongs to him."[42]

In 1920, the humorist Clarence Day published a little book called *This Simian World*, attributing human personality traits like gregariousness, prolixity, and a tendency to rowdiness to our primate heritage. Taking an inventory of various animals from which we might have evolved if chance events had worked out differently, Day imagined a variety of alternate possible versions of humanity. How much more sleek, elegant, and dignified we might have been, Day suggested, if we had descended from the great cats rather than the chattering monkeys. "Chattering monkeys" appeared often in popular references to human evolution. Jack London's *Before Adam* narrator complained of companions who chattered incessantly. The 1923 movie *Evolution* suggested that chattering in monkeys, far from being annoying, represented a great step in the progressive evolution of humans. Clarence Day agreed. Libraries were filled with the results of the human—and simian—tendency to chatter: "Books! Bottled chatter!"[43]

Popular images of cavemen had been shaped by ancient and tenacious visual traditions, and, as caricature, they responded to changing times. For example, by the 1920s an ironic new meaning entered the caveman cartoon lexicon. A variation on the brutal caveman lout theme emerged in the self-conscious new consumer society of the decade, combining the consumer theme with concerns about gender-role instability. Now the henpecking consumerist cave wife put her caveman husband in jeopardy by demanding that he use his club to acquire fashionable new furs, the skins of dangerous but attractive large cats.[44]

Notions of race, gender, and class hierarchy could all find convenient expression through images of cavemen, ape-men, and missing links. Critiques of these notions could draw on the same kinds of images. Ape-man and caveman humor came from all directions. It could be subversive, but it could also buttress the status quo, seeming to reaffirm traditional cultural values. And these metaphors reveled in their ability to make themselves visible or, in some cases, visually imaginable.

By the 1920s, then, ironic new meanings entered the caveman cartoon lexicon. Cavemen carried significant cultural resonance in the 1920s in part because ambivalence about civilization suffused the decade. That the Great War had vitiated modern claims for civilization became a truism of the time. What counted as civilized seemed, to many people, to be newly labile. Cavemen also acquired new cultural resonance with the revitalized evolution debates of the 1920s.

Images of cavemen unsettled many people, including some evolutionists; they also armed anti-evolutionists with persuasive rhetorical weapons. At this moment of general uneasiness about mass culture and a nascent "hieroglyphic civilization," images and objects spoke loudly, often eloquently, but not necessarily in predictable ways.

Scientists, Cavemen, and the Jungle Theory:
Picturing Evolution in the Jazz Age

The discovery of cave art in Europe beginning early in the twentieth century had added a new dimension to the caveman's appeal as a symbol. The cave art of the Cro-Magnon people stimulated enormous curiosity; these large-brained artists lent themselves especially well to romanticized versions of human antiquity, and they added a significant dimension to the debate about human origins that took such vigorous form in the 1920s. The appearance of art, for many people, signaled the real beginning of humanity. Cro-Magnon artists made it possible to imagine a noble human past.

The existence of cave artists, and of their subtle and sophisticated art, led some leading evolutionists to suggest that our history was not, after all, simian. It was human. We were artists, and we were quite distinct from all the other primates. The caveman was not a brute but an artist. Defenders of evolution emphasized the distance separating humans, including cavemen, from apes because, among other things, apes and monkeys carried a strong iconographic tradition, including depictions of simians as aberrant or degenerate forms of humans, as sinners, fools, and monsters.[45] These images had been used to discredit Darwinism in the late nineteenth century, but they predated Darwin and outlived him. Cavemen, too, carried metaphoric baggage.

Evolution, even for some scientists, was not simply an internal matter of arcane scientific theory. The debate was not simply about whether evolution itself was true. It was a debate about what such truth might imply. Just as the 1925 Scopes trial was not only about evolution, cavemen were not only about human prehistory. And the evolution debates took place in a context in which "the pictures in people's heads" carried complex associations.

In 1926 the Reverend Harry Emerson Fosdick wrote in his column in *Harper's* that recent developments in astronomy presented serious challenges to the "picturableness" of traditional faith, and he confessed he felt that "believing in God without considering how one shall picture him is deplorably unsatisfactory." The old cosmology had provided "a cozy stage," on which to imagine God. Modern science had destroyed that vision. Perhaps, Fosdick suggested, the radio offered a replacement model: the "living voice out of the unseen, the mystery of fellowship with the invisible." But no, this would not satisfy. And yet we had to face the truth: "The plain truth is that we cannot picture God at all. . . . The interesting fact is that, not only can we not imagine God, but science has brought us to the place where we cannot imagine the physical universe." Einstein's universe was "utterly unpicturable." Scientists had tried, in popularizations, to give us images to help us understand this universe, but they were "trying to picture the unpicturable!"[46]

Everyone, however, could picture human evolution in some form. It was the form those images would take that was contested. Scientists attempted to offer noble visions of how the human past might have looked. They tried to find visions of human evolution that could coexist, if not compete, with familiar mental pictures of Adam and Eve. They offered new visions of the human past and of possible human ancestors. But the field was already crowded.

Anti-evolutionists understood well the power of visual imagination. Monkeys appeared everywhere in the year of the Scopes trial—as they had in the years following Darwin's publication of *On the Origin of Species* in 1859. The more combative opponents of evolution, such as Billy Sunday and John Roach Straton, gleefully labeled evolution "the jungle theory" and hammered relentlessly on the monkey theme. Everyone knew how evocative these images were. Anti-evolutionists continued to use scientists' visual illustrations of evolution to challenge the credibility of scientific authority and, perhaps even more effectively, to ridicule evolution. A commercial slide company marketed some of Bryan's popular lectures, illustrated with lantern slides. Visual images and metaphors often revealed more than essays and lectures did.

Anti-evolutionists may have appropriated monkey and caveman images to suit their own rhetorical purposes—and to express their visceral distaste for evolution—but partisans of evolution and secular humorists also put monkey, caveman, and missing-link images and metaphors to use, and these "friendly" simian images may have undercut the images scientists tried to convey just as much as hostile ones did. That may not have been a bad thing, of course: scientists' use of images was not necessarily always benign. Everyone had an agenda, and the boundaries separating science from the extrascientific seem to have become particularly permeable during the course of the debate in the 1920s.

In 1925 the Baptist minister W. E. Dodd wrote: "I would say that a modernist in government is an anarchist and a Bolshevik; in science he is an evolutionist; in business he is a Communist; in art a futurist; in music his name is jazz; and in religion an atheist and infidel."[47] The forms and definitions of modernism were, of course, much more various than Dodd admitted, but he made an intriguing point. The evolution controversy was actually more complicated, culturally, than it would seem. It may seem odd to think of cavemen and jazz as linked together, but in the dialogue of the 1920s, they were. Often they were linked through that other pervasive image, the "jungle," a word redolent of many of the most disturbing preoccupations of the decade. In the confusion of modern and primitive, in the symbolic universe of that uneasy decade, when old verities seemed to slip into uncertainty, jazz was a potent symbol. Edmund Wilson wrote of "the excitement in the air of our time which gets its popular expression in jazz," but what Wilson called excitement others perceived as a dizzying and unpredictable pace of life. The term *jazz* itself may have originated as a slang term for speed.[48] The "jazz culture" theme resounded on all sides of the debate in similar ways.

Occasionally though, commentators suggested that the irreverence so many people associated with jazz might not be such a bad thing. In the aftermath of the Scopes trial, an editorial in the *Nation* titled "Jazzing the Scriptures" noted that an enterprising business had recently advertised "jazzed"-up versions—popularizations—of the Bible. Bemoaning the ignorance of the Bible on display at the trial, the editor suggested, "Now, jazz has a way of penetrating a good deal further than scholarship, and it may very well happen that the advertiser . . . will, all unknowingly, do more for the cause of liberalism in theology than all the experts who offered their services at Dayton."[49] It was an optimistic thought. But many people, including the editors of the *Nation,* understood the use of monkey, caveman, and jungle imagery to have distinctly clouded the issues not only at Dayton but increasingly during the long course of the evolution debates of the decade.

Anti-evolutionists used the word *jazz* as a code for a constellation of cultural concerns, but so did some defenders of evolution. In *The American Earthquake,* Edmund Wilson listed Prohibition Era slang terms for drunkenness, one of which was "jazzed."[50] The Reverend John Roach Straton, a New York fundamentalist, blamed the evils of modern life, including evolution, on "jazz emotions." In his sermons and essays attacking evolutionists, the evangelist Billy Sunday often dwelled less on theology than on the evils of working women, divorce, and decadent jazz.[51] But at least one prominent evolutionist also referred disparagingly to the "jazz mind" and ascribed the mentality he associated with jazz to modern forms of decadence.

The hostile reactions of many people to evolutionism were not simply intellectual; as historians have recently shown, anti-evolutionism grew out of many cultural dislocations of the period. One conspicuous plane of polarization, in a highly polarized society, many people suggested beginning in the 1920s, was between the city and the country, taken as symbols of the modern against the traditional. Yet the tensions so often attributed to the rift between the urban and the rural flourished within cities, too. And although newspapers and magazines published in the city often painted the fundamentalist movement as a relic of an obsolescent small-town way of life, the truth was not so simple. The tensions that fed the debate existed in all parts of the country, including New York, that epitome of modern urban life.

CHAPTER TWO

The Museum in the Modern Babylon

In 1935 the Reverend Henry Sloane Coffin told the story of his debate, some thirteen years earlier, with the Reverend John Roach Straton, recalling that Straton—railing against the rampant immorality of New York City—had asked rhetorically, "Who is responsible for this lewdness and this animalism?" Then, answering his own question, Straton thundered, "Henry Fairfield Osborn!"[1]

The occasion for Coffin's reminiscence was a memorial service for Osborn, and his audience, numerous friends, relatives, colleagues, employees, publishers, and protégés of Osborn's, would have warmly appreciated the irony of Straton's accusations—and they would have remembered the context. Ever since his arrival in the city as pastor of the Calvary Baptist Church, Straton had been famously tilting at the windmills of New York licentiousness as a vocal proponent of Sunday blue laws, an advocate of Prohibition, and a critic of modern styles of dancing, women's clothing, and theater—and of "jazz emotions." New York—a city he called the "modern Babylon"—represented for him all the sin, lack of discipline, and immorality of modern life.[2] Osborn might have seemed a peculiar target, however.

Born to great wealth, self-consciously patrician, Osborn counted among his friends and cohorts many of the most powerful people in the country. He was well connected. His father, William Henry Osborn, was a shipping magnate and founder of the Illinois Central Railroad, and from him Osborn inherited a fortune. The family included the banker J. P. Morgan, Osborn's "Uncle Pierpont." The Osborn family occupied an estate on the Hudson, next door to Olana, the famous Orientalist mansion of the artist Frederic Church, a close family friend. Theodore Roosevelt remained a lifelong friend. In 1907, planning a trip to the Fayum, Osborn carried a letter of introduction to the Egyptian government from his good friend the president; preparing for a European excursion in 1921, he requested that Charles Evans Hughes, then secretary of state, arrange for his passport and for his wife's, and Hughes complied. In 1928, featuring his picture on its cover, *Time* magazine called Osborn "a Zeus among Olympians."[3]

He was also a confirmed cultural conservative. An ardent proponent of eugenics and immigration restriction—an influential leader, in fact, in those movements—Osborn was as dismayed as Straton by modern women's clothing, suggestive theater, and the various manifestations of what he himself often called the "jazz mind." He disapproved of the funny pages and, indeed, of much that was published in the newspapers, and he took considerable pride in his own bearing and dignity as a gentleman. Although he certainly had his critics, his friends would have described him in much the same terms Edmund Wilson used to describe a character based on his brother William Church Osborn in a novel, *The Higher Jazz:* dignified, tasteful, and gracious, a portrait of old fashioned gentility.[4] Why would Straton blame such a man for the evils of the city and of modern life?

Osborn was an evolutionist. More important: he was an evolutionist with a museum. Henry Fairfield Osborn, president of the American Museum of Natural History in New York, was one of the most prominent scientists defending evolution against organized challenges from fundamentalists in the 1920s, and one whose views the interested public was likely to encounter frequently. His stature as a scientist and as museum president, along with his social prominence, ensured his influence and access to the media. From the time challenges to evolution began to resurface in the early years of the decade, Osborn took an active part in the defense of evolution before the tribunal of the public. Newspaper editors called on him for statements; he gave lectures and radio interviews; and authors of popular books and magazine articles drew on his work and on illustrations of evolution from his museum. In 1925, when John Thomas Scopes was indicted for breaking the Tennessee law against teaching evolution in public schools, the American Civil Liberties Union sent the members of the Scopes defense team to the museum for advice from Osborn—and to have their pictures taken with him. He worked energetically to try to mobilize scientific witnesses for the Scopes defense. Osborn's views of evolution did not by any means represent a consensus of scientific thought—indeed, no such consensus existed. Although virtually all biologists in the 1920s had long accepted evolution as established fact, they did not all leap to its defense. Osborn did. He held a secure place in the media as a representative of the scientific point of view. Osborn was a central figure in the debate not because he was typical of scientists but because of the ways in which he was anomalous. His were among the most available of available ideas.

And he made them visible. Osborn had a disproportionate influence for a variety of reasons. Most important, the museum gave him a medium for communication. In 1904 George T. Brett, an editor at the Macmillan Company, wrote to Osborn to urge him to write a book that would offer people a way to imagine the evolutionary past. People had many images in their minds, Brett wrote, "from the old illustrations, what the several steps in the process of creation looked like according to the Mosaic tradition." They knew how to conjure up mental images of Adam and Eve but did not have so many ways to imagine the past as described by evolutionists. People were hungry for this knowledge: "There is nothing outside of fiction in which the people would, in my judgment, be more heartily interested." Who but Osborn could supply such a thing? "I am persuaded that if you do not do the book we can hardly look for it from anyone else in this generation."[5] Of course, Brett wanted to flatter Osborn, but it was true that Osborn was well positioned to supply such images to the public. He wrote well, took a synthetic approach to paleontology, and had the resources, at the museum, to provide a wealth of illustrations. The resulting books, *The Age of Mammals in Europe, Asia, and North America*, published in 1910, followed in 1915 by *Men of the Old Stone Age*, were engagingly written and lavishly illustrated with images from the museum. They treated the science of paleontology, as Osborn put it in the arresting first sentence of *The Age of Mammals*, as "the zoology of the past."[6] At the museum as in his books, Osborn worked with visual artists to bring the past to life.

Art and the Zoology of the Past at the American Museum of Natural History

Scientists and artists at the American Museum of Natural History made the past imaginable in new ways, and many of the available images of what ancient life might have looked like came from this museum. Osborn's visions of the life of the past would help to shape the imaginations of generations of students—not only because the museum was such a popular destination for school trips and family outings but also because popularizers drew on images created by museum artists to illustrate books, articles, and movies. Images from the American Museum appeared regularly and often in newspaper and magazine articles, in books meant to explain evolution to the general public, and in films. Ultimately, however, Osborn found that he could not control the use of those images. His participation in the evolution debate of the decade illustrates the complexities of the situation of the scientist attempting to represent science for the public.

Straton had reason to fear Osborn. In a contest for the minds of young people, how was an opponent of evolution, speaking from the pulpit, to compete with the immense halls of a museum populated by dinosaurs, mammoths, and cavemen? Many people visited the American Museum of Natural History—according to the museum's annual report for 1926, there were 2,070,265 visitors in that year, with an additional 171,769 students attending lectures at the museum. The number of visitors increased steadily during the early 1920s, from 1,775,890 in 1920 to 2,292,265 in 1927.[7] Many of these visitors were schoolchildren; museum personnel developed active education programs both in the museum itself and for outreach to the schools, and by all accounts, the exhibits of fossil animals in particular were spectacular. Furthermore, the museum's influence extended far beyond New York. Smaller museums purchased plaster casts of reconstructions of early hominids on display at the American Museum. Teachers bought or borrowed lantern slides of museum diagrams and exhibits. The images created at the museum were widely distributed to schools, magazines, books, and newspapers across the country, and Osborn's status as a scientist and as a public figure meant that science teachers, journalists, and popularizers relied on him as an authority on the details of evolution.

Straton was not the only anti-evolutionist to criticize the museum. Exhibits at the American Museum drew fire from creationists partly because of the prominent role assumed by Osborn and his colleagues in the evolution debates but also because the museum provided so many of the illustrations in popular books and articles about evolution. Osborn and his colleagues took an active part in designing exhibits and, committed to the museum's educational mission, widely distributed images based on the exhibits. The museum was generously staffed with talented artists; trustees such as J. P. Morgan Jr. funded murals by the artist Charles R. Knight at considerable expense, and Knight's murals were extremely popular.[8] Museum personnel used lantern slides extensively in public presentations; they also provided slides to teachers around the country and supplied illustrations to writers of textbooks and popular books about science and to colleagues for their lectures before the public. Replicas of the famous Family Tree of Man on display in the Hall of the Age of Man were sold to several other museums, and Osborn had a photograph of it sent to Dayton, Tennessee, in 1925 to be used as evidence at the Scopes trial. Copies appeared across the country in newspaper accounts of the trial. Textbook publishers and authors of popular science books drew liberally from the visual repertoire of the American Museum; images from the museum appeared disproportionately in popular books about evolution— and in popular books not specifically about evolution but touching on evolutionary themes, such as H. G. Wells's successful *Outline of History*—in the 1920s.[9] The American Museum of Natural History did as much as any institution in the world to make the deep past imaginable.

In addition, the intellectual approach Osborn took to paleontology had fostered innovative techniques in museum display. Many passages in the draft of the artist Charles R. Knight's autobiography emphasize Osborn's role and the creative approach to exhibit design of curators at the museum. Osborn's design for the display of the carnivorous dinosaur *Allosaurus*, for example, vividly depicted predation, capturing an allosaur about to devour a brontosaur. The drama of these exhibits, Knight suggested, was a natural outgrowth of Osborn's lively sense of paleontology as the "zoology of the past" and of his genuine love of both art and nature. Osborn had always had a feel for art; he had organized a sketching club as a student at Princeton. And his artistic sensibilities, combined with his imaginative views of the fossil past, shaped the exhibits at the museum. A memorial to Osborn in one of the main halls of the museum is inscribed: "For him the dry bones came to life and giant forms of ages past rejoined the pageant of the living."

Knight, whose career as an artist grew in tandem with the museum, remembered the American Museum of Natural History as a pioneer of exhibits of mounted articulated fossil skeletons in lifelike poses, exhibits of animals in groups and in context. Recalling his early days at the museum at the beginning of the century, Knight reflected that "few museums in the world at the time could boast of more than a very few fossil creatures actually set up in approximately natural positions," and he credited Osborn with encouraging imaginative exhibit techniques that would have popular appeal. Osborn, Knight remembered, was "an artist at heart, and had a great appreciation of all things living and beautiful." The museum in those early days had been an exciting place, and Knight emphasized the conviction he and Osborn had shared that fossil animals should be portrayed as "at one time living breathing creatures existing in an atmosphere of color. . . . There had of course been many restorations made of prehistoric creatures, [but] unfortunately the men who made them were not artists," and the results had often been "dreary and lifeless renderings." Osborn, Knight reflected, was "first of all a scientist—yet he was not the dry as dust order of mind—but on the contrary reveled in the contemplation of all living things—seeking ever to correlate the present with the past." This marriage of art and science, Knight believed, had been a new thing and made the American Museum of Natural History a magic place to work. The excitement stimulated by lively exhibit techniques dovetailed with the museum's emphasis on public education.

Osborn's dinosaur exhibits, Knight recalled, had created a great stir in magazines and with the public: "These creatures were then absolutely new to the majority of the American public and they went wild over the new mounts, the Museum elected many new members, and got no end of publicity as a result of this well thought out campaign." The publicity, along with Osborn's eloquence, persuaded wealthy patrons to increase their support.[10]

Knight's memory of the novelty of mounting specimens in lifelike poses may have been somewhat exaggerated. In the late nineteenth century Ward's Scientific Establishment in Rochester, New York, a supplier of natural history specimens, advertised mounted articulated animal skeletons for sale, and as early as the London Crystal Palace Exhibition of 1851, the artist Waterhouse Hawkins had worked with the anatomist Richard Owen to create sculptures of the recently discovered and named dinosaurs.[11] In the late nineteenth and early twentieth centuries other American museums, including the Carnegie Museum in Pittsburgh and the U.S. National Museum of Natural History, of the Smithsonian Institution, also mounted spectacular displays of articulated dinosaur skeletons at considerable expense.[12] Nonetheless, the American Museum of Natural History in New York occupied the forefront of this movement, pioneering new methods of displaying extinct animals. Traditionally, many museums had emphasized taxonomically or randomly arranged cases of specimens with few explanatory labels; at the end of the nineteenth century the new emphasis on teaching through the display of "object lessons" had altered philosophies of display, and well-funded new museums put innovative exhibition techniques to work. Combined with the commitment of many museum scientists to the conservation of what they feared was a disappearing nature, and with the extraordinary financial resources available to museum directors like Osborn who were able to excite the interest of wealthy patrons, the new perspective revolutionized museum display in general. Osborn's social connections, his energy and charisma, along with his gift for publicity, and the museum's location in New York made the American Museum of Natural History a leader in this movement.

In addition, Osborn enjoyed an unusual ability to identify and attract to the museum especially gifted students, curators, and assistants. In particular his protégé, William King Gregory, and the geologist William Diller Matthew enlivened both the scientific life of the museum and the exhibits on display before the public. In looking back on those early days, the paleontologist George Gaylord Simpson, who came to work at the museum as a young man in the 1920s, emphatically noted the unusual creativity and scientific insight of Matthew and Gregory. He also acknowledged that Osborn, who had by then become a rather controversial figure because of his support for eugenics and because he had acquired a reputation for arrogance, had fostered an atmosphere in which such creativity could thrive. William King Gregory had, like Osborn, a significant artistic bent, and a dry wit often illuminated his scientific diagrams. Collaborating with the artist Helen Ziska, he illustrated his two-volume 1951 book *Evolution Emerging* with a wealth of remarkable and engaging evolutionary tree diagrams. Although they were curators—a status that in some museums insulates scientists from the demands of public education—they participated actively in exhibit design.[13]

These curators worked closely with artists; Knight warmly recalled his collaborations with them in his autobiography. Other artists at the museum, such as Erwin Christman, were similarly gifted and collaborated closely with scientists. The Columbia University biologist J. H. McGregor, remembered by his students as a talented artist—he enjoyed astonishing students by drawing with both hands simultaneously on the classroom blackboard—created reconstructions of human ancestors for the museum and worked with artists such as Knight in the design of paintings, sculpture, and drawings.

A report on the history of the museum, written in the 1940s by the anthropologist Clark Wissler for the board of trustees, underscores Osborn's emphases on display and on vertebrate paleontology: under Osborn, Wissler noted repeatedly, the museum had devoted a large share of its considerable resources to exhibits, especially of fossil animals. Correspondence within the museum confirms that Osborn and the scientific curators took active roles in designing and mounting exhibits. Memos and letters refer not only to concerns about the accuracy of exhibits and captions explaining them but also to such details as the arrangement of specimens and exhibits in space, lighting, and even the color and material of the letters on labels and the kind of adhesive used to attach them.[14]

In addition, Osborn's influence was extended through his authority among hinterland scientists—scientists and science teachers across the country whose institutional roles would not match his and who published less or not at all but who turned to him as an authority. Such people wrote often to Osborn asking questions about teaching and about current controversies in the interpretation of evolution. They sought his advice about how to handle classroom debates over the resurgence of Lamarckism, the doctrine that characteristics acquired by organisms could be inherited; about the meaning of disputed terms and concepts such as "mutation"; or about the pronunciation of names like "Cro-Magnon"; they also requested copies of museum photographs, casts of fossil specimens, and especially lantern slides for teaching.[15] And increasingly after 1922, they turned to Osborn for advice about how to handle challenges from anti-evolutionists and reported to him on their own activities in response to fundamentalists. They wrote letters to the editors of their local newspapers, gave speeches, and participated in debates, even though they thought such activities would have more influence coming from, as one of them wrote to Osborn, "some much bigger biologist and also better writer."[16] These were often people who were not officials in the American Association for the Advancement of Science (AAAS), not widely published, not on AAAS committees or offices, in the National Academy of Sciences, or at the most prestigious institutions. In some ways they were part of the public influenced by people like Osborn. And in this new age of specialization, even scientists were part of the lay public when it came to sciences outside their own fields of specialization.

Osborn's prominence was also extended through his own engagingly written and beautifully illustrated books. His 1915 history of Stone Age culture, *Men of the Old Stone Age,* enjoyed considerable success, both through its own very respectable sales and through its widespread and lasting influence on teachers and other writers. Osborn received many letters from readers telling him that they had purchased multiple copies of the book to give to libraries and acquaintances, that their children had read it more than once, and that their libraries never seemed to be able to keep it on the shelf. Several librarians wrote to him that the book was as popular as a novel, and the novelist Edith Wharton sent him a postcard in 1923 telling him that she was bringing it with her on a trip to look at cave art at Les Eyzies.[17] By 1922 *Men of the Old Stone Age* had sold nearly fifteen thousand copies, and Osborn noted that sales increased conspicuously after the heating up of the evolution debates in that year. The book sold well enough to be issued in thirteen printings during Osborn's lifetime; it was translated into many languages and remained in print into the 1940s.[18] Textbook writers drew liberally on both the text and the illustrations. A French writer, M. V. Forbin, published a Cro-Magnon romance, *Les Fiancées du soleil,* dedicated to Osborn and inspired, the author related, by *Men of the Old Stone Age.* According to a letter from Osborn to Charles Scribner, Forbin's novel was edited for use in teaching French language in American schools.[19] As Osborn often pointed out, H. G. Wells borrowed extensively from *Men of the Old Stone Age* in his phenomenally successful *Outline of History.* Wells's use of his work sometimes irked Osborn, but he recognized its value to him. In 1923 he wrote to Scribner: "Interest in my work has quadrupled since I wrote 'Men of the Old Stone Age,' entered into controversy with Bryan, and drew down the influence of Wells' 'Outlines.'"[20]

The lavish illustrations accounted for a significant part of the success of *Men of the Old Stone Age;* it was as much through visual art as through words that Osborn influenced popular notions about cave people, especially in his collaboration with the artist Charles R. Knight in the design of Knight's paintings, drawings, and murals of early hominid life. The success of the book and the familiarity of Osborn and Knight's collaborations and of exhibits from the museum in general also explain why the museum drew fire from anti-evolutionists.

Attacks on the Museum

Anti-evolutionists, understanding the power and complexity of visual images on evolutionary themes, targeted scientific illustrations and museum exhibits for criticism and ridicule. The American Museum was especially subject to attack because it was so effective. Osborn and his brilliant staff had created dramatic and innovative exhibits that made the museum a theater of ancient life. Not surprisingly, then, anti-evolutionists prominently targeted both Osborn and the museum, and their attacks were reported in newspapers around the country.

Many of the fundamentalists' attacks on the museum in the 1920s focused on the famous and controversial Hall of the Age of Man. Begun in 1916, the Hall of the Age of Man received considerable public attention at the time and has continued to be an object of fascination among historians.[21] In 1922 Alfred Watterson McCann, a Catholic lawyer, devoted an entire book, called *God—or Gorilla*, to excoriating Osborn and challenging the veracity of the Hall of the Age of Man. Osborn received numerous letters about that book, and clergymen in New York and Boston joined the assault on the museum.[22] In 1922 the *Sunday School Times* announced a series of articles that would expose the "amazing credulity of many who pride themselves to-day on being scientists." The series would focus especially on the "science falsely so-called" on display at the American Museum of Natural History.[23] The *Sunday School Times* accused the museum of fabricating fake "missing links" for display. In the same year Father Francis P. Le Buffe of Fordham University issued similar challenges, publishing pamphlets attacking evolution, denouncing Osborn, and engaging Gregory in a newspaper debate.[24]

In 1924 John Roach Straton delivered a series of sermons denouncing the Hall of the Age of Man at the museum because, he said, "It has been my terrible and woeful experience to witness thousands of little children flocking to the museum to have their juvenile minds poisoned by the foul miasma of evolution"[25] Straton's sermons were advertised well ahead of time, and on March 9 the *New York Times* announced: "At Calvary Baptist Church plans are being made to accommodate a record crowd tonight to hear Dr. John Roach Straton assail the American Museum of Natural History for teaching evolution."[26] In 1925, when lightning struck one of the museum's towers, he triumphantly announced that it was a sign from God. Newspapers published in places as far from New York as California covered the Osborn-Straton exchange, and Straton continued to cite his visit to the museum in debates with evolutionists—for example, in a debate with the Unitarian minister Charles Francis Potter at Carnegie Hall which attracted audiences estimated at around three thousand people and received considerable newspaper coverage.[27]

Osborn's responses to these attacks were adamant: "No one can point out either in the exhibition halls of the American Museum or in its lectures a single untruthful statement, because the lectures and the exhibition halls do not set forth theories."[28] The exhibits at the American Museum, Osborn wrote repeatedly, included only facts. He was acutely aware of the relationship between museum exhibition space and implied message; in requesting increased funding for museum exhibit space, he reasoned on several occasions that when collections were "jumbled together out of their natural order," they would inevitably convey "entirely erroneous impressions." The museum needed new space not for incidental aesthetic reasons but because in crowded display conditions the arrangement would be "less truthful and more misleading."[29] Adequate space to arrange specimens correctly was crucial, for the exhibits at the museum adhered scrupulously to facts. There was no speculation involved.

Nonetheless he wrote to his colleague J. H. McGregor, "The Museum is weathering Dr. Straton's attack, but I earnestly desire to make Case I more idealistic," and he notified curator George Sherwood that because "this exhibit is the subject of constant attack" he wanted to alter it in order "to make it absolutely beyond criticism in clearness and accuracy of arrangement."[30] In December 1924 he ordered the removal of a sequence of the "progression and progress from lemur to men" which had been used as a transition from the Hall of the Age of Mammals to the Hall of the Age of Man.

Despite these efforts, however, the museum continued to draw fire from anti-evolutionists who accused curators of devising misleading exhibits. Osborn's efforts notwithstanding, anti-evolutionists suggested that museum exhibits contradicted scientists' words. When Cardinal William Henry O'Connell of Boston repeatedly attacked Osborn and the museum for misrepresentations between 1922 and 1926, his focus was on the "lies" told in the exhibits; newspapers carried the story prominently.[31]

When three-time presidential candidate William Jennings Bryan, who continued to enjoy an enormous following in the 1920s, assumed leadership of the anti-evolution campaign, he incorporated anti-evolution lantern slides in his popular lectures; he, too, focused conspicuous attention—and ridicule—on the museum and its depictions of evolutionary themes. Anti-evolutionists understood that ideas about the human place in nature were implicit in images of evolution. Paintings and sculptures of human ancestors evoked emotion on their own, and as Straton noted, the arrangement of such figures also carried messages. But the messages Straton and Bryan saw so clearly—or used rhetorically—were not those that museum scientists intended to convey. Anti-evolutionists who criticized visual images of evolution appreciated the impact of such pictures: pictures of evolutionary ideas molded the debate. Visual images both shaped and expressed the way scientists conceived of evolution, but the pictures most available to the public could not have communicated in the straightforward way scientists expected them to.

Words and Pictures

The words published by scientists and science popularizers during the debates of the 1920s were often at odds with the messages implied in the illustrations that accompanied those words. This was so for several reasons. In his 1910 book *The Age of Mammals in Europe, Asia, and North America*, Osborn had written that evolutionary tree diagrams, among the most familiar ways to depict evolutionary relationships, had "fallen into disfavor" but that "the present reaction against these trees does not seem to be altogether wise, for we must remember that they are among the working hypotheses of this science, which serve to express most clearly the author's meaning."[32] That family trees and other kinds of diagrams were among the "working hypotheses" of evolutionary science was an astute observation on Osborn's part, but whether they "serve[d] to express clearly the author's meaning" was more problematic. For scientists, diagrams were not simply decorations but elements in visual languages with their own grammar and tacitly understood conventions. Scientists developed visual lexicons, sets of motifs that stood for ideas and assumptions familiar among colleagues.[33] Scientific illustrations functioned in a number of different ways, and these functions might not always be compatible.

First, as Osborn suggested, scientific diagrams were often "working hypotheses." Scientists—especially in certain fields including geology, anthropology, and paleontology—more than scholars in many other disciplines, used diagrams, restorations, and other visual images not only to communicate their ideas but also to form them and to test them. Writing to the artist Charles R. Knight of the "value of drawing as a great aid in 'discovering,'" the anatomist Adolph Schultz observed: "I enjoy [drawing] more than any other phase of my investigations, just because it leads my eyes to points which meant nothing to me before." Knight and the scientists at the American Museum commonly used drawings and sculptural restorations to test interpretations of incomplete fossil skeletons.[34] Restorations of living animals from the evidence of fossil skeletons did more than just illustrate the appearance of extinct animals. They allowed morphologists to test their predictions about missing bones and about how the animals might have functioned anatomically. Illustrations could, as one historian of science has written, "make visible some presumed invisible order."[35] Similarly, diagrams helped to visualize evolutionary patterns, to work out and test theories. They did function as working hypotheses, or thought experiments.

Second, scientists used illustrations—as Osborn also implied—to communicate, and to do so in at least three very different ways. Diagrams functioned, first, to communicate ideas and theories among scientists; second, they served as persuasive devices; and, finally, scientists also used them to teach and to communicate with a lay public. Because the same diagrams were used for all three purposes, they could mean different things to different audiences. Images meant primarily to convince colleagues of a theory often emphasized different patterns than did those intended for students or lay audiences. Diagrams designed to test hypotheses, to communicate with colleagues, or to persuade them might rely on visual vocabularies, lexicons not necessarily familiar to outsiders. The scientists who defended evolutionary theory in the 1920s belonged to a community that increasingly spoke a private language, and even the pictures they drew contained specialized professional vocabularies. Outsiders might well have misunderstood.

Scientists were not unaware of the ramifications of the conventions of scientific illustration. In this period of increasing professional specialization, and of the widespread influence of positivist philosophies of knowledge rejecting all metaphysical explanations as speculation and insisting on strictly empiricist methods, scientists' self-consciousness about the idea of objectivity was heightened—indeed, like many nonscientists, they sometimes used the terms *scientific* and *objective* as if they were synonyms. In the late nineteenth and early twentieth centuries many scientists sought methods of illustration, such as photography, which would demonstrate their objectivity by removing the scientist as an individual from the cognitive process of technical illustration.[36]

But when put on display in museum exhibits, which are expensive and difficult to mount and therefore long lived and reproduced in popular books and multiple editions of textbooks, diagrams conceived either as illustrations of a specific observation or pattern or in a spirit of exploration and hypothesis testing could become fixed, suggesting a misleading degree of certainty. A 1902 diagram by William Diller Matthew, for example, illustrating the evolution of horses remained on display at the American Museum and became one of the most widely published images in the science popularizations of the 1920s and after.[37] Diagrams designed for display at the American Museum were, in general, more likely than diagrams from most other sources to find their way into textbooks, newspaper and magazine articles, and books for the general public. Some of Knight's murals, painted before 1930, have remained on exhibit into the new century.[38]

Throughout the 1940s Knight continued to receive requests to reproduce the murals he had painted for the museum between 1900 and 1925, and several biologists have recently written of the profound impression Knight's restorations of the fossil past made on them as children.[39] Publication in popular books or museum display magnifies the influence and increases the longevity of such images. Their long-term impact is varied, complicated, and probably unpredictable. Furthermore, the designers of scientific illustrations may lose control over the presentation of their work. Osborn often lamented the use of museum images without permission, for example. And George Langford, the author of a book about Neanderthal life, *Pic the Weapon Maker*, written for children with the intention of presenting only the most scientifically accurate possible illustrations, found that the publisher insisted upon marketing the book wrapped in a dust jacket sporting a popularized and, in Langford's own opinion, thoroughly unscientific image of a caveman.[40] For a variety of reasons, then, visual images may become fixed in time, and both their intended messages and their hypothesis-testing functions may well be obscured. Expressing the author's meaning may be a complicated thing.

Finally, nonscientists who misinterpreted the intended messages of scientific diagrams were not always entirely mistaken. Creative misreading can tell us a good deal. Scientific images sometimes reveal extrascientific concerns on the part of scientists—assumptions, biases, or predilections of which they may be unaware but which may strike a chord with lay observers. Illustrations could convey messages that would be proscribed by the developing ideology of scientific objectivity. One historian of science has argued, for example, that nineteenth-century geological illustration subtly suggested that even though geology had demonstrated convincingly that the world was much older than strictly literal interpretations of the Bible might warrant, the geological history of the world still had an endpoint, with the present as the ultimate goal of the past—that, with the advent of humans, geological history had come to an end.[41] Illustrations of human evolution in the 1920s would often, without their authors saying it in so many words, imply something similar. And they would sometimes reflect origins in traditions of thought that might compromise their meanings.

In the 1920s, scientific diagrams often reflected ambivalence in the thinking of some evolutionists; scientists' diagrams sometimes included potentially confusing mixed messages. Biologists did not agree on the mechanisms of evolution, the pattern of the history of human evolution, the proper role of scientists in public controversy, or even the boundaries of science. Cultural preoccupations infused the conflicts among scientists. Questions about the mechanism of evolution, for example, were linked to concerns about determinism and human will. And the image of evolution as a neat, frictionless progress toward a goal in which inferior forms yielded to superiors reinforced beliefs that differences among humans—notably race—implied a hierarchy in which some were inferior to others.

Occupying positions of cultural prestige, some scientists felt compelled to take public stances in the evolution debates and did so actively. The visual images they published often made the extrascientific concerns and ambiguities in their thinking strikingly evident, and anti-evolutionists like William Jennings Bryan were quick to seize upon ambiguity in evolutionary metaphors.

For many observers, it also seemed that the objectivity of scientists who participated in the public debate about evolution had been compromised by their commitments to agendas beyond the boundaries of science itself. This was especially so in the case of those scientists, including Osborn, who participated in the eugenics movement advocating the "selective breeding" of humans. Osborn was a leader in this movement. The infamous Hall of the Age of Man has been criticized by historians who argue that it reflected Osborn's strong belief in racial and ethnic hierarchy; Osborn directed his staff to work hard to complete it in time to be used as a showcase when the museum hosted an International Eugenics Congress in 1921. A significant number of prominent scientists participated in this movement, including many who also took part in the evolution debate of the decade. But others objected sharply to Osborn's advocacy of eugenics and especially to his use of museum resources and prestige in service of the movement. Osborn drew bitter criticism for contributing the foreword to an extremely controversial eugenicist tract by his friend Madison Grant, *The Passing of the Great Race*. Like Osborn, Grant was a committed conservationist, and he was a trustee of the New York Zoological Society, which ran the Bronx Zoo. Unlike Osborn, however, Grant was no scientist, and many people criticized Osborn for sponsoring Grant and offering, by implication, scientific sanction for his work.

The boundaries of science could sometimes seem elastic in this decade when the disciplinary matrix had grown so complex. As a museum scientist and educator, Osborn extolled an interdisciplinary approach to paleontology, but in some of his work of the 1920s, other scientists thought they perceived far more speculation than modern definitions of science would allow. Definitions of the boundaries of science were very much at issue.

In 1925, in an attempt at an interdisciplinary synthesis that he hoped would suggest a new approach to explaining some of the mysteries of evolution, Osborn published a book called *The Origin and Evolution of Life: On the Theory of Action, Reaction and Interaction of Energy*. Acknowledging that much remained to be learned about the mechanisms of evolution, Osborn's introduction expressed a hope that the book would offer "some of the initial steps toward an Energy conception of Evolution and an energy concept of Heredity and away from the matter and form conceptions which have prevailed for over a century."[42] Much of the text was couched in the terms of physics, and many reviewers found it puzzling. Experimental scientists, especially the physiologist Jacques Loeb, and Osborn's Columbia colleague Thomas Hunt Morgan thought that the book was more metaphysical than scientific, an attempt to use the language of science to reject materialism or to reinstate the vitalist theory that neither life nor evolution could be entirely explained without reference to "forces" without physical substance. These were ideas that Morgan and Loeb considered not only unscientific but mystical.[43]

Selling Evolution to the Masses: Scientists and Publicity

In December 1925 the *New York World* published a review by the anthropologist George A. Dorsey of a book called *The Chain of Life*. "Another attempt to sell evolution to the masses," Dorsey complained, adding, "It can't be done. . . . Evolution is not a subject of 'universal interest,' as the jacket claims. . . . Nor can any book on evolution be 'thoroughly popular,' not even"—and here Dorsey came, perhaps, to the real source of his irritation—"if written by the wife of Henry Fairfield Osborn."[44]

Dorsey's annoyance with *The Chain of Life* was directed not so much at Lucretia Perry Osborn, the author of the book, as at the publisher and at the author's husband. The publisher's inflated claims for the book reflected confusion about the limits and purposes of science, as well as an unrealistic view of the definition of "popular" interest. "Why not a little more scientific honesty? To say, as do the publishers, that Mrs. Osborn's book 'explains the origin of life on earth,' is to make a false assertion. Science can not [*sic*] 'explain' anything. It is not the business of science to explain but to describe." Mrs. Osborn herself had not done a bad job of outlining evolution: "Considering the source of her materials [she] does it well. The reviewer's quarrel is primarily with the mystical energy complexes of her husband and the absurd claims made by her husband."

Quoting a particularly obscure passage from *The Chain of Life*, Dorsey exclaimed: "And the publishers call this a 'thoroughly popular book!' It would have been more popular had Mrs. Osborn told us more of the known facts of evolution and less of what the professor has said about life's mysteries and mystical energy complexes." Mrs. Osborn's book, perhaps because it was intended for a popular audience, invoked the same "mystical energy complexes" that Thomas Hunt Morgan had faulted in Osborn's *Origin and Evolution of Life*, and it emphatically reaffirmed Osborn's emphasis on "creative evolution."

Dorsey objected not only to the speculative flavor of Osborn's evolutionary ideas but also to his implied claims to authority beyond the boundaries of his own area of scientific expertise: "He is an 'authority' in certain restricted fields of paleontology; outside that field he is no more an 'authority' on evolution than Scopes of Dayton fame." Osborn's claim to amplified authority particularly galled Dorsey because it had important extrascientific implications and because these included Osborn's record on extrapolating from science to race theories: "On the 'Nordic' and other race questions he is the sponsor of Madison Grant!"[45]

Osborn and others in the eugenics movement treated moral and spiritual questions as if they lay securely within the purview of science. In a time of increasing attention to the distinction between science and nonscience, this inclination could make for inconsistency, and Osborn vacillated when it came to these issues. He was often invited to address audiences—including church congregations and Sunday school classes—on religious topics, and on principle he usually (but not always) declined, especially by the end of the decade, demurring that theology was not within his expertise as a scientist. In 1929 he wrote to the editor of a book about evolution: "For the rest of my life I expect to adhere closely to direct scientific observation and inductions which may be made from such observations but I shall not enter the field of philosophical interpretation."[46]

Where the field of philosophical speculation began, however, seemed in the eyes of some of his contemporaries, like Dorsey, to elude him. Both science and fame held special prestige, and like most scientists of renown, Osborn received frequent queries from people seeking general wisdom, advice, or help in getting their work published. Many lay people and aspiring writers asked to visit with him in order to discuss their ideas about such topics as "cosmic Christianity," the possibility of communicating with the dead, and "the true and great secret of life."[47]

Publishers, editors, and writers regularly sent requests for his ruminations upon extrascientific matters of general concern. Will Durant wrote to him to ask what he considered to be the meaning of life. Dorothy Giles, an associate editor at *McCall's*, requested his participation in "a symposium of statements from twenty of the world's most famous men and women in reply to the question: 'Do you believe in immortality? Why?'" Giles assured him that his words "would have tremendous import for the millions of McCall's readers."[48] Osborn had an assistant reply that he did not feel competent to participate in the symposium on immortality but that he would be glad to contribute an article "on the influence of fashion on human evolution."[49] The editor of the *Woman's Press*, the national magazine of the Young Women's Christian Association, wrote to ask his participation in "a symposium of outstanding and representative writers . . . on the meaning of the movement for world peace . . . for a larger spiritual unity of mankind in its search for a knowledge of God." She requested Osborn's contribution because she thought "such a symposium would be incomplete without a word from one of our leading scientists."[50] In this case, Osborn had his secretary respond that he did not "feel prepared" to contribute his thoughts on world peace but that he would be glad to contribute a short statement on the subject of "Conduct and Evolution."[51] For Osborn, world peace, immortality, and the meaning of life dwelled beyond the bounds of science, but the relationships of fashion, conduct, and evolution lay within his area of expertise as a paleontologist, as did government policy on immigration restriction.

Osborn's pronouncements in defense of evolution also regularly included assertions that science revealed religious truths. Though he claimed on principle to eschew any pretense to authority in matters theological, he routinely failed in practice to differentiate between scientific and philosophical inference nearly as carefully as did many of his critics, including those from outside the community of scientists. He saw himself as an objective scientist, taking care not to pontificate on extrascientific matters, but critics accused him frequently and with much justification of doing exactly that. Part of the complexity—or inconsistency—of his positions came from his ambivalence about his role not just as a "scientific man" but as a public man of science.[52]

Both scientists and nonscientist intellectuals worried about the responsibilities of scientists in public debates. An article in the *New York Evening Post*, reproduced in *Science* and excerpted and discussed in the *Literary Digest*, noted that "serious investigators" were understandably reluctant to publish their findings before presenting them in proper scientific contexts for criticism by colleagues. According to the *Post*: "In this day of horn-blowing, it is refreshing to come upon a group of men who are doing great things, yet who shun publicity as they would the plague." The editorialist opined, "There is a deep-seated prejudice against publicity . . . which plays its part in the suppression of information which should be presented to the public. One of the most important reasons for this feeling lies in the constant flood of 'claims' which second-rate scientists make of 'discoveries' and 'cures.'" But this scrupulousness could itself imply a certain peril: "The fear of publicity is carried too far when leading men are afraid to speak for publication . . . simply because they fear criticism from members of their own profession."[53]

The hesitancy of many scientists to engage in the evolution debate must have been reinforced by such pronouncements as the statement of the *Post* that "it is largely because of the reticence of the men best qualified to speak that those not nearly so well qualified occupy so much of the newspaper space devoted to science." According to the editorial, echoing a common sentiment, "the public never was so hungry for authentic information as it is to-day." But partly because of scientists' cautious reserve, newspapers turned to more sensationalistic sources. "It is also partly because of this reticence that publicists have conceived the notion that the public wants its science information jazzed up and distorted."

Scientists found themselves in something of a cultural bind. The professed tenets of their professional ethics included an obligation to educate the public, but the tenor of the time confused publicity with propaganda and publication with commercialism. Dorsey had commented on Mrs. Osborn's *The Chain of Life:* "It almost seems at times that she is primarily interested in selling the professor rather than evolution to the masses."[54]

Dorsey was no stranger to popularization. His own book, *Why We Behave Like Human Beings,* earned a place among the best sellers of the decade. Unlike many of his colleagues, however, he was firmly opposed to the use of biological language in support of racist and eugenic ideologies; for him, the "selling" of evolution was implicated in the use of the language of science to promote ideology.[55] But scientists who differed with him on the subject of race still shared his misgivings about selling science.

The theme of publicity, and ambivalence about it, wound its way continuously through the fugue of 1920s public discussion. Fundamentalists used publicity and public relations quite self-consciously, but with apparent trepidation about the delicacy of the balance required. In a report on a discussion about publicity during a meeting of the Chicago Area Council of the Methodist Episcopal Church, the head of the publicity department, the Reverend J. T. Brabner Smith, advised that "every church should spend money for advertising and purchasing space in the newspapers . . . and there should be cordial relationship and cooperation between the pulpit and the press." At the same meeting, however, Smith also complained, "Too many preachers are fond of and seek personal publicity."[56] Detractors of the Reverend John Roach Straton, believed to have been the model for Sinclair Lewis's Elmer Gantry, commonly accused him of being a "publicity seeker," a familiar phrase in the idiom of the decade. A former parishioner informed Osborn that Straton was "an agitator and a sensation seeker, the kind of fellow who is continually finding mare's nests . . . he would do almost anything to attract attention."[57] And many journalists would agree with the assessment of Charles Wood, who commented in the *Nation* that Billy Sunday had more to do with celebrity than theology: "It was his mannerisms, not his message, that made him good copy."[58]

Scientists had particular professional reasons for wariness about publicity. On the one hand, scientists after the Great War mounted organized efforts to institutionalize public education about and support for scientific research, and members of the American Association for the Advancement of Science called for the organization and similar scientific societies to form publicity departments. The founding of the Science Service in 1921 with the support of I. W. Scripps and the AAAS created a mechanism for generating stories about science and distributing them to newspapers across the country. The newly formed Science Service also marketed a pamphlet titled "Leaders in Science," which it described as "A Complete Photographic Library Comprising Photographs of 1200 Scientists . . . Ideal for Classroom or Home or Den."[59] On the other hand, the ascendant ideology of scientific objectivity and the specialization of disciplines, which made most subdisciplines increasingly arcane, militated against popularization. As David Starr Jordan, eminent zoologist and erstwhile president of Stanford University, wrote in a note to *Science,* "The greater the public interest in any branch of science, the more likely it is to attract the charlatan."[60]

Osborn himself was ambivalent about celebrity—at least for others—and he often voiced reluctance to have anything to do with any enterprise that might have a "commercial taint." It was all too easy for renown to slide into crass commercialism. And the effort to communicate with an unscientific audience in acceptably cautious scientific terms could be extremely trying, perhaps even futile. The much-discussed problem of the sensationalizing tendencies of newspapers exacerbated the difficulty. Complaining about the Hearst papers in particular, Princeton biologist Edwin Grant Conklin wrote to book editor Frances Mason, "I have several times written articles very carefully avoiding any sensation, and have been humiliated to find that they had been published with headlines that made me the target for criticism from all over the world."[61] Osborn and Conklin shared a concern for preserving their dignity, a preoccupation that reflected their sense of their professional status as "scientific men" and of the proper role they should adopt as public men of science. This was a concern commonly expressed by scientists of the time, at least partly related to their claims of disinterest, professional distance, and objectivity.

But Osborn also manipulated public relations effectively in building his remarkable museum. When the aspiring young explorer-naturalist Roy Chapman Andrews came to the museum from Wisconsin, for example, Osborn soon recognized that the young man's assets included a flair for publicity, supported by a gift for storytelling and a charismatic personality that made him both an extremely popular lecturer and a brilliant fund-raiser. Andrews declared his passion for expeditions—a passion matched by that of the newspapers—and understood well how perfectly a series of fossil-hunting trips to Mongolia would suit Osborn's research ambitions. Putting his charm and Osborn's connections to work, he made even the financing of the operation into a great story, claiming to believe, as he put it in his lively account, that successful businessmen were "adventurers at heart." The series of Mongolian expeditions Andrews led on behalf of the museum became wildly popular with newspaper editors and, judging by Andrews's success as a public speaker and popular writer, with the public as well.[62]

Museum publicity and newspapers collaborated to sell the romance of expeditions. In addition to helping to promote Andrews's expeditions, Osborn sponsored the African exploration-and-collecting trips of Carl Akeley, a naturalist who collected specimens and mounted exhibits for the museum and who published popular books about his expeditions. The biologist, writer, and Bronx Zoo curator William Beebe also published popular accounts of his biology fieldwork; his books and magazine essays were admired not only by the public but by the literary establishment at magazines including the New Yorker. Through his position at the head of the New York Zoological Society, Osborn helped to support Beebe's work and to cement close ties between the museum and the zoo in general. The museum's popular journal, Natural History, devoted an issue to "the romance of natural history." Although he was not much of a "field man" himself, Osborn was an imposing figure, well able to convey his enthusiasm for his work, and when he did visit field sites, he made sure to construct stories of the momentousness of his fossil finds and to have his photograph taken at such sites. An adulatory interview with him published in the World's Work gushed, "He is living proof that science is real romance."[63] In an interview for the introduction to a book called To the Ends of the World and Back, about the scientific fieldwork sponsored by the American Museum, he articulated in imaginative terms the exciting implications of Andrews's work in Mongolia: "I have a theory that this ancient terrain was the homeland and dispersal center of most of the mammalian life of the globe, including man. Somewhere hereabout may be the Garden of Eden!"[64] Comparison of the wording and details of many of his letters with interviews about him reveals a pattern of repetition that suggests a formidable ability to control and shape interviews, at least when dealing with friendly journalists, which he was careful to do.

Museum workers sometimes poked fun at Osborn's instinct for publicity, shared with the charismatic Andrews. A cartoon by one of the staff artists portrayed the scientific staff peering into a large box with the American Museum address printed on the side and "Roy Chapman Andrews, Mongolia" on the return address. A figure in evening dress leaned eagerly into the box. This was clearly intended to be Andrews, who had spent, he once wrote, nearly every evening in white tie and tails while drumming up support among Osborn's society friends for the Asian expeditions. A clearly recognizable cartoon version of Osborn gestured eagerly out of frame, and the caption read: "Camera! Camera!"

Self-consciously patrician, supremely confident—many have said arrogant—and endowed with an unremitting paternalist impulse toward subordinates, Osborn evoked complex and mixed feelings on the part of associates and employ-

"Camera! Camera!" This cartoon, drawn by a museum artist, depicts Osborn calling for a photographer to record the unpacking of fossil treasures from one of Roy Chapman Andrews's Asian expeditions. The figure in tails is probably Andrews, who, in raising funds for the expeditions, became a favorite of New York high society. The cartoon was found in the Christmas Luncheon folder in the Vertebrate Paleontology Library at the American Museum of Natural History. American Museum of Natural History Vertebrate Paleontology Library

ees. The anthropologist Margaret Mead, who worked at the museum beginning in the late 1920s and who voiced profound dislike for Osborn's racial attitudes, nevertheless later told an interviewer that she remembered Osborn as "a magnificent old devil." The museum, Mead added, "was his dream, and he built it. He was arbitrary and opinionated, but I got my first view of many things from his books and the exhibits he sponsored. . . . We would never have had the Museum without him."[65]

Osborn was very attentive to publicity in building the museum. The museum's publicity department grew in both absolute and relative size under his leadership.[66] He made a point of knowing which newspapers best served his various purposes and maintaining good relations with them. The museum's trustees and members generally read the *New York Times* and the *Herald-Tribune,* he wrote in a memo, but "papers like the World and the American should be flooded with news because their political support is extremely valuable." He knew, and made sure the museum's publicists knew, which papers New York's mayor and other powerful politicians in the city followed.[67] Even editors who disagreed with him in significant ways treated him with cautious respect. When he had a tiff with the editor of *Scientific American,* he and the editor were both very careful to make clear that maintaining friendly relations mattered a good deal to both of them. He closely monitored the newspaper articles that arrived from clipping services. Many of the newspaper clippings preserved in the archives of the museum include notes in Osborn's distinctive hand. He kept close track of newspaper references to the museum—often circling or underlining them.

He did not underline his own name, but this may not have been a sign of diffidence—he identified himself with the museum. And he made a stated principle of tasteful modesty, at least in theory. In a letter to his daughter Josephine he praised his wife, Loulu, who in her volunteer work, he boasted, accomplished a great deal behind the scenes without having her name appear—a well-bred aristocratic reticence he admired and which he claimed, rather improbably, he aspired to emulate himself at the museum.[68]

Osborn had grown up with a strong sense of his own rightful place as a man of influence—according to *Time* magazine, in Fairfield, Connecticut, where he was born, "an Osborn was decidedly an institution"—and made adept use of his position and his confidence in its naturalness.[69] But he found the appearance of bids for fame or publicity in others crass, vulgar, and generally distasteful. This could well be attributed in part to his consciousness of himself as a member of the patrician class.

But unlike most members of his social class, he became a scientist. And scientists, too, had a sense of their proper public dignity—part of a professional creed having to do with objectivity. Scientists who engaged in efforts to communicate with a larger public expressed continuing concern that such efforts would impair their credibility among colleagues. Even as established a figure as Edwin Grant Conklin, clearly collegial and well liked in addition to being well published, voiced such fears. And Roy Chapman Andrews, whose status as a scientist did not match his reputation as an explorer and public speaker, complained often and ruefully that the newspapers' insistence on labeling his famed series of Central Asiatic Expeditions in the early 1920s as "missing link expeditions" could cause the scientific community to look at him askance. Osborn's own self-confidence was always legendary, and he was capable of stunning hubris, but even he suffered occasional qualms and continuing ambivalence about the distinction between educating the public, drumming up support for the museum, and promoting himself. The importance of avoiding "any commercial taint" appeared regularly in his memos. His trepidation about publicity would intensify with the crescendo of the evolution debates, and with good reason. In attempting to balance his ardent defense of evolution, his promotion of the museum, his instinct for publicity, and his commitment to evolutionary theories of racial hierarchy with the dictates of scientific protocols, he ran into trouble.

A small number of scientists could command the attention of the news media, and their views would be those most visible to the public; they belonged to the category one analyst of scientific communication has labeled "visible scientists."[70] The "public" would also include many teachers across the country, who, uneasy about the disagreements among scientists about the details of evolutionary theory, would seek advice from better-known scientists.

Osborn, among the most visible, exercised a disproportionate influence for a variety of reasons. He holds special interest for historians, in part because of the ways in which he was *not* representative of scientists as a whole. Biology, in particular, had splintered into many pieces by the 1920s. Biological disciplines had proliferated, so that disciplinary infrastructures—research organizations, funding structures, and professional communities—had multiplied. Idioms and vocabularies had also multiplied and begun to diverge, often disrupting communication among scientists even in conceptually related fields. In addition, the very definition of what it meant to be scientific lost its status as a given. It was a difficult time for scientists to find themselves faced with a noisy public controversy.

The Scopes trial did not occur suddenly or unexpectedly; it signaled a crescendo in a longer public conversation. Anti-evolutionism gained momentum in the wake of the Great War, when William Jennings Bryan was drawn to a previously less well-publicized creationist movement. By 1922 it had become a crusade. Bryan's popularity and his ability to generate publicity acted as catalysts for the movement. Willa Cather's famous dictum that "the world broke in half in 1922 or thereabouts" resonates for science. Most newspapers in 1921 included relatively little discussion of science. In 1922 the evolution debate changed that situation dramatically.

CHAPTER THREE

Nineteen Twenty–two or Thereabouts

On Friday, February 24, 1922, Henry Fairfield Osborn recorded in his private diary: "Bryan in the Times!!!!" William Jennings Bryan's enlistment in the anti-evolution cause, and the publicity it generated, galvanized scientists—at least some of them. On hearing from the *New York Times* that Bryan's article would be featured on the following Sunday, Osborn flew into action. Invited by the paper of record to compose a response to Bryan's forthcoming article, he immediately (Osborn's letters and diaries show that "immediately" was among his favorite words, second only, perhaps, to "important") mobilized one of his secretaries and devoted a day to dictating his reply, which he delivered to the office of the *Times* in person.[1] In a book of essays about the controversy, Osborn pinpointed this as the moment when he first felt compelled to engage in the evolution debate: "Early in the year 1922 I was suddenly aroused from my reposeful researches in palaeontology" by Bryan's article in the *New York Times*.[2] The editors of the *Times* also invited responses to Bryan by the Princeton biologist Edwin Grant Conklin and the New York minister Harry Emerson Fosdick. Conklin's essay appeared on the same day as Osborn's, Fosdick's on the following Sunday. The three respondents made significantly similar cases on behalf of evolution. Fosdick was among the most renowned of liberal Protestant ministers, a focus of controversy and frequent contributor to magazines and newspapers. Osborn and Conklin would become leaders in the attempts by scientists to counter attacks by anti-evolutionists, and their role in the debate—especially that of Osborn, whose museum supplied so many images of evolutionary themes—illuminates significant differences between the responses of scientists and those of secular nonscientist defenders of evolution.

Initially scientists were caught off guard; although they continued to debate how evolution worked, they had assumed the question of its reality to have been settled. Though the renewed controversy surprised them, a number of them entered the public discussion, debating Bryan and other anti-evolutionists in newspapers, in books and magazines, in public speeches, and on the radio.

Many of the scientists most prominent in the controversy of the twenties had, like Osborn and Conklin, been young men in the midst of the nineteenth-century debates over Darwinism. They had successfully wrestled with the implications of evolution for their own religious faith. Most of them imagined that the issue had been more or less resolved well before the 1920s. James McKeen Cattell, the editor of *Science* and *Scientific Monthly*, wrote to Osborn in 1922: "We might have supposed that evolution had proceeded far enough to have made extinct any species that objected to the teaching of evolution in the public schools."[3] Winterton C. Curtis, who in 1925 traveled to Dayton, Tennessee, to testify as an expert witness at the Scopes trial, later remembered wishing, early in the twentieth century, that he had been old enough to participate in the "wars" of the nineteenth. "The fight for Evolution seemed so completely won even in the Gay Nineties that I recall, as a college student, almost wishing I had been born a generation earlier when the scrap was on." He thought he had missed out.[4]

Religious conservatives had for many years adjusted to the idea of an ancient earth, and as Curtis remembered, it had seemed by the end of the nineteenth century that the evolution controversy had faded. Even William Jennings Bryan had said in 1905 that he did not object if other people believed in evolution, at least in the evolution of animals.[5] The anti-evolution movement of the 1920s drew much of its momentum from the deepening of a theological rift within American Protestantism, between theological modernists and traditionalists, and from the myriad cultural tensions that overtook Americans in the 1920s.[6] The theological tensions manifested themselves in the publication of the pamphlets called *The Fundamentals*, published from 1910 to 1915. But even *The Fundamentals* had not unilaterally opposed evolution. The anti-evolution movement reached an inflection point with the formation of the World's Christian Fundamentals Association in 1919 and the enlistment of William Jennings Bryan. Bryan's involvement, energy, and eloquence helped anti-evolutionism to crescendo and become fervent in the twenties, in large part because anti-evolutionists struck culturally resonant chords.[7]

Many scientists attributed the resurgence of anti-evolution sentiment to the effects of the war. Conklin, for example, labeling it "this new Inquisition," blamed it on "the emotionalism let loose by the war."[8] Possible hyperbole aside, there was considerable truth in this assessment. Bryan began to rethink his earlier stance partially in response to his reading of Vernon Kellogg's *Headquarters Nights*.[9] Kellogg, a zoologist, had been stationed in Germany after the war, and his memoir had recounted conversations with German officers who seemed to collapse Darwinism into a crude social Darwinism. After reading Kellogg, Bryan came to associate evolutionism, in the wake of the Great War, with German militarism, with a not-very-subtle understanding of Nietzsche—in vogue in American literary circles at the time and introduced to a wider audience especially by the journalist H. L. Mencken—and with a doctrine he blamed for the war and associated with Germans, pithily summed up, as he was wont to do with complex ideas, as the notion that "might makes right."[10]

The war was not the only disorienting force, however. Bryan had also read a study by the Bryn Mawr psychologist James Leuba purporting to demonstrate that with increasing exposure to science students tended to stray from or even discard their childhood religious faith.[11] Many other people also recognized the spread of public education, the increasingly urban influences in American life, and the new roles of women as causes of psychological turbulence, disruption, and unease. And analysts regularly mentioned the spread of changing models of psychology and new patterns of consumerism, along with the manifold influences of technology, especially the automobile, as forces that had upended old customs and values.[12]

Beneath the bright surface of American culture in the 1920s, anxieties simmered. Whatever inconsistencies wrestled beneath the surface of Progressivism, the Progressives were largely united in being enamored of science. Science had been the byword of proponents of that great Progressive Era shibboleth efficiency. The social sciences were an invention of the Progressive Era, when newly minted Ph.D.'s, armed with questionnaires, had fanned out across American cities to collect information that would be used, scientifically, to solve the problems of the modern world.

By the 1920s, however, some people had begun to suspect that science might be one of the more corrosive of—in Walter Lippmann's memorable phrase—the "acids of modernity."[13] The jurist Roscoe Pound, also trained in botany and an early science enthusiast, told a graduating class at Wellesley in 1929 that science "has been teaching distrust of itself."[14] By the end of the decade, Joseph Wood Krutch's bleak evaluation of the psychological and spiritual impact of science, *The Modern Temper,* published in 1929 and partially serialized in the *Atlantic Monthly,* would be a best seller. Science, Krutch suggested, had cast us adrift in a meaningless universe. "Illusions have been lost one by one."[15] Krutch blamed science for having relentlessly exposed the essentially tragic predicament of humans, claiming that, especially since Darwin, god and poetry had lost their power.[16] When, at a meeting of the British Association for the Advancement of Science in 1927, the bishop of Ripon called for a ten-year moratorium on all scientific research, many people took his suggestion, widely reported in U.S. media, perfectly seriously. The *Chicago Evening Post,* for example, suggested that many people shared the bishop's sentiments, since "science has been leading rather a giddy chase for the last two or three decades."[17] So although many Progressives—especially those who defended evolution in the press—associated science with tolerance and enlightenment, by the end of the decade uneasiness about its effects was not confined to religious conservatives.

Anxieties over the perceived spread of "materialism" had long been implicit in debates about science and religion, as had uneasiness about the role Darwin had assigned to "chance" processes. The range of this literature shows that these things were widely contested, or at least that people fretted about them at a variety of levels. Two late nineteenth-century books, Andrew Dickson White's *A History of the Warfare of Science with Theology in Christendom* and John William Draper's *History of the Conflict between Religion and Science,* enjoyed new and widespread influence during the 1920s. Simplified historical accounts of what they portrayed as a long-term state of "warfare" between science and religion, these two books influenced many people and supplied some of the most often encountered terms and metaphors shaping the vocabulary of the evolution controversy. Scientists—like Conklin—often referred to the new fundamentalist crusade as a curious "recrudescence" of the anti-evolutionism of the late nineteenth century—and, in the longer term, of the repression of science by the Inquisition. Over and over again, defenders of evolution would refer to the efforts of anti-evolutionists as a "new Inquisition," invoking the names of Galileo and Copernicus as a kind of evocative code.[18]

The history of science had its beginnings as an academic discipline in this period and made frequent appearances in journals written by and for scientists, especially *Science* and *Scientific Monthly.* The history of science articles in these journals, usually written by scientists, former scientists, and science boosters, celebrated an inexorable progress. Yet at the same time philosophers of science, and many nonscientist intellectuals, had begun to adopt a far more skeptical epistemological position about scientific knowledge. Evolution was not an issue in isolation. It was also not the only science that might stir uneasiness. Psychology certainly did. As the historian Jon Roberts has shown, new theories of psychology, especially Freudian psychology, unsettled many Christians. Popularized versions of Freud and of behaviorism came into vogue as well.[19] The much-discussed and little-understood "Einstein theory" provoked theological and epistemological anxiety, though mostly of a vaguely literary sort.

Einstein's theory of relativity was big news in the 1920s. The media prominently reported the confirmation of relativity by the eclipse of 1919, and a period of widespread public fascination followed. Einstein's name has been said to have held an "incantatory power" in the twenties.[20] Magazines such as the *Atlantic,*

Harper's, and the *New Republic* carried articles speculating about the meaning for scientific method of what was then called the "Einstein theory" and about the ability of the public to have confidence in its understanding of scientific issues in the wake of Einstein, the uncertainty principle associated with Heisenberg, quantum theory, and non-Euclidean geometry. Even scientists who were not physicists were regularly asked to elucidate the theory of relativity. Osborn, a paleontologist, received frequent requests from schools and church groups to do so, and his wife, Loulu, wrote to the physicist Robert Millikan, a family friend, asking him to explain quantum theory to her. A 1923 movie, *Einstein's Theory of Relativity,* shows how easily nonscientists might make the leap from loose metaphor to epistemological generalization. The film attempted an explanation accessible to the general public but, not surprisingly, relied a good deal on metaphor and vaguely applied "common sense," conflating the distinction between the common understanding of the word *relative* (as in "everything's relative") and Einstein's technical meaning of the term *relativity.*[21]

Relativity might have generated religious protest if more people had been able to follow it in any detail, but because it was so arcane, it remained for the most part an object of curiosity, a little like a meteor or a visitor from some other world. This state of things was probably helped along by Einstein's iconic status—newspapers covered his 1921 visit to the United States prominently, and his face came to be one of the most famous of celebrity faces, the archetypal face of science in popular culture. And like its progenitor, the "Einstein theory," as it was then called, was itself famous. Many intellectuals found relativity unsettling; it attracted considerable press in relatively sophisticated magazines like the *Atlantic* and *Harper's,* and it undoubtedly caused some consternation as well as fascination among readers of those magazines. But it did not draw much fundamentalist wrath.[22]

Evolution, however, was very much of this world and was not, after all, so intellectually inapproachable. Indeed, many fundamentalists learned at least a good deal of the vocabulary of evolutionary theory, even if they rejected its logic. Evolution was intimately tied to theological modernism, the primary focus of fundamentalist concern. Theological modernists made much of the compatibility of their theology with science, and in turn, many biologists embraced modernist forms of Christianity. And, especially, evolution came with a long tradition of images and associations that made it potentially suspect. As in the 1860s and 1870s, images of brutish monkeys, apes, cavemen, and missing links crowded the common visual vocabulary. On all sides of the debate people mobilized phrases such as "from the slime," the "tadpole theory," and the "jungle theory" as part of an arsenal of humorous lexicon, and these terms carried long and complex connotations. Moreover, rifts among scientists over the mechanisms and details of evolutionary theory made it vulnerable to attack from outside. As Conklin suggested, the resurfacing of opposition to evolution came "partly from the fact that uncertainty among scientists as to the causes of evolution has been interpreted by many non-scientific persons as throwing doubt upon its truth."[23]

Scientists debated the mechanism of evolution among themselves, and anti-evolutionists followed these discussions, seizing on reports of dissension over natural selection among the scientific ranks. Although scientists were agreed as to the truth of evolution, and very concerned about threats to academic freedom, they did not agree on the mechanism and the details of evolutionary theory; nor did they all share a single perspective on the relationship of science to religion or on the important issue of scientific participation in the debate. Scientists in the 1920s disagreed not only about evolutionary theory but also about the proper response among "scientific men," as they generally called themselves, to the evolution controversy. They were divided in addition along disciplinary lines, by generation, and by location within the field. Newspaper coverage of the debate did not reflect these distinctions clearly. The culture of professional science and the culture of the media combined to ensure that the public would not be offered anything like a cross section of scientific opinion.

Challenges to the validity of evolutionary theory seemed to scientists to be tantamount to denials of reason itself. Scientists had fun ridiculing creationists, in an often distinctly elitist tone. But many of them were also genuinely alarmed. No bona fide scientist would entertain the suggestion that the pattern of evolution written in the fossil record could be an illusion; nor did they dispute any of the evidence from anatomy, biogeography, or embryology confirming the pattern of evolution. The debates among scientists had to do not with whether evolution had occurred but with how it had.

But how to make this clear to the public at a moment when scientists themselves were in disagreement as to the details of evolutionary theory was not so evident. Many popular accounts conflated the terms *Darwinism* and *evolution*, complicating the task of scientists who tried to communicate with a general public. And though many scientists avoided the public controversy altogether, others saw the need to explain evolution to the public as a duty.

Although there were scientists who wrote and lectured for the public before the 1920s, the evolution controversy of that decade contributed to a significant crescendo of popular writing about science. Popular science, disseminated in museums, magazines, books, newspapers, and Chautauqua lectures, thrived in the late nineteenth century. By around 1900, however, it languished. After the

Great War, scientists again took up the cause of educating the public. In part, they understood that, given the changing institutional organization of science, money for research would increasingly be supplied, in one form or another, by taxpayers. After the Great War, then, scientists renewed their efforts both to organize scientific research and to publicize its value. The 1921 establishment of the Science Service, with the support of the American Association for the Advancement of Science (AAAS), was a part of this effort. Scientists commented often in the pages of the AAAS journal *Science* on the importance of communicating with the public. In one typical essay a Cornell University scientist, Frank E. E. Germann, exhorted colleagues, "We must learn to advertise our wares," adding that "such books as 'Creative Chemistry' have done wonders in this regard, and Science Service is carrying the work on in an admirable way, but we must all do our share to help."[24] Popularizations of chemistry followed the war especially, but the decade of the twenties also saw an efflorescence of science journalism in general, including attempts by scientists themselves to communicate with the public.[25] The surfacing of the evolution controversy added considerable volume to the effort.

A prolific group of scientists and science popularizers worked to "sell" evolution to the public. The genre might be identified by the title of one of the more widely advertised of their books, *Evolution for John Doe*, by Henshaw Ward. During the 1920s proselytizers for evolution offered the public many works belonging to the category of "outline of" books in vogue in the decade. Books on evolution for a lay audience were so numerous that for several weeks during the summer of the Scopes trial, Brentano's Bookstore in New York devoted an entire window display to them. Reviewers commented regularly on the proliferation of popular explanations of evolution. Such books sold well enough that they continued to appear throughout the decade; a 1929 column in the *World's Work* called "Books for Babbitt" remarked on their continuing popularity, and several of them were included on a list of best-selling books as late as 1929.[26] According to an article in the *New York Times*, Edwin Grant Conklin's 1915 *Science and Heredity* had, by January 1926, sold twenty-five thousand copies in thirteen printings since publication.[27] Osborn's books also sold well, and George Dorsey's *Why We Behave Like Human Beings* ranked among the decade's best sellers.

The efforts of scientists and science popularizers to spread the good word reflected the climate of a decade that gave life to an expanding literature of popularization and self-improvement. Works in this category included the famous "Five-Foot Shelf of Dr. Eliot"—as the Harvard Classics were called in advertisements—and extremely popular compendia such as H. G. Wells's *Outline of History*, similar syntheses by Hendrik Willem Van Loon, and books recommended and marketed by the Literary Guild and the Book of the Month Club, both inventions of the decade.[28]

Although the decade of the 1920s spawned a remarkable increase in science popularization, its beginnings were patchy. Science Service notwithstanding, a survey of newspaper coverage of science reporting in 1920 and 1921 reveals a relative lack of serious discussion of science. There were exceptions, of course—some newspapers, such as the *New York Times*, covered science more seriously than others, and discussions of evolution appeared in books and magazines earlier and in greater detail than in newspapers—but in general, newspaper stories of those years covering science often dealt more with technology than with science; with superstition; with the comings and goings of such notables as Einstein and Madame Curie, both of whom made well-publicized visits to the United States in 1921; or with the pronouncements of Thomas Edison and Luther Burbank, the premier American science celebrities of the moment. New York newspapers frequently mentioned events at the American Museum of Natural History and the Bronx Zoo, and newspapers enthusiastically covered the exploits of scientific explorations, such as those of the popular writer and New York Zoological Society ornithologist William Beebe and the museum taxidermist, sculptor, and explorer Carl Akeley. Roy Chapman Andrews's famous Mongolian expeditions, sponsored by the American Museum, were particular favorites of the media, as was Andrews himself, and newspapers celebrated his trips in romantic terms, often in articles accompanied by dramatic photographs of the expeditions in their Gobi Desert setting. In general, however, journalists focused as much on the romance and color of such efforts as on their scientific content.

Scientists—at Least Some of Them—Join the Controversy

Discussion of scientific ideas took on renewed and sudden prominence with three events in late 1921 and early 1922. The 1922 exchange including Bryan, Osborn, Conklin, and Fosdick in the *New York Times* came just as the controversy began to heat up. Anti-evolutionism surfaced in newspapers generally with Bryan's vigorous engagement in the movement. Bryan, who had three times run for president and who remained enormously popular as an eloquent public speaker, could bring extraordinary publicity to any cause he embraced, and he embraced anti-evolutionism fervently. Two other events of late 1921 and early 1922 worried scientists as well. First, Bryan's support seemed to have helped tip the balance in favor of anti-evolution legislation in the state of Kentucky. Ultimately defeated, the measure proposed seemed a harbinger of ominous things to come, and scientists' journals and correspondence noted their concern.[29] The skirmish in Kentucky, everyone knew, drawing on the popular warfare metaphor, was only the first in what promised to be an extended campaign.

The third event that unsettled scientists followed from an address at the Toronto meeting of the AAAS late in 1921 and the responses of creationists. The British geneticist William Bateson, who had long believed that the rediscovery of Gregor Mendel's work on patterns of inheritance vitiated Darwin's theory of natural selection, delivered an address reiterating his conviction that science had failed to demonstrate the action and efficacy of Darwin's mechanism. Bateson explicitly denied that he was rejecting evolution itself: "Let us then proclaim in precise and unmistakable language that our faith in Evolution is unshaken." He continued, "The difficulties which weigh upon the professional biologist need not trouble the layman. Our doubts are not as to the reality of evolution, but as to the origin of 'species,' a technical, almost domestic, problem."[30]

But lay people did take notice—because Bryan did. Bryan had actually joined the AAAS, a cause of some levity among scientist members of the organization. A note in the association's journal *Science* joked that Bryan had conveniently failed to enclose his membership dues; *Time* magazine also made note of this incident, claiming that Bryan had enclosed an unsigned check. The laughter was uneasy among scientists, however; Bryan seemed to biologists to be monitoring their scientific pronouncements for ammunition to use against them. He had joined the AAAS to make a point: ordinary people were capable of understanding and criticizing science just as they were capable of understanding the Bible without the intercession of learned theologians.[31] This was not a point that scientists, in Bryan's case especially and in the face of attacks by fundamentalists, were necessarily willing to concede.

Bateson's remarks revealed tensions not far beneath the surface among the scientific community. Bateson had explicitly stated that he was not attacking evolution but only the Darwinian mechanism of natural selection, and he had acknowledged in his address that anti-evolutionists had capitalized on disagreements among scientists. Nonetheless, he insisted, his remarks at the AAAS meeting were a purely internal matter among scientists, a technical thing, of no concern to the public. But the public expressed concern. And scientists fretted among themselves about Bateson's remarks.

Osborn was furious. "Bateson has done a good deal of harm here," he wrote to Sir Arthur Shipley; to Conklin he hinted that he suspected Bateson was looking for publicity.[32] He thought such remarks were irresponsible because they gave ammunition to the fundamentalists, who, he predicted, would make noisy rhetorical use of them. In a note to the Columbia University biologist E. B. Wilson, he complained that Bateson had been "listened to with open mouth like a prophet." Reiterating the impact of the incident, he added, "He has seriously shaken thousands of people."[33] Osborn wrote an essay for *Science* maintaining that Bateson was behind the times and that he did not understand evolution because he was too much of a specialist to appreciate the big picture. The relationship between Osborn and Bateson had been adversarial for some time because Bateson, as an ardent proponent of experimentalism, dismissed Osborn's kind of science as outmoded. Osborn warned his colleagues that narrow specialization would lead only to scientific dead ends. Now he added the criticism that public statements on matters of narrow scientific interest added fuel to the fundamentalist fire: "The separation between the laboratory men and the systematists already imperils the work, I might almost say the sanity of both. . . . When such confessions are made the enemies of science see their chance."[34] Bryan, he noted, had made adroit rhetorical use, in his *New York Times* salvo, of the ammunition Bateson supplied: "The force of the article lay in his clever citation of the wide differences of opinion existing among evolutionists as to the *causes* of evolution, and especially his citation of the hopeless attitude of the distinguished Cambridge evolutionist, William Bateson."[35] No matter how difficult questions about evolutionary mechanisms might be, it was imperative that scientists exercise caution "especially in our popular addresses, which are eagerly read by the public. . . . When we give voice to our own opinions we should clearly indicate them as our opinions and not as facts."[36] Cautionary notes of this kind would echo more and more frequently as the controversy continued, and from all directions.

Bateson denied having spoken irresponsibly. He wrote to Winterton Curtis: "I have looked through my Toronto address again. I see nothing in it which can be construed as expressing doubt as to the main fact of Evolution." He implied that if there was blame to be placed, it should be directed not at him but at some of his colleagues. At Toronto, he said, he had been addressing an audience of professionals and was trying "to call the attention of my colleagues to the loose thinking and unproven assumptions which pass current as to the actual processes of evolution."[37]

The energetic reactions to Bateson's address at Toronto demonstrated the awkwardness for scientists of the timing of the fundamentalist anti-evolution initiatives. As Conklin predicted, anti-evolutionists would persistently conflate Darwinism and evolution while scientists sought to make that crucial distinction clear. When scientists used the term *Darwinism*, they customarily referred to natural selection, but in the public debates Darwin had attained iconic status. Darwin's name—and even his very familiar photograph—had become inextricably associated with anything having to do with evolution, so that a challenge to Darwin's theory on the part of any scientist could potentially confuse the interested public. And the public soon became more interested. Conklin wrote to David Starr Jordan, a paleontologist and former president of Stanford University: "I have had dozens if not hundreds of letters from men all over the country assuring me on the authority of Bateson that evolution was a back number."[38] In 1926 the zoologist Edwin Linton remembered that many people had perceived the controversy ignited by Bateson as "the Sarajevo shot which precipitated war."[39]

Many scientists responded with concern to the Kentucky initiatives, the public response to Bateson, and Bryan's involvement. Some of them called on their professional organization, the AAAS, to take measures. One especially long and persuasive letter from a member, Ira D. Cardiff, of Yakima, Washington, circulated among officers of the organization. Dated March 1922, Cardiff's letter to Burton E. Livingston, permanent secretary of the organization, went also to Conklin at Princeton and to Osborn at the American Museum. Referring to the situation in Kentucky, which he attributed to "the pernicious activity of a more or less discredited politician," Cardiff took issue with the tendency he perceived on the part of the newspapers to minimize the significance of the anti-evolution movement. Cardiff cited a recent editorial in the *New York World* belittling Bryan and assuring readers that "no harm can come from his efforts." Cardiff disagreed. Too many people were susceptible to such "propaganda" as Bryan's.[40]

Cardiff blamed his scientific colleagues. Scientific men, he wrote, generally responded that "evolution is an accepted fact, and it needs no more special attention than the teaching of the rotundity of the earth," but they failed to appreciate the lack of science education on the part of the public. Cardiff contrasted the contemporary state of affairs with that of his own youth. "I recall as a boy, that it was not an uncommon thing for noted scientific men to be giving public and popular lectures over the country upon the subject of evolution. This is a practice unheard of at present, which to my mind is unfortunate."

Evolution, Cardiff claimed, was "probably the greatest generalization the human mind has ever made and its establishment as a working scientific principle has probably contributed more toward the general mental and physical improvement of humanity than any other single factor in civilized society." Yet it was relatively little understood, partly because of its difficulty and partly because "it is constantly and systematically being misrepresented and misinterpreted by certain classes of selfish propagandists." Although "the Scientist, of course, should not be put to the necessity of being a propagandist," the need for action was clear. Scientists "have only themselves to blame" for the ignorance of the public. The AAAS, Cardiff urged, should establish a publicity department to furnish up-to-date and understandable reports on scientific progress to the press, the movies, lecture bureaus, and magazines. In addition, the organization as a whole should pass a strong resolution opposing measures such as the one before the Kentucky legislature attempting to "place legal restrictions on scientific work and freedom of thought." An official pronouncement from the AAAS would carry much weight. "It is my opinion that if the layman understands that evolution is a doctrine, accepted by all Scientists, and has the unqualified endorsement of the American Association, it would have great effect, especially if this matter were given wide publicity." Scientists often remarked that the public wanted to hear from the AAAS as the official representative of science, and they expressed confidence that endorsements from that august body would carry significant weight with public opinion. In practice, however, publicity would come to seem like something of a conundrum for scientists; the connotations of the term *propaganda* seemed significantly labile in this decade, and the distinction between *publicity* and *propaganda* would appear increasingly vexatious.

The Executive Committee of the AAAS, in good bureaucratic fashion, discussed Cardiff's letter and decided to take no action but to refer the matter to a committee. The committee, made up of Conklin, Osborn, and the geneticist and eugenics advocate Charles B. Davenport, was more inclined to action, or at least activity. A flurry of letters between Osborn, Livingston, and Conklin ensued. Osborn wrote to Livingston that "the first thing to do, is for the Committee to make a very careful selection of thoroughly sane and sound reference works to which inquirers may be referred . . . a delicate task, because we cannot have any commercial taint attached to it."[41] This cautionary note about commercialism would, like the general uneasiness over publicity and propaganda, permeate subsequent discussions.

Osborn and Conklin discussed other potential actions, including the possibility of collecting and publishing scientific responses to Bryan in book or pamphlet form, before deciding that other organizations would supply this need, allowing them to return to the demands of their own research. The active publication program of the American Institute of Sacred Literature, located at the Divinity School of the University of Chicago, a center of theological modernism, in particular seemed to absolve them of the responsibility. The American Institute of Sacred Literature intended to distribute some 300,000 pamphlets proving the compatibility of science and religion, including 30,000 to be sent, free of charge, to every member of the clergy in the country. Among the first pamphlets published in this series were reprints of Conklin's response to Bryan in the *New York Times*, as well as that of Harry Emerson Fosdick. Fosdick contributed another, as did several other prominent scientists, and these essays were distributed not only to churches but also to schools, libraries, and colleges.[42] Shailer Mathews, a modernist University of Chicago theologian, edited the series; the participation of both theologians and scientists reflected the close relationship between religious modernists and religious scientists fostered by the evolution debate of the decade.

Although the AAAS committee's initial flurry of letters did not result in a strong institutional reaction, it did mobilize this core group of scientists in defense of evolution, and the existence of the committee shaped the response of the AAAS to later outbursts of anti-evolutionism, including the advent of the Scopes trial. A committee was in place, and it was reactivated when the Tennessee case arose. The central roles of Osborn and Conklin meant that public defenses of evolution would emphasize the religious respectability of the scientists most in evidence in the press. Osborn's unusual access to the media, his eminence in the scientific community, his concern with education and communication with the public through museum exhibits, the influence of his museum in providing visual images of evolution, his large and efficient staff of assistants at the museum, and his role on the AAAS committee all positioned him to take a leading role in the public defense of evolution. In addition, he and Conklin were religious men, concerned with defending evolution specifically as consonant with religious faith. And they worked to enlist other scientists who shared that perspective.

The scientists most active in the defense of evolution not only argued that evolutionary theory, properly understood, was no threat to traditional religious (by which they meant Christian, and for the most part Protestant) values; they claimed that a true understanding of evolution actually reinforced religious faith. Part of the strategy of a significant number of the defenders of evolution was to try to weave God into the fabric of evolution. This argument reflected, in part, the nineteenth-century tradition in which many of them had been educated. It was also possible in part because biologists did not agree on the mechanism of evolution, especially on natural selection—and because definitions of the boundaries of science and scientific authority often seemed unusually labile—in the 1920s. It was a rhetorical strategy that would have profound effects, fragmenting the scientific community and confusing or alienating segments of the public.

Although a number of scientists dissented from the rhetorical stances of these leaders in defense of evolution, those most active in the debate shared and reinforced their positions, adopting an argument that emphasized the compatibility of evolution and religion. Many scientists insisted that science could—and should—speak on issues of spiritual import. Michael I. Pupin, a colleague of Osborn's at Columbia University, had achieved a place in the public limelight through the publication of his popular autobiography, *From Immigrant to Inventor,* a Horatio Alger tale of the Serbian immigrant who rose to success as an engineer—a book that won a Pulitzer Prize and became a best seller. Religious sentiment suffused the conclusion of Pupin's memoir. "The physical facts of science are not cold, unless your soul and your heart are cold," he wrote. In effusive and flowery terms he declared that science illuminated the beauties of life and of the universe, adding depth to our understanding of Genesis and revealing the divine power at the heart of all physical phenomena: "I am sure that many loyal members of the National Research Council believe that scientific research will bring us closer to this divinity than any theology invented by man ever did."[43]

Some scientists, of course, would object to such statements, often strenuously and occasionally bitterly; but those who expressed ideas similar to Pupin's appeared more often and more prominently in newspapers. In 1924 the physicist Robert Andrews Millikan drafted a statement that became known as the "Millikan Manifesto," asserting that, far from casting doubt on religious faith, evolution actually "furnished . . . a sublime conception of God."[44] Signed by business and religious leaders and some fifteen prominent scientists, including Osborn and Conklin, Millikan's statement was published in *Science* and widely reported and printed in newspapers. These three were probably the best-known and most energetic defenders of this position, but others also participated in the public defense of evolution. Other signers of the Millikan Manifesto, including William Patten, Kirtley Mather, Charles Walcott, Edward Murray East, John Merle Coulter, J. Arthur Thomson, William Henry Welch, James Rowland Angell, and J. C. Merriam, wrote and spoke for lay audiences about evolution. William Patten assigned American Institute of Sacred Literature pamphlets in his evolution course at Dartmouth University.[45] Kirtley Mather, a Harvard geologist, wrote prolifically in books and essays about the compatibility of evolution and religion, appeared in debates and on radio programs, and authored one of the pamphlets distributed by the American Institute of Sacred Literature. In 1925 he would volunteer to travel to Dayton, Tennessee, to testify at the Scopes trial.

E. A. Birge, president of the University of Wisconsin, ignited considerable debate when he lectured and wrote similar defenses of evolution, also asserting its compatibility with religion. Birge, a zoologist and a Congregationalist Sunday school teacher who had studied with the modernist theologian Shailer Mathews at the University of Chicago, engaged in a series of exchanges with Bryan over evolution in 1921 and 1922.[46] The zoologist H. H. Newman of the University of Chicago, who wrote and edited books about evolution intended for a relatively wide audience, also agreed to testify at the Scopes trial in 1925, affirming the defense claim that evolution posed no threat to religion. Less well-known scientists and teachers participated locally, writing for local newspapers, lecturing at schools and churches, and corresponding actively with Osborn and Conklin to keep them apprised of local situations and to seek their advice. Walter C. Kraatz, an assistant professor of zoology at Miami University in Oxford, Ohio, sent Osborn a collection of letters and articles from the *Ohio State Journal* and the *Columbia Dispatch,* noting that these papers had "given space lavishly" to the evolution debate, especially in publishing letters to editors on the subject. Kraatz worried that most people never read scientific books, relying instead on newspapers for knowledge of this important issue. Though he feared that he lacked the stature to have as great an impact as he would like, he labored to do his part, seeing it as something of a duty and one that most scientists neglected, for an understandable reason: "Very few seem prone to write in such an arena. A gladiatorial arena is just about what these objectors try to make of it."[47]

The arena certainly came to seem gladiatorial to some of the people defending evolution, especially those whose positions did not insulate them as Osborn was insulated by his power and status. H. H. Lane, who in 1923 published *Evolution and Christian Faith,* for example, wrote to Conklin that his efforts to educate students about evolution had jeopardized his job. Others seemed to enjoy the sense that they were engaged in battle. An erstwhile student of Osborn's, Arthur Miller, who taught in Kentucky, penned a series of lengthy newspaper essays explaining evolution, and he remained in close touch with Osborn, sending him newspaper clippings and seeking his advice.[48]

Although many scientists rallied to the defense of evolution, many did not, as Kraatz and others complained. Furthermore, the defenders most visible in newspapers and in popularizations did not represent a broad or random cross section of scientists. Not all scientists were equally likely to be asked by the *New York Times,* or the Hearst papers, for example, to comment on matters of public concern. As one persistent journalist protested when Osborn suggested another, less busy scientist for an interview, newspaper chains did not want to interview scientists who were not famous. The formation of the Science Service represented an attempt to promote credible science and a worthy public image of science, but it also implied an acknowledgment that not all scientists could expect to gain access to the media. The question of who should speak for science seemed increasingly troublesome as the evolution debates unfolded. Moreover, not all scientists were equally willing to comment publicly.

Exacerbating the tensions, some scientists and science popularizers displayed every bit as much intolerance as the most radical of the fundamentalists. In a popular book of the 1920s, *The New Decalogue of Science,* the eugenicist and science popularizer Albert Edward Wiggam wrote that science "lies at the basis of all morals. You cannot be truly righteous until you find out how. Science alone can teach you how."[49] Several defenders of science sounded similar themes. Most of the scientists who came to the defense of evolution emphasized their Protestant faith, claiming that science could affirm religious truths. Frequent statements of this kind by proponents of evolution, like Wiggam's claim that science could teach righteousness, suggested that the boundaries of science were very much at issue in the debate.

Many biologists—active proponents of racialist interpretations of evolutionary theories—enlisted in the eugenics movement in the 1920s. Among the scientists who participated most vocally in the evolution controversy were a large number associated with eugenics organizations. Scientists such as Osborn and Davenport played prominent roles in the eugenics movement, and they believed public support of eugenics to be as much a civic responsibility as the education of the public about evolution. Other biologists, less active in eugenics organizations than Osborn and Davenport, were nevertheless involved to some extent. This was true of Conklin, for example, whose friend Thomas Hunt Morgan gently chided him for it. The biologists who avoided or disapproved of eugenics may well have been in the minority. The scientific community was beginning to be divided on the subject, as on so many others. Anthropologists following Franz Boas dissented from eugenics, as did some younger biologists, but biology textbooks regularly included eugenics messages in their discussions of evolution. Across the country, eugenics organizations sponsored Fitter Families and Better Babies contests, eugenics messages appeared in movies, magazines, and popular science books: it is probably fair to call eugenics part of the orthodoxy of 1920s American culture. The implications of their support for eugenics led a number of scientists to support the movement for immigration restriction which culminated in the Immigration Restriction Act of 1924. For dissenters from this orthodoxy, however, scientists' activism in the eugenics cause represented an overreaching of their authority as scientists. Eugenicists claimed to be mounting scientific arguments; dissenters saw those claims as relying upon a dangerously plastic definition of the boundaries of science.[50]

Not Everyone Joined the Fray

Some young scientists who joined in the defense of evolutionism, such as the geologists Carroll Lane Fenton and Kirtley Mather, were political liberals, as were the most vocal of the nonscientist defenders, including many liberal members of the clergy and journalists.[51] And though many of them were willing to testify as to the compatibility of science and religion, they may have been more likely, ultimately, to draw sharp boundaries between science and nonscience. But the scientists whose voices were loudest in the popular media were generally older, more famous, and more conservative politically. These men (they were all men) worked to align with the cause scientists who could establish firm credentials as religious believers. And some established scientists who opposed scientific racism, such as Franz Boas and Thomas Hunt Morgan, deliberately avoided the public debate about evolution. The conflict was not simply a case of repression of the enlightened by the ignorant.

The scientists who eschewed the public debate altogether did so for a variety of reasons. The New York Zoological Society ornithologist, explorer, and popular writer William Beebe, for example, responded to a question from a reporter for the *New York Times* by saying that he thought the whole argument was trivial. He was too busy studying evolution to waste time defending it. Raymond Pearl, an ecologist at the Johns Hopkins University, also protected his research time. When asked by a journal editor for a popular article, he wrote back that he didn't have much time and that whatever popular writing he did went to his friend Henry Louis Mencken for publication in Mencken's *American Mercury*.[52] Although Pearl enjoyed joking about fundamentalists with Mencken, he did not involve himself a great deal in the controversy, in part, undoubtedly, in order to protect his time. But it is also true that Pearl could not have made a religious argument for evolution— he and Mencken routinely enjoyed an extremely acid idiom in their private correspondence, making fun not only of religious people but even of the Deity: Pearl certainly belonged to the same professional societies as the scientists who tried to establish the harmony of evolution and religion, and his standing in the community of scientists was assured, but he was a far more iconoclastic character.

Pearl probably resisted participation for reasons similar to those of the Columbia University geneticist Thomas Hunt Morgan: like many other scientists, Morgan believed that such public controversies were antithetical to the real mission of science. An advocate for the newer disciplinary matrix emphasizing experimentalism, empiricism, and focus on research, Morgan scoffed at suggestions that he participate in the public evolution debate.[53] Morgan and Conklin enjoyed a warm friendship but did not share quite the same perspective on the relationship of science and larger social issues; Morgan could also be quite critical of Osborn, his colleague, technically, at Columbia.

Not all the scientists who spoke out took blatantly religious positions, of course. In an essay in the *New Republic* in 1923, the entomologist and zoologist Vernon Kellogg suggested, "Perhaps it is not a bad thing that Mr. Bryan and the Fundamentalists are stirring up matters about evolution, and hence stirring up the evolutionists to interrupt for a moment their evolutionary research in order to take stock of their present knowledge, and to tell the public, in more or less intelligible language, just where evolution stands now."[54] Magazines like the *New Republic* and the *Atlantic* offered scientists space for relatively detailed explanations of the evidence for evolution.

Kellogg, raised in Emporia, Kansas, and a close friend of the journalist William Allen White, had been a member of the faculty at Stanford University, where he co-taught a course on Darwinism with David Starr Jordan. Both men were committed to teaching. During the war, Kellogg worked under Herbert Hoover in European relief efforts, and after the war his commitment to public education brought him to the National Research Council, where he served as permanent secretary until 1931. Kellogg had a good deal to say about evolution in the public arena. In the summer of 1925 he debated the formidable Princeton University theologian J. Gresham Machen in the *New York Times*, and in 1927 he participated, "for the scientists," also in the *Times*, in a forum including Arthur Lovejoy, "for the college professors," and John Roach Straton, "for the Fundamentalists." Kellogg argued, essentially, that evolution and religion occupied different spheres and that accepting evolution need not imply a rejection of religion.

Theistic evolution did not represent a consensus view in the 1920s. It was, however, the view of evolution that would be most often presented to the newspaper-reading public. The scientists who chose to participate in the anti-evolution debates of the 1920s and who had the means to command public attention were frequently men of an older generation, influenced by ideas current in the late nineteenth century. Though their attitudes toward the issue of science and religion were not typical of the ideas of most scientists, and though they did not speak for all scientists when it came to the details of evolutionary theory, they were the scientists the public would most often hear. Their ideas about what was at stake were quite different from those of the defenders of evolution from outside science—and by 1925 there was a distinct dichotomy between those inside and outside science.

The arguments mounted by scientists differed significantly from those espoused by nonscientist defenders of evolution. By 1925 the category "intellectual" could not comfortably contain both scientists and nonscientists without qualification. Different definitions of science and varying understandings of the boundaries of scientific authority necessarily simmered just beneath the surface of the debate. And nonscientist secular intellectuals often understood those boundaries in ways that would put them essentially at odds with their scientific allies in the defense of evolution.

Nonscientists in the Defense of Evolution

Although defenses of evolution by nonscientists came from several quarters, in many cases they took approaches very different from those articulated by scientists and sometimes drew different lessons from the debate. Some of the most active defenders of evolution were liberal and modernist ministers, such as Harry Emerson Fosdick and Charles Francis Potter. Like the scientist signers of the Millikan Manifesto, these people argued for the harmony of theistic evolution and liberal Protestant theology, and a number of liberal ministers corresponded actively with scientists. Newspapers' announcements of Sunday sermon topics included as many affirmations as denunciations of evolution—a lot, in both cases.

Charles Francis Potter, minister of the First Unitarian Church on Manhattan's Upper West Side, called scientists "the prophets of today." He explained, "Science, through evolution, has given us an infinitely higher idea of God."[55] In a well-publicized and well-attended event, Potter debated John Roach Straton at Carnegie Hall. The debate attracted a large crowd and was broadcast on the radio. Potter also sponsored an "Evolution Day" at his Upper West Side church and traveled to Dayton to serve as an expert witness at the Scopes trial.[56]

Others defended evolution from a very different perspective, especially from the secular political left. For many of these people the evolution debate carried considerable import beyond the content of the science itself. In 1923, a twelve-year-old California girl by the name of Queen Silver, a socialist, delivered a lecture called "Evolution from Monkey to Bryan." The lecture and the lecturer made quite an impression, and the lecture was repeated, published, praised by Luther Burbank, and reprinted in pamphlet form. Queen Silver, who billed herself as "the Godless girl," adopted the familiar humorous idiom of monkeys, apes, and tadpoles, ridiculed Bryan, and studded her speech with references to the writings of nineteenth-century evolutionists. Humans, she said, were very obviously ape-like, even though some scientists had minimized the similarities "in an attempt to placate the enemies of Darwinism." The real point of her colorful argument, however, was this: "Bryan's ideas, if carried out in laws, will deprive the children of America of the right to learn the truth."[57]

William Green, the editor of the *American Federationist,* the magazine of the American Federation of Labor, wrote to Osborn that "organized labor was one of the most influential factors in establishing as our national policy our system of free public schools. Because we and our children are educated chiefly through public schools, we are much concerned that educational policies should be broad and constructive." He was interested, therefore, in offering readers reliable articles on evolution. The reason he approached Osborn in particular, Green wrote, was because "You are one of our foremost scientists and at the same time have a profound appreciation of religion."[58]

Emmanuel Haldeman-Julius, a socialist autodidact, created a legend around his idea that ordinary Americans were hungry for good literature at reasonable prices, books in cheap pamphlet form that could be carried in a pocket and read on one's lunch hour. Using innovative methods of advertising and marketing, he eventually ran the largest mail-order book-publishing house in the world, and his series of Little Blue Books—he called them a "University in print"—reached extraordinary numbers of people: he is estimated to have printed some five hundred million copies of more than two thousand different titles. They are especially interesting in the context of the evolution debates because they reflected Haldeman-Julius's extreme antipathy to organized religion.[59] For people like Emmanuel Haldeman-Julius and many of the authors publishing with him, the evolution controversy had, most of all, to do with political liberties. These authors did include some scientists, such as the young paleontologist Carroll Lane Fenton. Haldeman-Julius published a number of books specifically dealing with the evolution controversy, including a pamphlet called *Poems of Evolution.* In 1925 he and his wife brought a collection of Little Blue Books—all pro-evolution and a number of them hostile to religion—to Dayton, Tennessee, to sell at the Scopes trial. Crusaders from the left like Haldeman-Julius incorporated the evolution controversy into a general push for freedom of thought and speech, in many cases directed at religious conservatism, which they saw as a threat to larger freedoms.

From this perspective, the association of the Ku Klux Klan with the rhetoric of religion loomed ominously over the debates. Several African American editorialists extended the association of fundamentalism with the Inquisition to include conservative Christian support for slavery and participation in lynching. The syndicated communist Ernest Rice McKinney argued that fundamentalists had put Christ to death, persecuted Galileo, and "were the backbone of slavery in the south." The anti-evolutionism of the twenties, McKinney thought, was just another instance of an all-too-familiar bigotry; he asserted, in an image meant to conjure up an association with the Ku Klux Klan: "A Fundamentalist is nothing more than the same old reactionary strutting forth in a brand new robe."[60] William Pickens ridiculed Bryan, commenting, "Bryan and Tennessee are attacking the law of evolution—as if they thought that natural laws yielded to attacks . . . but no law of nature can be broken." Scientists and journalists alike invoked the notion of evolution as natural law, but Pickens added a sobering dimension to the familiar theme: "As for Tennessee—the law of evolution cannot be overthrown by the wisdom of a state where they burn men alive."[61]

The most active—indefatigable, it might seem—and possibly most effective opponent of anti-evolutionism was Maynard Shipley, founder of the Science League of America. Shipley was a great admirer of science; for him, it represented the casting aside of prejudices and superstitions that had caused bigotry, intolerance, and suffering. Convinced that the Inquisition served as a general model for the historical relationship between science and religion, he believed that religious fundamentalists threatened social regression to the attitudes of the Dark Ages. For him, the anti-evolution movement was at least as much political as it was religious; it was dangerous, and he took it quite seriously as a threat to liberty. Fundamentalists did not just oppose evolution, he insisted. They rejected the scientific method itself, which he viewed as tantamount to a rejection of reason and therefore of the values of the Enlightenment. Declaring that evolution was only the initial foray in a larger battle, Shipley warned anyone who would listen that the ultimate goal of the fundamentalists was theocracy. He debated fundamentalists, monitored their activities, and worked energetically to dissuade legislators contemplating anti-evolution laws. He was not alone. His language was that of a crusade, and his efforts really amounted to a campaign against anti-evolutionism. Working with his wife, the writer Miriam Allen de Ford, who wrote articles about the Scopes trial for the *Nation,* he worked tirelessly to publicize and combat the legislative initiatives of anti-evolutionists. He corresponded with liberal members of the clergy and with newspaper editors who provided him with information from all over the country and who, like him, saw themselves as engaged in a struggle against a repressive political agenda.[62]

Many scientists shared his concern about the dangers of anti-evolutionism and allowed their names to appear on the masthead of Science League publications, but they did not for the most part share his political convictions. Some of them were quite ambivalent about him and about the Science League, and the AAAS stopped short of official alliances with him or endorsements of his work.

Several prominent members of the Honorary Advisory Board, including Herbert Spencer Jennings, Raymond Pearl, and Edwin Grant Conklin, either resigned or threatened to resign in protest against the inclusion on the board of people whose scientific credentials they questioned. They did not wish to see their names associated with scientific outsiders or upstarts. Nor did they entirely approve of Shipley, who was a socialist and a good friend of Eugene Debs's.[63] Although scientists applauded his efforts on behalf of science, some of them suspected, as Jennings put it, "that the people who are running the League know nothing about science, and that they are likely to let us in for other things that we don't wish to stand for."[64] They suggested privately to one another that he would not confine his efforts to a simple defense of science, that his political views would intrude, and that his politics were suspect. In the end, the influence of such prominent skeptics meant that the AAAS would not officially support Shipley's efforts, although some individual scientists would continue to lend their names to his masthead.

In general, a number of the scientists defending evolution fretted about the politics of some nonscientist allies. One of the most prominent of these allies, even before the Scopes trial, was the Chicago lawyer Clarence Darrow, long an active proponent of progressive political positions and a frequent defender of labor leaders and political radicals. Like many of the defenders of evolution whose views of religion had been influenced by the popular books of John William Draper and Andrew Dickson White chronicling a long history of antagonism and even warfare between science and religion, Darrow conflated organized religion with intolerance and repression, with a superstition culminating in witch trials and persecution of people espousing new ideas. His rhetoric resonated with terms common to the defenders of evolution—fundamentalism was analogous to medievalism, linked with the Inquisition. For Darrow, science meant liberation from ancient superstitions, tyrannies, and bigotry. Science, the avatar of enlightenment, the antithesis of superstition, represented a way out of centuries of oppression and cruelty, all of which was irrational. Science was the symbol and the method of all that was best in the modern world, the vehicle of progress.[65]

Darrow's remarkable summary at the 1924 trial of the young Chicago thrill-killers Nathan Leopold and Richard Loeb eloquently espresses his commitment to science and what he meant by it. Darrow viewed human history in evolutionary terms, casting religion as a response on the part of primitive humans to the great dilemma of consciousness of mortality—like other animals, humans responded most viscerally to an overwhelming survival instinct; religion began as a first, crude attempt to reconcile that instinct with the paradoxes of the human condition. But religion, like all primitive things, was shot through with brutality.

The Bible, Darrow claimed in a 1929 debate with representatives of the Protestant, Catholic, and Jewish faiths, "defies every principle of morality, as man conceives morality."[66] The notion of an anthropomorphic god was a primitive relic of a time before science had revealed the insignificant place of our earth in the universe. Organized religion had been responsible for centuries of violence, repression, and brutality. Reason and science had cleared a path through this bloody history and offered the only real salvation.

Many people criticized Darrow for defending Leopold and Loeb, accusing him of "selling out" for a large fee and objecting to the reasoning he used in their defense. The trial was meant not to establish the guilt of the two killers, however, since they had confessed to the crime, but to decide their fate, and Darrow ardently opposed the death penalty. In arguing against executing Leopold and Loeb he constructed an evolutionary theory of responsibility. According to "the old theory of responsibility," the product of a primitive superstitious worldview, "if a man does something it is because he willfully, purposely, maliciously, and with a malignant heart sees fit to do it." Our modern civilization had transcended such ideas: "Science has been at work, humanity has been at work, scholarship has been at work, and intelligent people now know that every human being is the product of the endless humanity back of him and the infinite environment around him."[67] Cries for the death penalty were but the atavistic remnants of the instinct for revenge, which "roots back to the hyena; it roots back to the hissing serpent; it roots back to the beast and the jungle."[68] In an eloquent appeal to the judge for mercy, Darrow concluded, "Your honor stands between the past and the future. . . . I am pleading for the future; I am pleading for a time when hatred and cruelty will not control the hearts of men, when we can learn by reason and judgment and understanding and faith that all life is worth saving, and that mercy is the highest attribute of man."[69] Darrow's plea had lasted more than twelve hours. The two killers' lives were spared: they were sentenced to life in prison. Darrow edited the argument for publication in pamphlets, and newspapers printed long excerpts of it. His eloquent essay based on the plea in the case would come back to frustrate him at the Scopes trial, however, when Bryan would accuse him of having made, in the case of Leopold and Loeb, an argument fundamentally inconsistent with the claims he made on behalf of Scopes and evolution. For many people, his attitude toward religion would also make his role at the Scopes trial extremely controversial.

Probably no one denounced religion in as acid a tone as did H. L. Mencken, notorious reporter for the *Baltimore Sun* and editor of the *Smart Set*, then founder and editor of the *American Mercury*. Mencken could be positively vitriolic, both in his published writing and in his private correspondence. In letters to his friend Raymond Pearl, Mencken relentlessly ridiculed religion and extolled the virtues of alcohol. Both publicly and privately he took delight in referring to anti-evolutionists as "gaping primates," "simians," and "Neanderthals." He denounced Bryan relentlessly, sarcastically, and mercilessly: even his obituary of Bryan rang with contempt. An influential journalist and a friend of Clarence Darrow's, Mencken would come to exercise considerable influence in shaping the tone of journalistic accounts of the Scopes trial and perhaps even more influence in coloring later popular histories of the anti-evolutionists of the decade.

Although there were certainly anti-evolution papers in some parts of the country, many newspapers and magazines—especially the large New York papers, the newspaper chains, and the wire services—assumed the truth of evolution and the ignorance of anti-evolutionists. Morrill Goddard, the Sunday editor of the Hearst Sunday newspapers, wrote to Osborn in 1923, claiming a circulation of nearly five million papers every Sunday and extrapolating from that to estimate a readership of perhaps twenty million. He wrote to ask Osborn for a "straightforward, clean cut, scientifically accurate statement of the evidence of evolution, which will be written in such language that a school-child or day-laborer will be able to follow it and understand it." Goddard made clear where his sympathies belonged: identifying himself as a "college man," he commented: "I feel as if I have a duty to perform, possibly, in controverting Mr. Bryan's ignorant assertions as to 'Darwinism.'"[70] A statement published in the *Detroit Free Press* called the anti-evolution movement "essentially a backwash from the stream of progress and enlightenment."[71] The *New York Times* called upon Osborn regularly and drew on American Museum images to illustrate many of its articles about the controversy. Reconstructions of human ancestors executed by Charles R. Knight and J. H. McGregor for the museum would become familiar figures on the front page of the *Times* and in newspapers generally.

The *Times* seemed to aim for a moderate and even tone most of the time, but humor weighed heavily in newspaper coverage of the controversy. Even before the push for anti-evolution measures, and even before the caustic H. L. Mencken entered the fray, newspapers made fun of fundamentalists. Before the evolution controversy percolated to the top of the stew, fundamentalists had conspicuously pushed for local blue laws, as well as for Prohibition. This was as true in New York City as it was in the rural South or the Midwest. New York newspapers in the early twenties crammed together stories describing sermons preaching the sinfulness of dancing on Sundays with articles making fun of such sermons and editorials and letters about the inconveniences and practical conundrums raised by the prospect of laws prohibiting activity on New York Sundays. These measures provoked a good deal of ridicule in the press. *Puritan* became a common term of opprobrium. Cartoons about Puritans were common, as were Mencken's snide references to Puritans. The term *Puritan* acquired a particular resonance in the twenties, much like the term *Victorian*.[72] The new texture of these terms reflected the proliferation of laws around the country regulating even trivial behavior.

Laws governing behavior became so common that the magazine *Judge* featured a recurring cartoon figure called "Aunty Everything." Such laws inspired a good deal of indignation as well as ridicule, indignation couched in terms of liberty and tolerance. An editorial in the *Nation* suggested that the "campaign against evolution—against the scientific method—is only part of a larger movement . . . to regulate all human existence by law—including personal habits and opinions."[73] The ridicule of anti-evolutionists so prominent in the press drew on the same vocabulary of verbal and visual references that supplied parodies of the widely maligned outbreak of twentieth-century Puritanism, the Inquisition, and the Dark Ages and cartoons featuring Inquisition and Puritanism references. These continuing themes were firmly in place for the advent of the Scopes trial, and the anti-evolution movement added to the mix the jokes that would so annoy scientists like Osborn: monkey jokes, ape and caveman jokes, and jungle references. In the cartoon world, Williams Jennings Bryan was harassed, badgered, and congratulated by all manner of simians. Cartoons regularly portrayed Bryan as a buffoon and as an Inquisitor and a Crusader.

The long tradition of invoking past episodes of repression associated with Galileo, Copernicus, and Giordano Bruno resurfaced with the renewed popularity of books by Andrew Dickson White and John William Draper. The language of the Inquisition was readily available, and both published accounts and private letters attest to the common vocabulary used to describe the situation. Scientists and science advocates regularly referred to the "medieval superstitions" motivating religious conservatives. The examples of Giordano Bruno, burned at the stake in 1600—though for theological heresies, not, as was often claimed in the 1920s, for his Copernicanism—and Galileo, forced to recant his assertions that the earth moved around the sun, came easily to hand.

Evolution acquired an important symbolic significance for journalists and nonscientist intellectuals of the period. Not that these people agreed at all points either, of course. They certainly did not. But consistent themes emerge. There were important differences between the responses of the scientists who took up the public crusade and the preponderance of journalists and intellectuals from outside science who wrote and spoke about evolution and its significance. In many newspapers and magazines evolution stood for science, which meant modernity; it meant enlightenment, rationality, and liberation from "medieval" intolerance and "superstition."

This was so even in periodicals taking humorous positions in response to the controversy. The *New Yorker* and *Vanity Fair*, for example, were not really concerned with demonstrating or explaining evolutionary theory—they accepted its truth as a given of science, and they accepted science as a preeminent source of authority. Science, for many literary nonscientists, epitomized the modern. Adopting self-consciously sophisticated stances, editors of such magazines ridiculed anti-evolutionists; to be anti-evolution, they implied, was to resist modern science. Anyone who did that must be retrograde and foolish. Wits of the decade, in company with H. L. Mencken, found the impulse to draw analogies between fundamentalists and simians, Neanderthals, and Cro-Magnons irresistible. Throughout 1926 the *New Yorker* continued to recommend T. S. Stribling's novel *Teeftallow*, a book that caricatured ignorant southerners as "primitives" and devised a plot based on the repressive potential of their ignorance. The language, including the visual lexicon, of the "primitive" served varied purposes.

The editors of the *Nation* took a less flippant approach than those at the *New Yorker*, but they agreed that the substance of science was not the essential point. Rollin Lynde Hartt declared that "evolution is only superficially the issue at Dayton. At bottom, the issue is a vastly older and vastly more important question— the question as to whether the separation of church and state shall be maintained."[74] According to another editorial, anti-evolution laws mattered because they represented the general proliferation of laws seeking to control individual behavior and even belief which imperiled American liberties: "The important question is not whether the doctrines or theories which these laws seek to suppress are true; it is whether it is ever possible or desirable to determine the content of teaching, to attempt to define truth and suppress error, by law."[75] Scientists sometimes claimed that the truth of evolution was precisely what was at issue; more often they ridiculed the proposition that a "law of nature could be determined by popular vote." This was not quite the same objection as the one made by the editors of the *Nation* and other civil libertarians.

Some nonscientist evolutionists, like Clarence Darrow, celebrated the dislocating of tradition by science; some, such as Maynard Shipley and W. E. B. Du Bois, agreed that science could be a force for liberation. Others, including Walter Lippmann, saw the transition to a science-dominated world as inevitable but painful. In general, though, these people spoke in a rather different idiom than the one most often represented in newspapers as the voice of science.

Some of the scientists actively engaged in the defense of evolution did not, to be sure, conform in essential ways to the model established by Millikan, Osborn, and Conklin. Winterton Curtis, the anthropologist Fay-Cooper Cole, and the young Harvard University geologist Kirtley Mather, all of whom went to Dayton in 1925 hoping to testify at the Scopes trial, were in some ways not at all typical of the more famous "first string" scientists whose names would be associated so prominently in the newspapers with the scientific defense of evolution. Mather certainly did, like Osborn and Conklin, argue that evolution was compatible with Christian faith, but unlike them he always opposed eugenics and scientific and other forms of racism. He remained committed to communication with the public and to progressive causes throughout the rest of his long life. Fay-Cooper Cole fell into the anthropological camp that denounced the theories of "Nordic" superiority so prominent among eugenics advocates such as Osborn. Although Curtis dwelled on the "spiritual" aspects of evolution in his Scopes trial testimony, his attitude toward religion ultimately had more in common with Clarence Darrow's than with Osborn's and Conklin's; indeed, he admired Darrow deeply, and the two struck up a friendship that lasted long after the end of the trial.[76]

Although they were asked to give lectures on evolution after the Scopes trial, these men were not, until then, as consistently sought or quoted by newspapers as were the people Walter Kraatz called the "bigger" biologists. And the issues for scientists were not precisely the same as those concerning most nonscientists. At this moment when the details of evolutionary theory were very much in dispute, scientists engaged in the debate labored to make the substance of evolutionary theory clear. Many of them also worked to illuminate its larger significance.

There were four separate things at issue: the truth of evolution; the interpretation of the trajectory it had taken, or the patterns of the history of life; the mechanism of natural selection; and the philosophical, metaphysical, and religious implications of evolution. These four themes would frequently be conflated in the media and by leading fundamentalists, but it was the last of them that made for the most passionate discussion and which made the role of symbols in the debate so telling and so crucial.

Some scientists, including Osborn, would themselves compound the confusion by mixing evolution with philosophy or metaphysics. Attempts to reconcile two themes—questions about the role of chance events and the idea of progress—had long been intertwined with evolutionary thinking. Although scientists had accepted evolution and natural selection in the late nineteenth century, relatively few of them had believed that natural selection could be the sole mechanism of evolution, and late nineteenth-century evolutionary theory incorporated a variety of non-Darwinian modifications of Darwin's theory. Reluctance to accept natural selection as the exclusive or even primary mechanism for evolution reflected strong philosophical, religious, and aesthetic predilections, but it also reflected a body of scientific observations and of traditions in scientific thought and a number of persistent scientific conundrums. These themes would reappear in the debates of the 1920s as scientists attempted to craft a reassuring perspective of evolution for the public. Osborn reflected, in this respect, the nineteenth-century traditions in which he had been educated. Theistic evolutionists, especially Osborn, were among the most visible—and vocal—of the visible scientists.

CHAPTER FOUR

Saving the Phenomena

Scientists found themselves in a difficult position in the 1920s. The fossil evidence for human evolution remained fragmentary, and evolutionary theory was in disarray. Although scientists accepted the fact of evolution, they did not agree on natural selection. And natural selection had implications that turned their attention to the philosophical dimensions of evolution. The scientists who defended evolution in the 1920s had come to terms with these things in the late nineteenth century; their views of evolution had been formed by accommodations that allowed for direction and purpose in evolution, striking a precarious balance between increasing pressure to exclude extrascientific non-naturalistic explanations of natural phenomena and a conviction that evolutionary history did have direction and meaning. Scientists who participated in the debates drew heavily on those traditions; in order to understand their perspectives, it is necessary to understand the tradition of late nineteenth-century evolutionary theory. Without that perspective, scientists' statements about evolution in the 1920s often seem puzzling—and did, to many nonscientists of the time.

Late-Nineteenth-Century Theories of Evolution: Saving the Phenomena

Well before Darwin published *On the Origin of Species* in 1859, geologists and paleontologists had been accumulating evidence of the progression of life from simple beginnings over immense periods of time. Revelations of the history of life extending back into "deep time" had not proved incommensurable with Christian faith; in fact, many of the naturalists uncovering this history perceived it as proof of the goodness of the Creator. At least as early as the Middle Ages, European naturalists had argued that God had revealed himself to humanity through two complementary texts, the books of the Bible and the "book of nature." In an influential 1691 work called *The Wisdom of God Manifested in the Works of the Creation*, the naturalist John Ray developed this theme, and Ray's work became part of the canon of a tradition elaborated in William Paley's influential 1802 book *Natural*

Theology. Among the most lasting tenets of the tradition of natural theology was the notion of design. The organisms we see in the world around us, Ray and Paley maintained, were so beautifully and precisely adapted to their functions and their environments that they could have come to be that way only by design. And design implied a designer. This notion of design in the biological realm extended the clockwork universe idea that had allowed religious thinkers to assimilate Newton. If the universe worked as if on its own, like a mechanism such as a clock, there must have been an original clockmaker. Naturalists following Ray and Paley compiled an impressive body of evidence of the exquisite "perfection" of organisms, attesting, as Ray had put it, to "the wisdom of the Creator manifested in the works of the creation." Evidence of perfect—or imperfect—adaptations among organisms would become central to arguments about evolution before and after Darwin and in a variety of intriguing permutations.[1]

Darwin's theory of natural selection has sometimes been said to have "stood the argument from design on its head." This is so only in a very complicated way, however. Darwin postulated that the appearance of perfection of design could be achieved through the process he called "natural selection" by analogy with the "artificial selection" practiced by animal and plant breeders. Just as animal breeders selectively paired individuals displaying desirable traits, nature might favor individuals possessing particular advantages that enhanced their chances of survival and reproduction. Drawing on the insight of Thomas Malthus,[2] who worried about the possibility of the human population outstripping the available resources, Darwin noted that, as Malthus had observed, populations increase geometrically but resources do not. The inevitable result would be competition. Noting that animal and plant breeders achieved desired results by taking advantage of small variations among individuals, allowing the individuals who exhibited favored variations to breed, Darwin suggested that competition among organisms might act to select favored variations in a similar way. Those organisms best suited to the environments in which they found themselves would be more likely to produce offspring similarly well adapted, and the favorable variations would multiply over long periods of time.

Darwin made an essentially ecological argument—favored variations were "perfect" not in any abstract or absolute sense but relative to their environmental context. It was also a population argument. Though not comfortable with mathematics, Darwin was thoroughly at home with nature. His thinking was shaped by a profound appreciation of the natural world, and his appreciation of the variety among individual organisms allowed him to imagine a pool of variation that would provide the material for natural selection. He did not use statistical or mathematical language or illustrations or test his arguments mathematically, but his theory had statistical underpinnings. He thought in terms of populations of organisms, as many of his contemporaries did not. His experience as a field naturalist, as well as his discussions with breeders, convinced him that variations—the raw material for natural selection—existed in abundance. And Charles Lyell's influential three-volume *Principles of Geology*, published between 1830 and 1833, which he had read during his formative voyage on the *Beagle*, demonstrated that the time available for evolution was sufficient to amplify small variations over time.[3]

Concerned with the controversial implications of his work, Darwin ruminated on the nature of scientific method. In order to be convincing—and scientific—his argument could not rely on the teleological element in the argument from design, the argument that adaptations existed to fulfill a purpose. Within the next several decades the argument laid out so painstakingly in *On the Origin of Species* convinced the majority of scientists that evolution had indeed occurred. Many people remained skeptical about natural selection, however. The role of "chance" or random variation presented a particularly stubborn obstacle.[4]

Laypeople were not alone in responding apprehensively to the element of chance implicit in Darwin. In the debate following publication of the *Origin of Species* in 1859, the English philosopher John Herschel had famously referred to the theory as "the law of higgledy piggledy." It was impossible, Herschel wrote, to accept "the principle of arbitrary and casual variation and natural selection." It would be different if Darwin would just admit the necessity for "intelligent direction" in evolution.[5] The sticking point was most often whether evolutionism could be made to include progress, which for many people was incompatible with the notion of "chance" variation, or contingency.

Chance was a stumbling block for theologians as well. Yet through the 1870s and 1880s, many thinkers had accommodated evolution by incorporating teleological views and grafting them onto their evolutionary theories. The fossil record, considered by Darwin rather sparse in 1859, had become too convincing to allow anyone to dismiss evolution without considerable difficulty. Scientists and theologians such as James McCosh, president of Princeton University and Osborn's early mentor, found ways to reconcile evolution with their Christian faith by effecting compromises that accommodated their belief in progress and meaning. Those scientists who accepted evolutionism in the late nineteenth century more often than not did so by circumventing Darwin's efforts to banish teleology.[6]

Teleological explanations—explanations attributing immanent purpose to natural phenomena—had a long and venerable history among European naturalists, traceable ultimately from Aristotle's final causes through Paley's natural theology, and they were not easily forsaken. In the late nineteenth century, naturalists and philosophers resorted to several variations on non-Darwinian evolutionary mechanisms in order to make Darwin more palatable, especially to downplay the role of chance; these variations on the evolutionary theme would reappear in the literature of the 1920s.

Evolution as Development

Many people in the late nineteenth century found solace in the development metaphor, the idea that the evolution of species is "just like" the development of the individual. Development, of course, implies a predetermined, predictable direction and a goal; it was common in the late nineteenth century to use the terms *evolution* and *development* interchangeably.[7] The development metaphor was pressed into service to explain social evolution and to postulate an anthropological theory of the stages of development of civilization from primitive to advanced.

In biology the development metaphor preceded Darwinism and evolutionism. In the 1830s German *Naturphilosophie* contributed the notion that the animal embryo passed through stages reflecting the hierarchy of nature. At the end of the century Ernst Haeckel cast this idea in evolutionary terms: the stages of development (ontogeny) of the embryo, he claimed, reflected the whole evolutionary history of the species (phylogeny). The threefold parallel of form through embryology, paleontology, and morphology both confirmed the fact of evolution and provided a method for mapping evolutionary history. Although embryologists eventually challenged and discarded much of Haeckel's "biogenetic law," his pithy phrase "ontogeny recapitulates phylogeny" assumed a place in the pantheon of widely disseminated, if little understood, truisms. The idea that growth of all kinds could be seen as similar to the development of the individual organism resonated powerfully in Victorian descriptions of the world. It became axiomatic in anthropology and psychology. In biology and in social science it became theory and method as well as metaphor. Beyond the realms of science, it appeared as metaphor, and as metaphor it shaped ideas as if it were method there, too. By the end of the nineteenth century the metaphor was ubiquitous.

The theory that ontogeny recapitulated phylogeny guided scientific practices and theories in very explicit ways. It was especially attractive as a theory because it seemed to confirm evolution at a time when the fate of Darwinism itself was by no means assured. It fit comfortably along the experimental track taken by embryology. Because the history of the embryo was supposed to have mirrored the history of life, the development analogy complemented paleontological observations. It also helped ease many people into acceptance of evolution by implying progress and meaning.

The analogy found its way into the standard college biology curriculum. The paleontologists Joseph Le Conte and David Starr Jordan, for example, offered students a form of evolutionism mediated by teleology. Both men, students of the Harvard University anti-evolutionist Louis Agassiz, converted to evolutionism by adapting the embryological teleology they absorbed from German romanticism through Agassiz to an evolutionary model. Le Conte's widely used textbook took the development analogy as a virtual leitmotif. David Starr Jordan's 1895 course syllabus similarly emphasized Haeckel's biogenetic law. There was, by these accounts, a predictable, meaningful direction in the evolutionary process. If evolution worked in this way, as Le Conte repeatedly insisted it did, then it was not nearly as radical a departure from the comforting and familiar theory of design as the unmediated theory of natural selection might imply.[8] The analogy of evolution with development confirmed the purposeful unfolding of historical progress. It was as much a theory about history as about biology, but it offered a comforting perspective on biological evolution.

Unlike Darwinian evolution, the development metaphor necessarily implies a teleology, progress toward a predetermined end. For this reason, many people who were uncomfortable with the element of chance implicit in Darwin were able to accept the reality of evolution by conceiving it as analogous to development.

Lamarckism

Teleological impulses also found expression in the form of neo-Lamarckism. In the late eighteenth century the French naturalist Jean Baptiste Pierre Antoine de Monet de Lamarck had postulated that individual organisms could pass on to their offspring the useful characteristics acquired during their lifetimes in response to environmental challenges and that this could account for the changes observed over time within lineages. The accumulation of such changes would, by definition, constitute progress. Though emphasizing his own explanation of evolution by natural selection, Darwin had not ruled out the possibility that the Lamarckian mechanism could be at work in evolution, and many of his followers preferred it to natural selection. In the late nineteenth century an amended version of Lamarck's theory was so influential in the United States that it was sometimes called the "American school."[9]

Paleontologists in particular found modified Lamarckian explanations convincing because the discoveries of more and more fossils, especially in the American West, did seem to reveal an unmistakable pattern of progress in the history of life. Life appeared to evolve into more and more complex patterns, and organisms seemed to become, on average, bigger, more complex, and ultimately, in the case of our own lineage, more intelligent. If organisms could bequeath to their offspring the characteristics acquired during their lifetimes, paleontologists thought, progress in a discernible direction could result; there seemed to be no way to account for such progress using the theory of natural selection.

The paleontologist Edward Drinker Cope was among the most outspoken and influential founders of the American school. An energetic collector at a time when the rich fossil fields—called "fossiliferous" by paleontologists—of the American West were just being explored, Cope was among the first to discover, describe, and name many of the animals in the history of life. The founder and editor of an important journal, the *American Naturalist*, Cope was absolutely committed to a belief in the inevitable progress of the history of life and, furthermore, to a belief in mind as the agent of progress. Lamarckian inheritance, for Cope, meant that organisms actively chose their destinies. Natural selection, by contrast, meant "chance" variation, an untenable notion. Cope believed that variation had to be directed, and directed by mind, by consciousness. This was not a radical perspective in the 1870s and 1880s; it had, on the contrary, broad appeal and was included in the training of many students. Joseph Le Conte's widely used textbook *Evolution: Its Nature, Its Evidences, and Its Relation to Religious Thought*, for example, ascribed a similar role to mind as a directing force in evolution.[10]

There were both scientific reasons—the observation of trends by paleontologists and the fact that the proliferation of these observations was a recent and extremely impressive development—and metaphysical predilections for taking Lamarckism seriously. It offered an explanation of evolution's progressive trend while at the same time avoiding determinism, satisfying both the optimistic faith in progress so prevalent among the Victorians without, for many Christians, challenging a theological commitment to the notion of free will.

Lamarckian inheritance had not been a great problem for Darwin, whose flexible approach had allowed for the inheritance of acquired characters. And in the years immediately following publication of the *Origin of Species*, many theistic evolutionists such as the Harvard botanist Asa Gray and the geologist Joseph Le Conte flexibly adapted evolutionism to their own strongly held religious beliefs. But by the 1890s an increasingly hostile debate between the British sociologist and polymath Herbert Spencer and the German biologist August Weismann had polarized the thinking of evolutionists. Weismann had performed experiments testing Lamarckian inheritance, persuading him and many others that such a pattern did not occur. Furthermore, he joined an increasing number of apostles of positivism, arguing that acceptance of such an unproven theory of inheritance in the absence of good evidence was simply not scientific.

In this time of protean disciplinary boundaries, Herbert Spencer—a social scientist but not an experimental or natural scientist—counted himself a firm evolutionist. He was, after all, the author of the phrase "survival of the fittest." Though his famous phrase came to be used to denote natural selection, Spencer increasingly committed himself to Lamarckism, largely for philosophical reasons. He had a lot at stake in the debate about Lamarckism. Progress was one of the key elements in his all-encompassing system of thought, and he was convinced that if individual organisms could not pass their achievements along genetically to their offspring, progress would not be an inevitable feature of evolution or of social evolution. Lamarckian inheritance allowed organisms—including people—to choose their own evolutionary pathways; natural selection left them passive victims of chance.

Spencer grounded his evolutionism in a laissez-faire liberal metaphysics rather than in religion, but his objections to natural selection were congruent with those of religious thinkers such as the Catholic evolutionist St. George Jackson Mivart and the novelist Samuel Butler, both of whom mounted energetic challenges to natural selection for similar religious reasons. Natural selection, for these people, meant evolution guided only by chance, a world governed by accident. This was not the world they wanted to live in. Furthermore, natural selection seemed harsh, even cruel, mandating the triumph of brute force. For many theistic evolutionists, the rule of brute force—or the "survival of the fittest"—seemed disturbingly un-Christian.[11]

By the 1890s, the experiments of August Weismann had cast doubt on Lamarck's notion of the inheritance of acquired characteristics, but teleology lingered, if only beneath the surface. Nevertheless, it would be an oversimplification to ascribe to metaphysics or religious commitments alone the conviction of paleontologists that the history of life moved along an inexorable upward path. Patterns like the recently uncovered record of the evolution of the horse from the tiny four-toed, more or less omnivorous Eocene "dawn horse" to the modern single-hoofed, long-legged, grass-eating *Equus*, for example, or like the apparently almost universal trend toward larger brains among the mammals, made a remarkable impression. For paleontologists, especially those who studied the patterns of mammal evolution, the evidence for direction in evolution seemed incontrovertible. Real evidence of progress had mounted as fossil hunters sent specimens from the field to museums and universities for analysis. Progress stood out among the aesthetic dispositions of late Victorianism and dovetailed nicely with the ancient tradition of a hierarchical chain of being, from single-celled organism to humans, of course, but the fossil record seemed to bear it out.[12]

Orthogenesis

Osborn, like many paleontologists, was struck by the prominence of long-term trends in the history of life. "Marsh's law" described the observation of the Yale paleontologist Othniel Charles Marsh that the brain sizes of most success-ful mammal taxa had increased steadily during the Tertiary period, often called "the Age of Mammals." Similarly, "Cope's law" referred to a tendency for ani-mals' overall size to increase in evolution. Not surprisingly, then, although they were by no means alone in this respect, paleontologists were among the most en-thusiastic proponents of orthogenetic theories, theories that postulated various kinds of forces inherent in organisms propelling them in the linear evolutionary trajectories that scientists observed in the fossil record and which seemed to de-mand interpretation as evolutionary progress.[13]

From the beginning of his career as a paleontologist, Osborn noted trends in the fossil record that did not seem to him explicable by natural selection. Such characters as the horns that developed in great variety among the Titanotheres—an extinct group of mammals Osborn studied throughout his career—first ap-peared as rudimentary, even "infinitesimal" features with no apparent survival value and grew gradually and progressively, often for a very long time, before they could possibly have any use. Such features could not be explained by any mech-anism that depended upon chance, Osborn concluded: "The main trend of evo-lution is direct and definite throughout, according to certain unknown laws and not according to fortuity."[14]

More problematic were trends in apparently neutral or even deleterious fea-tures, the most often cited being the enormous antlers of the now extinct Irish elk. Why would natural selection have favored traits that must have carried dis-tinct or even lethal disadvantages? Before the assimilation of Darwin's notion of sexual selection—that some traits might flourish not because they enhanced an individual's chances of survival but because they increased its chances of repro-ducing—such characteristics seemed to many evolutionists to be impossible to explain by any mechanism other than a kind of evolutionary momentum gone haywire. The notion of ultimately fatal evolutionary momentum seemed to confirm the idea of mysterious orthogenetic forces.

Still, the major pattern of evolution seemed to be upward and therefore difficult to explain by, as Osborn put it, "fortuity." How, paleontologists asked, could anyone account for the observed progress of life through evolutionary time—an ascent (as they were inclined to put it) that, since it led from simple organisms to a culmination in human beings, obviously made sense—by the kinds of stochastic processes that seemed to lie at the heart of Darwin's theory? Natural selection, many of them thought, could act as a sieve, eliminating unfavorable variations, but how could it account for slowly accreting change in a positive direction?

Such progress simply could not have happened by chance. Paleontologists struggled to find ways to explain it without recourse to "vitalistic" ideas—attributing nonmaterial "vital forces" to living things—which scientists increasingly viewed as more mystical than scientific. These trends, which came to be described but not explained by the term *orthogenesis*, resisted explanation in terms of natural selection because most scientists, unlike Darwin, conceived of natural selection as a purely negative force. The psychologist James McKeen Cattell, editor of the important journal *Science* and later of the *Scientific Monthly*, wrote in 1896 that natural selection—which he placed inside quotations marks—was "no cause of the origin of species, but may be the cause of the annihilation of the unfit species."[15] Many committed evolutionists were skeptical about the adequacy of natural selection, and their arguments against it carried a lot of weight. Even Darwin himself, in *Descent of Man*, confessed that he was sobered by much of the criticism and feared that he might have placed too much emphasis on natural selection in the first edition of the *Origin*. And as Lamarckian explanations came to seem less convincing, paleontologists increasingly invoked "unknown factors" such as Cope's "bathmism" or Osborn's "growth forces." Until he died, Osborn remained convinced that unknown forces were required to explain the trends he saw in the fossil record. He gave names to the trends—for example, "aristogenesis"—but he never found a convincing explanation for them.

In 1896 Osborn proposed a mechanism he called "organic selection" which seemed under certain circumstances to be more satisfactory than natural selection. Intriguingly, three different scientists proposed the same mechanism in the same year—and two of those scientists, the British animal behaviorist Conwy Lloyd Morgan and the American psychologist James Mark Baldwin, did so at the same scientific meeting.[16]

Organic Selection and the Role of Mind
in Evolutionary Progress

In its barest outlines, the theory of organic selection, also sometimes called the "Baldwin effect," postulated that adaptations acquired during organisms' lives could direct their evolution without being inherited. Flexibility of response in the face of environmental challenges could allow animals, for example, to escape elimination by natural selection. This would buy the animals enough time for genetic variations in a similar direction to arise and be selected, adapting them in a more permanent way. For example, if the habitat of a bird species were permanently flooded, those individuals who could learn to wade and forage in and around the water would not be eliminated by natural selection. Their offspring would probably learn from them, giving morphological and instinctive behaviors in the same direction time to accumulate through natural selection. Such a response would mimic Lamarckian evolution but would actually have occurred without the inheritance of acquired characters.

Later generations of biologists would dismiss this idea as unnecessary—in the case of the birds in flooded habitats, natural selection gave the advantage to behavioral flexibility; it was not necessary to invoke any other kind of selection. But the late nineteenth century saw a proliferation of proposals for various alternatives to natural selection. Organic selection was one of a plethora of posited alternatives or supplements to natural selection, many of them called some other sort of selection. The ninth edition of the *Encyclopedia Britannica* described something called "subjective selection" in an article on psychology; even the adamantly selectionist August Weismann offered up "germinal selection"; and the immediate precursor to organic selection, thought by Lloyd Morgan and Weismann to be identical to organic selection, was Weismann's "intra-selection."

There were reasons for these ideas, and for resistance to natural selection, which were grounded in metaphysics, religion, or aesthetics, but they also grew out of observations of nature and explanatory traditions in science. These things cannot be so easily separated. For example, "intra-selection," a physiological version of organic selection, was an attempt to solve the puzzle posed by the coordination of parts: if natural selection favored the alteration of an organ, how did the other organs, so perfectly balanced to fit into a harmonious whole, adjust? The tradition in physiology beginning with the experiments of Claude Bernard emphasizing the perfect homeostasis of the living organism combined with paleontologists' use of the comparative anatomist Georges Cuvier's principle of the coordination of all parts of the animal skeleton made it very difficult to see how natural selection could, for example, alter an animal's dentition without simultaneously adjusting its digestive system. Weismann's intra-selection did not represent simple philosophical predilections alone; nor did Osborn's observations of the trends in fossil horns. Indeed, these two joined the trend among their colleagues of trying to base science on naturalistic nonmetaphysical explanations. This was a central goal.

But at the same time, uneasiness about the notions of "chance" and "fortuity" remained, in most cases in latent form, to be revived by the challenges of anti-evolutionists and the need to defend evolution in the 1920s. This was a problem especially because the scientists who participated in the public debate were disproportionately those who chose to defend evolution as consistent with Christian faith and because many of them were old enough to have been shaped by the debates of the late nineteenth century.[17]

The Evolution Debates

Explanations aimed at the public were difficult because the status of natural selection among scientists remained tenuous through the decade of the Scopes trial. For many scientists of the 1920s, natural selection still seemed to be at best necessary but not sufficient as an explanation of the observed patterns. Though Darwin and other field naturalists had an intuitive sense of the kind of population thinking that made it possible to see how natural selection could work, many other scientists, not accustomed to thinking in statistical or probabilistic terms, did not.

The introduction of Mendelism beginning around 1900 complicated matters even further.[18] In 1866, Johann Gregor Mendel, an Austrian monk, had published results of a long study of patterns of inheritance in pea plants. The results suggested a way in which natural selection might work by postulating what was at the time a novel interpretation of the pattern of heredity. Mendel's experiments went unnoticed for many years, however. Then in 1900, biologists from several different disciplines—Hugo de Vries, Erich von Tschermak, and Carl Correns—within several months of one another, "rediscovered" Mendel's neglected experiments, and the conceptual obstacle presented by Darwin's and other dominant nineteenth-century ideas about heredity seemed at last to have been bridged. Darwin had, along with most of his contemporaries, imagined that the heredity material, whatever it was, acted like a fluid. When the genetic material of two parents combined to create an offspring, the contributions of each parent would flow together, something like the blending of two liquid solutions. This analogy made it hard to imagine how natural selection could overcome the dilution effect: any useful evolutionary advantage enjoyed by one parent would, it seemed, necessarily be diluted by reproductive combination. Mendel's experiments on hereditary patterns in pea plants suggested, however, that inheritance could be particulate, rather than fluid; hence, an evolutionary innovation could be preserved and could accumulate through the generations. But Mendel raised other difficulties for Darwinism. For many early twentieth-century scientists, Mendel's observations seemed to leave the hypothesis of natural selection superfluous: mutations alone, some biologists speculated, might account for evolutionary change.

The doubts of scientists about the efficacy of natural selection were ultimately resolved by the development in the 1940s of the cross-disciplinary synthesis biologists refer to as the "modern synthesis," after Julian Huxley's 1942 book by that name; the synthesis relied heavily on statistical reasoning and on communication across the disciplinary boundaries that divided scientists in the 1920s.[19] This statistical analysis had begun by the 1920s but had not made its way across disciplinary boundaries. Lacking the statistical framework that eventually made the synthesis possible, many biologists of the 1920s (and earlier) saw natural selection as a force that could work negatively, like a filter, to eliminate "unfit" strains but could not account for progress in a "positive" direction. The undeniable pattern of progress and direction seemed especially evident to those scientists who focused their attention on the fossil record.

Furthermore, the professionalization of many subdisciplines in biology involved a proliferation of journals, professional societies, and new disciplinary languages and perspectives. Historians have argued about the degree to which, with the ascendancy of experimentalism, such older traditions as morphology were left behind;[20] what is certain is that new lexicons, methods, and sets of practices coalesced in the various disciplines.

The fragmentation of the biological sciences was not just an intellectual transformation; the institutional and funding contexts of the biological sciences changed as well. As biology splintered into a variety of disciplines and each of these disciplines constructed professional identities and apparatus, professional idioms diversified along with the proliferation of new and specialized journals. In 1927 an article in *Science*, the journal of the American Association for the Advancement of Science and one of the journals that continued to be read across the emerging disciplinary boundaries, announced that Oxford University Press had just published "a list of all periodicals containing the results of scientific research in existence between the years 1900 and 1921." The editor marveled that this list "contains the stupendous number of just over 24,000 separate periodicals."[21]

With the fragmentation of the biological disciplines complicating communication among scientists, exacerbated by the debates with fundamentalists in the 1920s, Lamarckism and other teleological impulses enjoyed a brief resurgence. Lamarckism, though not quite respectable in the 1920s, flowered again, and teleology, still seductive, was more likely to surface in explanations of evolution aimed at the public than in the strictly scientific literature.[22] And it became a question of lively debate in college classrooms. As one Iowa correspondent, describing a passionate classroom debate over Lamarckism and seeking advice from Osborn, concluded, "The lecture room, politely speaking, is a field of battle."[23]

When the experimental biologist Paul Kammerer traveled to the United States in the early 1920s to lecture on evidence purporting to confirm the inheritance of acquired characters, his claims drew wide attention in the press—and cautious skepticism among scientists. Edwin Grant Conklin wrote to colleagues that he suspected that Kammerer had deliberately made reports of his experiments too complicated and difficult to follow so that no one would be able to test his work. Furthermore, Conklin added, noting Kammerer's popular American lecture tour and a feature article in *Time* magazine, Kammerer seemed entirely too eager for publicity: "The manner in which he has been exploited in this country and his published interviews in the papers have been most offensive to good taste and good sense."[24] Kammerer's audience did, nevertheless, include scientists, however skeptical they might be. Osborn wrote to his Columbia University colleague Thomas Hunt Morgan that he had heard a "very agreeable" lecture by Kammerer; although he added that "very few of us agreed with his interpretations," he did muse about it. A few months later he still pondered the question, telling Morgan that although he understood that acquired characteristics were not immediately inherited, they were "by and by inherited through organic selection," and that in that sense "Lamarck was right."[25] Even scientists who firmly rejected Lamarckism saw the question as unresolved, so much so that it required public responses. Thomas Hunt Morgan, who disagreed with Osborn, took the question seriously enough to write a reasoned refutation for the *Yale Review*.

Osborn had rejected Lamarckism in the 1890s, but he began to toy with it again in the 1920s, though he initially did so very tentatively. As he had in his correspondence with Morgan, he increasingly revived his earlier theory of organic selection. He certainly doubted that natural selection alone accounted for the patterns of evolution as revealed in the fossil record. While many anti-evolutionists of the 1920s, following William Jennings Bryan, were more concerned with the deterministic implications of natural selection than with the notion of chance,[26] many evolutionists, including religious scientists such as Osborn, and some non-scientist supporters of evolution did find the role of chance implied in natural selection troubling, and their desire to defend evolution in the face of challenges by anti-evolutionists often focused such evolutionists' attention on the potential of Lamarckian explanations to alleviate the problem. Lamarckism, in these circumstances, resurfaced as a temptation, if not as a conviction.

Furthermore, the process of disciplinary fragmentation in the biological sciences meant that, increasingly, people in laboratory sciences like genetics, for example, were trained differently, worked within different communities of colleagues, and developed verbal and visual lexicons that diverged from those of museum scientists like Osborn. Osborn and the geneticist Thomas Hunt Morgan, for example, exchanged letters that suggest they were asking radically different kinds of questions and speaking different technical languages.

Technically, Osborn, who held an endowed professorship at Columbia as well as the presidency of the American Museum of Natural History, straddled the worlds of the museum and the university, although the museum demanded most of his time. But the roles of the university and the museum had changed considerably. Laboratory scientists like Morgan saw themselves as working at the cutting edge of new and more scientific disciplines. The culture of science changed along with the content.[27]

Morgan criticized Osborn's 1915 book *Origin and Evolution of Life* in some detail, calling it speculative and protesting that Osborn invoked theories of evolutionary mechanisms in terms of a language of "energy," for example, which was borrowed from late nineteenth-century physics and which had no real meaning in Osborn's context. Other reviews leveled the same criticism of this book. Osborn tried to persuade Morgan in letters and discussions, but without success. Part of the problem with this attempt at a dialogue was that Osborn was trying to raise questions about the mechanics of embryology which could not be tested or even explored empirically at the time, but he did resort to a language derived from nineteenth-century physics which was not grounded in real data or method for biology. It was not uncommon for biologists at the time to draw on the language of nineteenth-century physics and to use it in a necessarily metaphoric fashion. This sort of thing, however, sounded vacuous and unscientific to the proudly empiricist Morgan. And his criticisms stung Osborn, who complained as late as 1932 to Nicholas Murray Butler, the president of Columbia, that Morgan's remarks about *Origin and Evolution of Life* had been patronizing. It became increasingly difficult for scientists to speak audibly (at least in great detail) to one another across disciplinary boundaries.[28]

These disagreements involved interdisciplinary competition among colleagues; changing disciplinary questions and perspectives exacerbated them. When Os-

born and Morgan discussed the role of "mutations," for example, they were referring to essentially different things. And some disgruntled naturalists and zoologists objected to the willingness of many of their colleagues to base theories on
experimental studies of a single organism, *Drosophila,* pioneered in Morgan's lab
at Columbia. Why, annoyed zoologists sometimes asked, should we assume that
the genetics of fruit flies can tell us anything about human evolution? Some experimentalists, on the other hand, complained that paleontologists and other naturalists allowed their imaginations far too much play for the purposes of science.

Disagreements about evolutionary theory, especially natural selection, were
based on real differences about scientific method and language, but they also involved philosophical predilections. Osborn was committed to an essentially teleological, progressive view of evolution that scientists like Morgan found more
mystical than scientific. Teleological themes surfaced conspicuously in debates
about the evolution of humans, in which objections to the idea of chance took on
enhanced significance. The closer an animal taxon was to humans, the more explicit the question of inherent direction in its evolution became.

The evolution of the mammals did seem to some people to suggest a progressive pattern, and philosophically there was more at stake in interpretations of
mammal evolution than in the evolution of animals less closely related to humans. Osborn did seem to imply that the rules were different for mammals, but
this was in the context of a general lack of consensus about how evolution
worked. In a disagreement about method, Morgan accused him of exempting the
mammals from the rules of organic evolution: "I am sorry to hear," Morgan remarked dryly, "that the mammals have not evolved by mutation. It would be too
bad to leave them out of the general scheme . . . and I cannot but hope that you
will relent some day and let us have the mammals back."[29] Osborn, Morgan
claimed, had implied that the methods of inference used to examine mammal
evolution should differ from those of other animals because mammal evolution
worked according to mysteriously different rules.

Morgan was not being entirely fair because he and Osborn differed fundamentally about the very meaning of the term *mutation* and about its role in evolution; nevertheless, Osborn did come to draw conclusions about human evolution in essentially different ways than he would about other animals' histories,
and that would cause qualms among some of his colleagues.

And scientists and admirers of science increasingly amplified a philosophical
emphasis on empiricism and objectivity—identifying these things as the essence
of what it meant to be scientific. Principled voices denounced "mere speculation."
The combination of disciplinary fragmentation, the growth of new disciplinary

literatures, and the continuing conceptual disarray of evolutionary theory made the 1920s an awkward time for biologists to defend themselves and their work against attack by anti-evolutionists.

Yet at this very moment of disciplinary restructuring and conceptual disarray, scientists were called upon to defend both their theories and their authority in public school classrooms and in the popular press. Many of them rallied to the defense of evolution, but the defenders did not represent a broad or random cross section of scientists. Fragmentation among the sciences would not be evident to the public, however; if scientists such as Osborn or Albert Einstein or Luther Burbank appeared frequently in the newspapers as authorities, they appeared as authorities on science generally, not just as experts in paleontology, physics, or horticulture.[30] Nor were all scientists equally likely to be asked by the *New York Times,* for example, to comment on matters of public concern; nor were all scientists equally willing to comment publicly.

Scientists continued to debate the merits of Lamarckism, natural selection, and other mechanisms, and these debates spilled over into the public defenses of evolution, despite scientists' best efforts to sequester them. Anti-evolutionists would persistently conflate Darwinism and evolution while scientists sought to make a crucial distinction between them. When scientists used the term *Darwinism,* they customarily referred to natural selection, but in the public debates Darwin was associated inextricably with anything having to do with evolution, so that a challenge to Darwin's theory on the part of any scientist could potentially confuse the interested public.

In addition, although many defenders of evolution argued that science and religion occupied separate realms, so that there need be no conflict between them, it was not always easy in practice to tell where science ended and religion began. Some very prominent scientists attempted to solve the problem of potential religious conflicts with evolution by bringing religion into the domain of science, a strategy that, while sincere, had profound implications for the idea of the boundaries of science and its authority. By the end of the decade the claim that science revealed truths about things like morality or the meaning of life would provoke increasing skepticism. It would also cause uneasiness among scientists, ultimately leaving many of them more reluctant to engage in dialogues with the public.

CHAPTER FIVE

Unlikely Infidels

Even John Roach Straton must have been impressed by the movie *Evolution*, judging by the vehemence of his denunciation of it. An August 1925 editorial in the *New Republic* described the movie, being shown "in all the better theaters" during that summer of the Scopes trial, as "quite remarkable in its effects."[1] The movie began—in the beginning—with the evolution of planets, then of the earth. It included early shots of single-celled organisms, followed by invertebrate animals—including an aquarium octopus pressed against the side of its tank and delicate and difficult-to-photograph coelenterates—swimming; it progressed along a sort of evolutionary chain of being. It had dinosaurs. They were cleverly arranged dinosaurs, too. At first, artistically positioned sculpture dinosaurs peered out from living plant habitats—plants waving in a gentle breeze gave the inert dinosaurs a feeling of movement. Then dinosaurs actually moved. Through brief stop-frame animation sequences, dinosaurs did things. A *Triceratops* munched on plants. A Tyrannosaur harassed a *Triceratops*. The great graphic artist Winsor McKay had produced a clever dinosaur cartoon animation, *Gertie the Dinosaur,* in 1914; there were also dinosaurs in D. W. Griffith's 1914 film *Brute Force* and more engineered by Willis O'Brien for the popular 1925 movie *The Lost World*. The dinosaurs in *The Lost World* were far more elaborate, more numerous, and more active than those in the didactic *Evolution* film; nonetheless, stop-frame animation dinosaurs were still novel in 1925, and many viewers of the film must have agreed that the movie's effects were remarkable. According to *Variety, Evolution* was "roundly applauded" in New York, but that may have been a sign of approbation of its message and an indication of the kind of audience it attracted.[2] After moving through a sequence from invertebrates to fish to terrestrial quadrupeds, the film featured some charismatic and photogenic mammals, including a sloth, an anteater, and a coatimundi; more than any other mammal, however, there were, significantly, primates. The evolution of the primates, the movie made clear, represented a singular evolutionary development, one with portents of greatness. A caption accompanying a monkey announced an implicitly significant innovation: "Mingled with the shrieks, cries and roars of the jungle, a new sound is heard. It is a *distinct chatter.*" Chattering, by implication an augury of speech, evolved with monkeys, according to the film—a familiar notion by 1925. Primate chattering had held a conspicuous place in such Stone Age novels as Jack London's *Before Adam,* and Clarence Day used it to humorous effect in his 1925 book *This Simian World.* Chattering, Day suggested, was the precursor to literature.

Like the movie about relativity, *Evolution* highlighted human technological progress. It arranged its animals in a trajectory of evolutionary ascent, from single-celled organisms through fish, amphibians, reptiles, and mammals to primates, culminating not only in humans but in human achievement in the form of trains, tractors, airplanes, cities, factories, and, of course, laboratories. "Man— a mere atom that could be whisked away like a leaf in the winds—" the movie's caption rather dramatically asserted, "has—by the power of his mind, harnessed the elements and, at his command, cliffs of stone and steel rise from the earth." Here was evolutionary progress on a grand scale. Acknowledgment (or a boast) of scientific guidance from the American Museum of Natural History appeared prominently in the opening credits, and according to the *New York Evening Post,* the movie had been vetted by scientists, but the vertebrate paleontologists and anatomists at the museum could not have been minutely involved with the final version: it included several odd errors that they would never have allowed.[3] Still, the emphases on progress and ascent, culminating in the final message of the film, were those sanctioned by Osborn and the scientists most prominent in the debate.

The film's implications disturbed the editor of the *New Republic,* however. The evolutionary sequence of aquarium, museum, and zoo faunas concluded with a frequently encountered piece of doggerel: "Some call it Evolution. Others call it the work of God." For the editor of the *New Republic,* the suggestion that science could guide us to theological conclusions was both ominous and false. Naming Osborn as one of the scientists broadcasting this sentiment in the popular media, the editor complained, "It can do no good to point out that 'god' is not a scientific conception, that scientific researches reveal nothing but material facts, that spiritual principles are as irrelevant to biological evolution as jabberwocky. Every scientist knows this." Emphasizing Osborn's stature explicitly, he went on, "No one understands better than the president of the American Museum of Natural History that in the process of biological evolution the one test of fitness is the fact of survival. . . . Nothing could be further from any 'spiritual principle' than biological evolution."[4]

But Osborn appeared to have understood no such thing. He was not being, as the editorial implied, disingenuous. More and more tenaciously during the course of the evolution debates of the 1920s, he advocated a form of theistic evolution that he believed would comfort those people who worried about the possible implications of evolution for their religious faith. In particular, he invoked a notion of "creative evolution" that avoided the materialistic implications of Darwinism by postulating an inherent teleological mechanism guiding the history of life. The term *creative evolution* would have been familiar to readers of the French philosopher Henri Bergson, whose book of that name enjoyed something of a vogue in the United States, and Osborn often cited Bergson. His notion of creative evolution had also grown out of his accommodation to evolution in the late nineteenth century as a student of James McCosh at Princeton and as a friend and admirer of the neo-Lamarckian paleontologist Edward Drinker Cope. It blended religious predilection with paleontological observations, including the observation of long-term, apparently directional trends in fossil lineages. These apparently directional patterns in the history of life made orthogeneticists of many—some said most—paleontologists. In the context of the evolution debates of the 1920s, this directionality seemed to Osborn to be especially germane: if only the public could be made to understand it, they would be reassured. Osborn and other signers of the Millikan Manifesto continued to argue that evolution was no threat to Christian values because it provided a reassuring message about the meaning inherent in human life. In this respect the popular evolution movie reflected the views of scientists most often represented in newspaper and magazine accounts of the debate. As the debate heated up, this kind of argument would cause confusion and misunderstanding.

Intriguingly, nonscientist intellectuals, like the editor of the *New Republic,* often assumed the centrality of natural selection far more confidently than did many scientists, who, though convinced of the reality of evolution, were tentative about the role of natural selection. This reluctance among scientists to accept natural selection as the mechanism of evolution did not in general mean that they rejected natural selection altogether—only that they did not see how it could be a sufficient explanation by itself for the evolutionary patterns they observed, especially in the fossil record. It was a complicated thing. There were scientific reasons for skepticism, but in some cases it was also associated with accommodations of evolution to avoid what a number of people perceived as disturbingly materialistic implications of pure Darwinism. Many of the defenders of evolution in the 1920s had, like Osborn, made this accommodation between evolution and religion as students in the nineteenth century. Adopting forms of theistic evolutionism had eased their own way earlier on, and in the 1920s they offered it to a public they saw as turning to them for guidance. Evolution, they argued, need not imply, as Bryan claimed it did, the doctrine that "might makes right," or, in Herbert Spencer's phrase, "the survival of the fittest." Rather, they argued, evolution revealed the glory of God. And it suggested a hopeful pattern of ascent, of progress.

Secular nonscientist intellectuals in general were also—perhaps ironically—more likely than the scientists associated with Osborn and the Millikan Manifesto to insist that science and religion occupied entirely different realms. Some of them even agreed with fundamentalists about the radically unsettling cultural and metaphysical implications of evolution. Although there were scientists who shared these perspectives, the public voices of science—the rhetorical positions of the most visible scientists—would try to soothe such fears. The number of scientists and science popularizers arguing for the compatibility of science and religion increased so conspicuously during the course of the controversy that the book review section of one science journal, the *Quarterly Review of Biology*, began referring to such works as a genre: the reconciliation books. Reconciliation books would turn out to be problematic as vehicles for the messages of science.

Scientists' attempts to solve the problem of potential religious conflicts with evolution by bringing religion into the domain of science had implications for the idea of the boundaries of science and of its authority. By the end of the decade this effort implicitly to extend the boundaries of science to the realms of morality would cause some intellectuals outside the community of science to become skeptical of scientific authority. It would also cause uneasiness within the community.

The Earth Speaks to Bryan, or, The Ascent of Man

In the summer of 1925, just in time for the trial of John Thomas Scopes for teaching evolution in Tennessee, Osborn published a book of essays addressed to William Jennings Bryan. Osborn called the book *The Earth Speaks to Bryan*, citing a line from the Book of Job (12: 2): "Speak to the Earth and it shall teach thee." The earth, Osborn quipped, had been speaking clearly to William Jennings Bryan for many years, but Bryan refused to listen.[5]

Citing a line from Bryan, "truth is truth and must prevail," Osborn insisted that the evidence of paleontology—written on the rocks and in the earth—demonstrated the truth of evolution so irrefutably that even Bryan would be convinced if only he would pay attention. Bryan denied it. Scientists like Osborn, he claimed, peddled mere guesswork wrapped in the pretentious jargon of science. Darwin's guesses, Bryan wrote, "would not have survived for a year" if they had not been "buoyed up by the inflated word 'hypothesis.'"[6] Bryan and other anti-evolutionists, such as Alfred Watterson McCann and Straton, hammered this ac-

cusation relentlessly: evolutionists made guesses and tried to sell them as "hypotheses." Osborn perhaps necessarily had to adapt his defenses of evolution in
response to this accusation, an imperative that would turn out to be frustratingly
difficult.

The Earth Speaks to Bryan was an anomaly among Osborn's books in several
ways. The title, which the publisher, Charles Scribner, disliked, was a departure
for Osborn, who normally prided himself on avoiding ad hominem attacks. Scribner suggested "The Ascent of Man," but Osborn vetoed that as too "highbrow"
for the readers he hoped to influence. "If I write this book," he told Scribner,
"I stoop to conquer." He envisioned the audience for the book as consisting of
"people who are traveling in railway trains and going off to their summer homes,"
which appeared, oddly enough, to be his idea of a middle-class audience.[7]

Unlike such books as *Men of the Old Stone Age,* Osborn's well-illustrated and
detailed history of human ancestry, *The Earth Speaks to Bryan* left out the real data
of evolution. Most notable, it included no illustrations—and Osborn, of course,
had long been adept at the use of images in communicating scientific ideas. Yet
he saw the *Earth Speaks to Bryan* as having a positive didactic effect, despite the
mixed reviews it received. He wrote to friends that his *Evolution and Religion*—an
expanded collection of similar essays—had played a critical role in staving off the
threat of anti-evolution legislation in several states. The mixed reviews of both
books, however, along with their sluggish sales in contrast to the very robust sales
of the much more technical and expensive *Men of the Old Stone Age,* suggest that
in reality, when he saw himself as stooping to conquer, he was addressing an audience he did not know or understand.[8] It also suggests, perhaps, the enhanced
appeal of illustrated texts.

Nevertheless, *The Earth Speaks to Bryan,* one of the more widely publicized of
the decade's attempts to popularize evolution, remains a fascinating document
in several respects. In it, Osborn insisted that the crucial issue at the Scopes trial
was not—as many people would argue—academic freedom or freedom of speech.
The important question at the trial was whether, in teaching about evolution,
John Thomas Scopes had been teaching his students the truth. Noting that lawyers on both sides of the case would undoubtedly raise such issues as "personal
rights, rights of opinion, rights of free speech, constitutional rights, [and] educational liberty," Osborn predicted that these problems might "for a time befuddle
the minds of the jurors." Such concerns were, he maintained, "mere temporary
side issues and will fade into insignificance in comparison with the supreme
issue." The essence of the case boiled down to this: "If Scopes has been teaching
the truth to his students he will win; if he has been teaching untruths he will lose,

and will deserve to lose." Though he was "a great believer in educational liberty," Osborn wrote, he insisted that the teacher "is at liberty only to teach truths which are well and soundly established."[9]

For the reader at the beginning of the twenty-first century, Osborn's dismissal of the civil liberties issues involved in the Scopes trial may seem startling. A number of recent historical revisions of the Scopes trial literature have insightfully focused on just the issues that Osborn so peremptorily dismissed.[10] For many people at the heart of the controversy, the truth of evolution mattered primarily because it highlighted the indispensability of academic freedom and freedom of speech. And scientists—including Osborn—were actually quite concerned with academic freedom and freedom of inquiry.

Osborn received many letters describing the fears of junior colleagues that they might lose their jobs for teaching evolution or for speaking out publicly in favor of it. Some of these fears appear to have been well founded; for example, his colleague, the embryologist Edwin Grant Conklin, wrote to him that a former student, H. H. Lane, had been threatened with losing his position at the University of Kansas if he were to participate in the Scopes trial defense.[11] Even Conklin, who was Osborn's most active collaborator in the defense of evolution, worried about the possible repercussions. Conklin, a professor at Princeton University and a highly respected scientist whose reputation and place in the academic hierarchy would seem to have been secure, worried that writing about evolution and religion might "raise a storm, especially in Blue Presbyterian circles." When the evolution debate heated up, he wrote to friends that personal attacks on him by anti-evolutionists made him apprehensive about the effects on the Princeton trustees and about the potential effect on his status among professional colleagues.[12] Osborn himself expressed concerns—not for his own position, certainly, but on behalf of others. In 1922 he had written to the publisher Arthur Scribner, for example, "College professors dare not mention the word evolution for fear of being turned out of their chairs."[13] And in 1925 he created a stir when he was quoted in the *New York Times* complaining he had learned during a visit to Nebraska that teachers there were afraid to include evolution in their lectures on biology.[14] His statement in *The Earth Speaks to Bryan* dismissing academic freedom as a central issue in the Scopes case makes sense only as an example of significant hyperbole, an intriguing rhetorical flourish, undoubtedly in large part shaped by the necessity of responding to relentless charges that his museum was putting nothing but fanciful guesswork on display, dressed up in pretentious words like *hypotheses* and *theories*.

The Earth Speaks to Bryan presents a striking counterpoint to the complaint of the editor of the *New Republic*. In this decade of magazine popularizations of the "Einstein theory" and of Heisenberg's uncertainty principle, arcane and difficult theories, radically simplified in popular accounts, enjoyed a certain vogue, often as metaphors taken to mean that the comfortable old notion of truth had, in the modern world, become unexpectedly tenuous. Surely a man as sophisticated as the president of the American Museum must be, the *New Republic* editor assumed, could not possibly be serious when he claimed that there were no theories, only facts, on exhibit in the museum. This claim, however, Osborn made most sincerely, not only in *The Earth Speaks to Bryan* but more and more frequently in essays, addresses, and interviews. Increasingly prominent in newspaper accounts of the evolution debate, it complicated things a good deal.

Evolutionary theory, for one thing, was not so untroubled as Osborn's confident pronouncements made it sound. Osborn's confidence in the transparent truth of scientific ideas is perhaps even more intriguing than his dismissal of the question of academic freedom. He was not being disingenuous, precisely, but in one sense the *New Republic* editorial was quite correct: Osborn certainly did understand the contingency of scientific theory. In 1910, in *The Age of Mammals*, he had referred to scientific diagrams, like those on display at the museum, as "the working hypotheses of science."[15] In museum guide books he had repeatedly cautioned visitors that as our knowledge improved the exhibits would change to reflect our growing understanding. He had himself long wrestled with the details of evolutionary pattern and mechanism and continued to do so. This was especially true in the case of human evolution. The evidence he defended so staunchly, on display at the American Museum, included some very fragmentary fossil specimens, including casts of the Piltdown skull, of which he and many others had initially been skeptical. Discovered in England in 1912, the Piltdown specimens combined a large modern brain with a "primitive" chimpanzee-like jaw in a semblance of "the missing link" that seemed made to order—too good to be true. Indeed, not all of the curators working with Osborn at the museum had been able to shake their doubts about Piltdown; those who suspected a hoax were ultimately proved correct.[16] And although he remained a formidable figure in the scientific community, some of Osborn's interpretations of the pattern of evolutionary history met with increasing skepticism during the decade, in large part because of the exaggerated positions he took during the course of public debates with fundamentalists.[17] The leitmotif of *The Earth Speaks to Bryan*, however, echoed and amplified the rhetorical strategy he and other scientists prominent in the debate had adopted to ease the fears of the devout.

The Earth Speaks to Bryan argued that evolution, properly understood, did not represent a challenge to anyone's religious faith; indeed, it confirmed religious hopes. "We naturalists take as transcendent," Osborn generalized, "the teaching that the universe is by no means the result of accident or chance, but of an omnipresent beauty and order, attributed in the Old Testament to Jehovah, in our language to God."[18] Quoting the Millikan Manifesto, he wrote that "it is a sublime conception of God which is furnished by science" of evolution "culminating in man with his spiritual nature and all his Godlike powers."[19]

These two arguments—that evolution was well and simply true and indisputable and that it was compatible with, and even reinforced, Christian faith—provided the scaffolding on which theistic evolutionists constructed responses to the anti-evolution crusade, in part because Osborn and Conklin led in the effort to defend evolution and because they enlisted like-minded colleagues in the cause. In addition, these very prominent scientists exercised considerable influence over hinterland science teachers and popularizers. Writers and educators looked to the American Museum of Natural History for vivid illustrations of evolutionary concepts. Teachers accustomed to turning to the American Museum for lantern slides, casts of fossils, reconstructions of human ancestors, diagrams, and wall charts now also wrote to the museum to ask for advice about how to handle the current controversy and to report on their own local efforts to do so. This was true also of interested laypeople, such as Osborn's old friend W. W. Keen, a Philadelphia doctor, who wrote essays and a book defending the compatibility of evolution and religion, and of many liberal and modernist ministers. These people wrote to Osborn because they were familiar with the American Museum, and they wrote to the scientists whose names they knew and to scientists they understood to embrace a view of evolution compatible with religion. Newspapers turned especially often to the same group of scientists for authoritative statements.

Scientists' statements of theistic evolution vastly oversimplified the positions of the larger community of scientists and even of some of the other scientists who participated in the debate. Such statements were, however, overwhelmingly predominant in popular media representations of evolution, including the movie *Evolution*. Newspaper and magazine accounts of the evolution controversy often adopted as obvious, sometimes with a something of a superior shrug, the assertion that evolution and religion were compatible—a truism of modern science. Yet it was not at all obvious to some dissenters, such as the *New Republic* editor, or to all scientists. It was, indeed, a line of argument that many scientists who avoided the public debate came to find troubling.

In most respects, it would have been impossible for one person to represent the views of scientists at large on the evolution controversy because scientists disagreed in important ways. The category "scientists" was far from monolithic or unified. As the thoughtful Harvard University geologist Kirtley Mather wrote, "No individual man of science can speak for all scientists, nor can he represent science. As a matter of fact, there is no such entity as 'Science.' There is rather a large group of individual scientists." Mather was, like Osborn, a public defender of theistic evolution, but his public statements tended to remain more cautious than Osborn's became. "All knowledge is merely an approximation to the truth" because "we observe a process of which we are a part."[20] Osborn had often voiced similar epistemological cautionary notes, but in leading the charge against fundamentalists, he began to downplay them in favor of confident assertions of indisputable truths established by science. Reconcilers' avowals of the meaning of evolution would be jumbled together in popular books and the press with discussions of the kinds of evidence on which scientists based their acceptance of evolution as established fact, and the distinction between these kinds of arguments would often be lost.

Defending Evolution: The Arguments

Defenses of evolution most often encountered in popularizations of the decade fit into several categories. First, scientists used the *Origin of Species* as a template, citing evidence from biogeography, embryology, anatomy, and paleontology. For example, the embryological arguments of the late nineteenth century, revived in the 1920s especially by Edwin Grant Conklin, combined the evidence of recapitulation of evolutionary pattern in the development of the embryo with the metaphoric solace that evolutionary change is really no different from development. One essay, by a former student of Osborn's, suggested that intelligent beings visiting from some other planet might be incredulous if they were informed that adult humans had originally been children.[21] Scientists and popularizers, naturally, sought metaphors and figures of speech that they hoped would pique the popular imagination. One of the staples of the evolutionary literature, Darwin's evidence from homologous organs—different organs that had originated from the same evolutionary origins, such as the wing of a bird and the arm of a person—became, in the 1930 children's book *The Earth for Sam*, the story of how "a fin becomes a foot."[22]

In an intriguing set of variations on Darwin, scientists cited the perfection of animal adaptation, with its echoes of the argument from design, but so was it common to find the argument that evolution must be inferred because of the imperfections of organisms.[23] If God had designed the kangaroo, why would he have made it so silly, or made the tinamou so stupid or the sloth so improbable? Why make the turtle so clumsy—surely the turtle could not be cited as an example of optimization in nature? The answer was that evolution, unlike the notion of divine plan, left room for historical oddness. This line of reasoning was similar to Darwin's argument that vestigial organs such as the human appendix could be explained by evolution, but it was perhaps, in the context of the 1920s, spiced with a more metaphysical flavor.

The second category of defenses had to do with scientific method. Scientists invoked the idea that the evidence for evolution was cumulative and corroborative across disciplines. They further argued that only evolution explained all the evidence in this way—as J. Arthur Thomson put it, all the lines of evidence met in evolution.[24] Many of the arguments made by scientists were, like Thomson's, verbal descriptions of images, illustrations of ideas about how science worked.

A related inference insisted that evolutionary theory was therefore necessary, that it was "the unifying principle that binds all science together," as Arthur Miller put it, "the 'open sesame' to which nature yields her secrets." This implied too that—as most of the scientific witnesses at the Scopes trial would attempt to testify—biology simply could not be taught without evolution. Many scientists making this argument added that an evolutionary perspective was essential to advances in modern medicine.[25] Advances in technology, so notable in the decade of the 1920s, seemed to bolster the claims of science in general, and defenders of evolution cited medicine and technology in support of the efficacy of scientific method. The airplanes, trains, factories, and laboratories depicted in the movie *Evolution* took on an iconic status that made the very images or mention of such things persuasive. Technology itself, like its representatives Thomas Edison, George Eastman, and Luther Burbank, had become a kind of celebrity, and popularizers recruited the star power of technology in service of evolution even as intellectuals debated the connection between science and invention.

Like other kinds of scientific theories—especially the laws of physics—evolution had predictive ability. Just as the law of gravity allowed physicists to predict the patterns of physical phenomena, and in the same way that astronomers predicted the existence of a planet where Neptune was before they could actually see Neptune, understanding evolution allowed biologists to predict "missing links" and then find them. Scientists had predicted—and Ernst Haeckel named—a human progenitor similar to *Pithecanthropus* before the fossil remains were actually found. The name, assigned to a hypothetical ancestor not yet encountered in the fossil record, amounted to a prediction, a hypothesis. And when the Dutch anatomist Eugène Dubois found the fossils he named *Pithecanthropus* in Java, in 1891, the "Java man" seemed to provide a confirmation of Haeckel's prediction. Prediction of such "missing links" was for biologists tantamount to scientific hypothesis testing.

Scientists emphasized that these predictions were not "mere guesses." They felt considerable pressure to make this point emphatically because of the relentless barrage of speeches and newspaper articles in which Bryan accused them of publishing and exhibiting guesses, of broadcasting speculations disguised as science: "The word 'hypothesis,' though euphonious, dignified and high-sounding, is merely a scientific synonym for the old-fashioned word 'guess.'"[26] Other anti-evolutionists vigorously took up this theme. To some extent, anti-evolutionists' relentless repetition of this charge forced the kind of adamant response Osborn made in the *Earth Speaks to Bryan*. Scientists were at considerable pains to distinguish between guesses and hypotheses. Anti-evolutionists' challenges, especially this rhetorical strategy, put scientists in a difficult position: they were pressured into invoking their experience, training, and authority and into insisting on the truth of scientific theories even as they agreed that the best way to teach science was not by demanding the rote learning of established facts but by active participation in the work of science in the field and in the lab.

Scientists and science popularizers often asserted that students should learn science not as a body of facts but as a method. Many popularizers pointed out that facts were not the essence of science: science was a method, an approach. In essays about science education, scientists routinely advocated teaching science by encouraging students to do experiments, to learn science by doing science. One of the most prolific popular science writers, Benjamin C. Gruenberg, a New York City writer on education issues, author of *The Story of Evolution*, wrote in an essay in *School and Society* in 1925 that science was "an instrument for restating truth, for changing opinions, for breaking up prejudices, for challenging convictions." When students were taught to memorize the "facts" of science, or even scientific laws, they were actually being trained to be *un*scientific because "scientific generalizations or laws are subject to change with little or no notice." The way to teach science was to teach experimental method, to instill skeptical habits. A belief in absolute truth, Gruenberg warned his readers, fostered bigotry. Science should be taught as "a way of looking at the world," one that "lives day by day on the basis of verities that are subject to change."[27] Gruenberg's themes must have resonated with scientists who encountered them; similar discussions occurred regularly in the pages of the journal *Science*. But the attacks of anti-evolutionists complicated this effort. Anti-evolutionists' accusations that scientists inflicted mere "theories" on innocent schoolchildren, and their insistence that theories amounted to no more than guesses, put scientists defending evolution in a frustrating position. Even though scientists often embraced views like Gruenberg's, the controversy forced them to fall back on assertions of authority. It was their experience as scientists that qualified them as arbiters of scientific truth; it was his lack of experience—or knowledge—of the practice of science that disqualified Bryan. Invoking authority could be perilous though, and sometimes people did it in a coded way, for example, through reference to technology, medicine, and physics—and, implicitly through references to physics, to a history of science as victim of intolerance and superstition.

Analogies with physics were common, such as the claim that evolution was as well established as Newton's law of gravitation or the common comparison with the straw man of the round earth. Physics, after all, enjoyed considerable prestige, and the idea of scientific "laws" held a lot of cachet—this is probably why Gruenberg made such a point of warning that such laws should not be taken as absolutes. Arguments in defense of evolution routinely associated the names, faces, and personalities of great scientists with scientific laws. Analogies of Darwin with Copernicus, Galileo, or Newton stood out, used as shorthand or code for invoking, implicitly, analogies with the heliocentric theory or with gravity—and with the very idea of physical laws. In addition, allusions to gravity called to mind the fashionable concept of the historical warfare between science and theology made popular in the books by Draper and White. Evolutionists, the association with gravity might remind readers, could be subject to the kind of persecution Galileo was widely understood to have unjustly suffered. So the very mention of gravitation suggested that scientists were in danger of being persecuted as advocates of truth and reason.

The Authority of Experts

Scientists often saw themselves as avatars of reason and as under attack by people who were irrational. And it often turned out to be difficult to explain evolution without recourse to their years of training and experience. Scientists and science popularizers regularly cited the authority of scientists and their years of practical experience, experiment, and observation. All scientists "in good standing" agreed, according to this argument, and therefore there could not be any reasonable doubt. Rejoinders to anti-evolutionists regularly included phrases such as "I know of no scientist of any standing who doubts the status of evolution." Nonscientist defenders of evolution, such as the indefatigable Maynard Shipley, made this argument as faithfully as did scientists themselves. The status and authority of the scientists defending it, these arguments implied, counted as evidence in favor of evolution. Biologists often claimed that no one could entirely comprehend evolution without firsthand experience in the field or the laboratory, and the weight of years of experience added substance to a scientist's credibility. Hence Osborn routinely cited his "fifty years of experience," and others said the same thing in their own ways. This was among the most frequent themes of scientists. Perhaps it had to be, given the kinds of challenges anti-evolutionists mounted.

As a corollary, scientists scoffed at anti-evolutionists, especially Bryan, who challenged evolution in the language of science, on the grounds that he and his allies had no scientific acumen and therefore lacked the authority to be taken seriously. Edwin Grant Conklin, for example, wrote: "It is not on record that Mr. Bryan has ever made any discoveries with regard to evolution or that he has made any careful study of the subject, even at second hand."[28] And when a popular figure like Luther Burbank remarked, "Those who would legislate against the teaching of evolution should also legislate against gravity, electricity and the unreasonable speed of light," he implicitly combined the status of physics with the concept of scientific law and the authority of science to make fun of anti-evolutionists as absurd, irrational and, especially, hopelessly behind the times.[29] Charges of being "behind the times" carried a special resonance in the decade of the 1920s and implied that the opinions of those who were not current with modern knowledge carried no weight—were even laughable.

Scientists often implicitly invoked their own authority—and challenged that of anti-evolutionists—by ridiculing fundamentalists. For example, Conklin quipped, "Apparently Mr. Bryan demands to see a monkey or an ass transformed into a man, though he must be familiar enough with the reverse process."[30] Droll references to Bryan and other fundamentalists became familiar to readers of *Science*, the journal of the American Association for the Advancement of Science (AAAS), and to readers of most newspapers. Conklin referred to the antics of "Billy Sundae," and a New York chemist, identifying himself as "one who is steeped in the 'slime' of science," wrote to Osborn making fun of "funnymentalists."[31] Scientists enjoyed cartoons of Bryan every bit as much as anyone else: the archives of the American Museum include whimsical Bryan parodies by museum artists, circulated among the staff, and the journal *Scientific Monthly* republished especially pointed cartoons of Bryan as an Inquisitor or a Crusader. Luther Burbank told an interviewer, "Mr. Bryan is an honored friend of mine, yet this need not prevent the observation that the skull with which nature endowed him visibly approaches the Neanderthal type." The *New York World* printed Burbank's comment beneath a photo of Bryan juxtaposed with a bust of a Neanderthal from the American Museum.[32] Humor, as Burbank understood, could be a potent rhetorical weapon.

Humorous images held particular force. Cartoons poked fun at almost everyone and everything having to do with evolution. People on all sides of the evolution question used humor as a weapon, however. Bryan himself certainly did. John Roach Straton, derided as a "funnymentalist," in turn dubbed evolutionists "funnymonkeyists." As the Scopes trial drama played itself out, many thoughtful people complained that the relentless monkey jokes obscured the serious issues at stake. Humor made a dangerous, if effective, weapon.

Arguments from scientific authority, however implicit, could also—like humor as a weapon—be somewhat risky, since scientists were at pains to reject arguments based on religious authority. It was an odd twist on the argument from scientific authority, in part, that led Osborn and the other scientists who rallied to the defense of evolution to emphasize the religious, churchgoing credentials of the scientists engaged in the debate. Anti-evolutionists routinely accused scientists in general of atheism. Osborn was astonished, and somewhat bemused, to learn that Bryan had lumped him in the same category with Voltaire, Thomas Paine, and the notorious agnostic Robert Ingersoll.[33] Although his approach to evolution had always taken a vaguely religious cast, it was probably in this context that he began to emphasize his devout upbringing. "I was brought up in a religious atmosphere by a wonderful Christian mother who taught me, above all things, to tell the truth," he wrote in a letter to the Reverend John Roach Straton; at the outset of the Scopes trial he sent a similar letter to John Washington Butler, the author of the Tennessee anti-evolution law.[34]

An account in the *New York Times* in December 1925 of a gathering of scientists and other luminaries to dedicate the new Peabody Museum at Yale University included a subhead: "All Believers in Religion." A majority of the scientists present, the paper noted, "are church members or active in church work, and all of them are believers in evolution. Close inquiry today did not reveal an atheist among the several hundred scientific men."[35] A few months later, the *Literary Digest* published a report, from the *Kansas City Central Christian Advocate,* on a Kansas City meeting of the AAAS which noted that "there was not one of those men, eminent in scientific research and, of course, an evolutionist, who was an atheist; and so far as we checked them up they were Christians, many of them workers in their local churches."[36] The argument for the reconciliation of evolution and religion had clearly made an impression.

Creation by Evolution

The editor of one of the most successful of the reconciliation books, the pro-vocatively titled *Creation by Evolution,* published in 1928, determinedly set out to recruit scientific contributors who had established sturdy credentials as religious believers. A fascinating, assertive, and tenacious woman, Frances Mason was not a scientist but was committed to popularizing the reassuring notion that evolu-tion reaffirmed the happiest kind of Christian theology. For several years she ded-icated remarkable energy to recruiting like-minded scientists to contribute essays for a book that would, she hoped, spread the message of theistic evolution to the lay public. She was careful to approach only the most prominent scientists and to choose scientists who shared a religious perspective of evolution; and she en-listed their aid through a tireless campaign of flattery, badgering, and cajoling.[37]

It would take her a great deal of effort to persuade these busy scientists to write chapters for her book, and it would take even more work to coax them to revise their essays into a form that could be understood by an audience of nonscientists. Her correspondence with Edwin Grant Conklin demonstrates how hard it could be for scientists to write for general audiences. Most scientists could not really imagine a popular audience very accurately. Conklin's idea of "non-technical lan-guage that all may understand," as the subtitle of *Creation of Evolution* put it, was profoundly unrealistic, as Frances Mason well understood, to her chagrin. They exchanged many frustrated notes. Mrs. Mason asked for "about 3500 *interesting* words ... Simple and non-technical—even romantic.*"[38] Conklin replied that he did not know whether it was possible to make his essay "simple and attractive" and at the same time "really a paper worthy of a scientific man." He feared that any essay "sufficiently accurate to be worthy of a scientist" was "apt to be less pop-ular or sensational than may be desirable in a book that is addressed to the un-scientific public."[39] In trying to persuade him to revise, Mason wrote that almost all the authors had rewritten their essays extensively, and sometimes entirely, in order to be comprehensible to outsiders: "Dear Prof. C. Lloyd Morgan wrote his paper all over three times! . . . I think nearly everyone has done his paper over and over! It *is very hard to come down to the anti-evolutionists standpoint!"*[40] Finally Con-klin, exasperated, declared that he could not possibly simplify more than he had without compromising his scientific integrity, and he suggested that his contri-bution to the book be eliminated: "I suggest that you omit my article as the surest and best way of improving it."[41] Eventually Conklin did revise his essay, although its vocabulary remained relatively arcane. And the book did find an audience, or at least sold well. But it became clear to many scientists that the balance between comprehensibility and misrepresentation of complex ideas was difficult to achieve, and not all of them continued to think it was worth the sacrifice of time spent away from their own research.

Ultimately, though, Frances Mason published a book that was widely lauded as one of the best books of its kind, a book that sold very well, was recommended by the Book of the Month Club, and excerpted for broadcast over the radio.[42] "Stories" from the book were broadcast on NBC radio, and an advertising flier featured a recommendation by Ida Tarbell, expressing a hope that "study groups, women's clubs and reading circles" would adopt it. One reviewer thought it "should be on the shelf of every library, on the table in every home from enlightened Massachusetts to benighted Tennessee," and several others suggested that Osborn's preface would, as one of them put it, "ensure it a warm welcome."[43] It did apparently receive a warm welcome. Mason's book packaged the varieties of theistic evolution that became increasingly familiar to newspaper and magazine readers throughout the decade. Her title, *Creation by Evolution,* sounded strikingly similar to the phrase Osborn used to describe his own version of theistic evolution, "creative evolution."

This perspective continued to echo in books and articles addressed to the public throughout the decade. The packaging of evolution as a latter-day argument from design belonged to what was ultimately among the most prominent—and most problematic—of the rhetorical strategies of the defenders of evolution: that evolution was a satisfying explanation and not irreligious. The visible scientists endlessly reiterated the claim Osborn made in *The Earth Speaks to Bryan,* that evolution reaffirmed Christian values. This was really not a scientific argument at all but a vaguely theological one, and although it was often cited in newspapers, it aggravated many secular nonscientist intellectuals and some scientists.

An entomologist sent Osborn a letter objecting to the attempts of scientists to reconcile evolution and religion, saying, "Many scientific men are hypocritical and have (for various reasons of their own) side-stepped the issue." He cited the Millikan Manifesto in particular. Bryan was correct in denying the compatibility of religion and science, and though he himself was a scientist, he was "disgusted with many that call themselves scientists." Echoing a charge others had lodged against Osborn, he explained, "I can't help but feel that many of them are not frank and that they are playing to the galleries." He enclosed two newspaper clippings "that would lead many to believe that you and Mr. Bryan were quite in accord and he may think you have gone back up the tree."[44] Some nonscientist book reviewers, responding to *The Earth Speaks to Bryan,* repeated the accusation that Osborn was "playing to the galleries."[45]

Edward Berry, of the Johns Hopkins University, wrote in *Science* that special-ization had left many scientists very narrowly educated and that "in these days of uneducated scientific experts" it was necessary to "stop our ears to the lure of the sirens of speculation, and the imaginary spiritualistic voices of arm chair phi-losophy."[46] The prominent zoologist Vernon Kellogg, writing in *World's Work*, agreed with the position that science and religion were not antagonistic but com-plementary, but he drew back from any claim on the part of science to reveal re-ligious truths: "The cause of things may be called God; the manner of things, sci-ence. Science has never explained ultimate causes. . . . Primal being and ultimate becoming are beyond the purview of science. They are truly something science doesn't know, and I doubt very much will ever know."[47]

Some of the most successful popularizers were actually much less likely to embrace theistic evolution, and because they surveyed biology across disciplines in a way that few practicing biologists did, they could be quite synthetic, incorpo-rating wide reading in the most recent literature across the range of biological disciplines. Henshaw Ward, author of *Evolution for John Doe*, for example, read widely among the popular works of scientists and thought and wrote very clearly. Employing a clever rhetorical device, Ward recounted an imagined tutorial ses-sion with a scientist about the argument from design, pretending to have be-lieved in this idea—which was, he said, a natural thing to believe—then relating the patient, clear arguments of the scientist, who explained to him why the de-sign idea is mistaken. Ward accepted natural selection, and he had read Thomas Hunt Morgan, whom he praised for having made a "wonderful" contribution to evolutionary theory by explaining the role of chromosomes and how they worked. In some ways *Evolution for John Doe* was much closer to the spirit of science as de-scribed by Gruenberg than were the reconciler arguments: Ward wrote respect-fully of scientists, even deferentially, but he was not afraid to respectfully dissent when he thought they were mistaken. Above all, he investigated the nature of the authority scientists held, carefully separating their authority on matters of sci-ence from any claims to special wisdom beyond their realms of expertise.[48]

Not everyone was so circumspect. The physicist Michael Pupin said in an in-terview that science had demonstrated that "back of everything there is a definite guiding principle, which leads from chaos to cosmos." Science proved "the beau-tiful law and order" of the universe, that the cosmos was not "simply the result of haphazard happenings." Pupin concluded, therefore: "Today the average man understands the structure of the universe incomparably better than did the prophets of three thousand years ago. We scientists, therefore, ought to be able to teach him more about the Divinity."[49] Statements of this kind worried people who believed that the boundaries between science and religion should not be so permeable. Some scientists thought that rhetoric like Pupin's and Osborn's called into question the very integrity of science. For some scientists who were evolutionists but not theistic evolutionists and for many secular nonscientist defenders of evolution, such as Henshaw Ward, Benjamin Gruenberg, and editors of such magazines as the *New Republic,* a defense of evolution in religious terms constituted a serious danger to science—and to the authority of scientists.

Ira Cardiff, the scientist whose letter to the AAAS had done so much to stimulate the organization to install Osborn, Conklin, and Davenport on a committee to study the problem of anti-evolution initiatives, ultimately expressed disgust with the result, setting off something of a debate in the pages of *Science.* Citing attempts to reconcile science and the Bible, Cardiff wrote to the journal, "I wish to protest against such methods. The cause of science is in deplorable straits when it must be defended by such so-called scientists who would attempt to reconcile it with primitive Jewish folk lore." Without naming names, he especially singled out "certain scientists of high station, who state that there is no conflict between science and religion."[50]

By 1925 the idea of purpose in evolution was hotly contested among scientists; although it appeared prominently in the rhetoric of defenders of evolution among the public, it left a number of their colleagues uncomfortable. It also failed to persuade not only Bryan and his followers but also many other thoughtful people, including more secular-minded members of the public. As one letter to the *New Republic* pointed out, "It is difficult to imagine anything more terrible than the laws of nature with purpose read into them."[51] An editorial in the *Truth Seeker* took a similar position, more acerbically: "Besides being imbecile, the creative evolutionary plan is cruel and without mercy, except in death. We can endure with composure the indifference of an insensate nature, but put a creator into the scheme and we rebel against the atrocity. Evolution is a mindless process, but add to it creative intelligence and it becomes murder aforethought."[52]

Blurring the lines between science and religion was anathema to those nonscientist defenders of evolution who saw science as a part of the advance they took the Enlightenment to have been, a great leap forward in human civilization, responsible for a general increase in tolerance, reason, and justice. Some insisted with the editor of the *New Republic* that scientists espousing religious views were being disingenuous; others charged them with hypocrisy or even cowardice. For these people, so given to characterizing the anti-evolution campaign in the language of the Inquisition, any compromise of science with religion was a compromise of the values of the Enlightenment.

In an editorial in the *Truth Seeker* Jacob Benjamin bitterly mocked the "pious pedants" who tried to reconcile science and religion and the complacency of New Yorkers who refused to face the reality of the conflict: "Everyone immediately became interested in evolution." Benjamin complained. "Booksellers reaped a rich harvest—even moving picture theatres gathered a few shekels by exhibiting films depicting the evolutionary phases of life." New Yorkers were smug. "The glorious city of New York was surprised, so the claim went, at the abysmal ignorance of the South. Here evolution was in style; therefore it was respectable." But this was an illusion: "Laid away under the dust of time, the pioneer thinkers, those men of real ability and manly courage who fought for evolution when it had on swaddling clothes instead of evening dress, were conveniently forgotten."[53]

People on all sides of the debate about religion and science in the 1920s tossed the word *truth* about, each side asserting an exclusive claim on that problematic quantity. But beneath it all there appears to have been considerable uneasiness about what exactly truth would be in an era dominated by science. Science advocates like Maynard Shipley, founder of the Science League of America, insisted that it was not simply evolution that was at stake but the scientific method, even rationalism itself.[54]

Joseph McCabe, a former priest and a writer of a popular book about evolution, objected strenuously to the scientists who signed the Millikan Manifesto. Estimating that there were approximately five thousand active scientists in the United States, McCabe pointed out that the number who signed the manifesto was actually quite small, fewer than twenty out of five thousand: "The silence of the others is eloquent, for we may be sure that at least all the more distinguished men of science were asked to sign it."[55] Those who—he assumed—refused to do so were entirely correct: "It is not the business of either science or history to talk about God. Many scientific men throw as much dust in the eyes of people as the Fundamentalist preachers do. 'Science is not opposed to religion,' they say, pompously. But the good men have no more right to talk about religion than the preacher has to talk about science."[56] Naming Millikan and Pupin, McCabe objected that they were not even biologists and therefore "they both speak in the name of sciences they do not know." Calling Osborn the "self-constituted loud speaker of American science" and the "high priest of the little tribe of Aaron in the American scientific world," he protested of the reconcilers and modernists: "The mischief is that they tell us so very little and say it so very loud."[57] The main thrust of McCabe's quarrel with the reconcilers was that they had overstepped the boundaries of science, that they had used the mantle of authority conferred on them by their status as scientists to pontificate in areas in which they were not entitled to claim competence. It was true that neither Robert Millikan nor Michael Pupin were biologists: the controversies in popular media seemed to foster a tendency for scientists to be asked to stretch their authority as scientists and, in response, to indulge the temptation to do so. McCabe, among others, objected. Professors Osborn and Millikan "have every right to tell any person whom it may interest . . . that they believe in both evolution and religion. But they have no right whatever to say this in the name of science." Humans did evolve from an apelike ancestor, McCabe insisted, and he claimed that no matter what the reconcilers said, "science as such is never concerned with religion. . . . Yet there is a deadly conflict, because science tells us a large number of truths which, in the opinion of highly educated people, are inconsistent with the belief in God and the soul."[58]

McCabe was not mistaken in inferring that the theistic evolution defense mounted in the Millikan Manifesto did not represent the views of all scientists. Many scientists and religious modernists argued that science and religion did not necessarily conflict precisely because they occupied different realms of knowledge. Vernon Kellogg, for example, reiterated this theme consistently. Science was not concerned with Final Causes. This was also the position taken by the editors of the *New Republic* and the *Nation*.

But other people, such as Walter Lippmann and the Princeton theologian J. Gresham Machen, responded that for those who believed that revelation had priority over reason as a way of knowing, the conflict was very real. The poet and critic John Crowe Ransom complained that the amiable theology of the modernist clergy defending evolution stripped religion of its majesty and its meaning—it was like "Hamlet without the Prince of Denmark," or, as he titled a book on the problem, like *God without Thunder*.[59] Though many scientists agreed that science and religion could be effectively compartmentalized, this was simply not true for fundamentalists.[60] But the reconcilers' suggestion that evolution could contain comforting religious implications provided no comfort at all to religious conservatives. Like Bryan, who saw theistic evolution as more dangerous than atheistic evolution because the former was more insidious, conservative Christian theologians objected to the quasi-religious arguments of theistic evolutionists.

From the point of view of religious conservatives, the reconcilers were making a theological argument—they were preaching. And they were preaching an essentially modernist sermon, one that could not possibly answer the objections of fundamentalists. J. Gresham Machen went further: modernists like Harry Emerson Fosdick and the theistic evolutionists were not, according to Machen, really Christians at all. Whatever it was, the Christianity of reconcilers like Osborn and Conklin was not the Christianity of conservative believers. Among other things, evolutionists urged their audiences to take comfort in the notion that evolution implied progress, describing the trajectory of human evolution as a story not of descent but of ascent. For many conservative Christians, evolutionists' references to "ascent" only made matters worse: for them, the fall of Adam and Eve from grace in the Garden of Eden was essential to Christian faith. It was inseparable from the Christian doctrine of redemption, and both required free will. Like Bryan, they found the determinism implied by Darwinism—and social Darwinism, which Bryan conflated with Darwinism—impossible to reconcile with Christianity. Invoking progress as the great moral lesson implicit in evolution could hardly mollify them.

The anti-evolutionist Francis P. Le Buffe, S.J., of Fordham University, among Osborn's and Conklin's more persistent critics, declared in 1923 that "evolutionists are now on the defensive and feel the need of justifying themselves against the accusations of irreligion frequently leveled against them both of late and in the past." Evolution teachers, Le Buffe claimed, were "materialistic," and he brushed aside the claims of scientists who published reassurances of the compatibility of science and religion, citing Edwin Grant Conklin in particular. Scientists like Professor Conklin, Le Buffe had shrugged, either were pantheistic "or hold to a symbolic God."[61] Conklin, along with Osborn among the most strenuous of the reconcilers, eventually acknowledged in a letter to Le Buffe, "Of course I do not believe in a God who is a 'Big man in the Skies,' and I presume no enlightened person holds such a view today."[62] But as Harry Emerson Fosdick pointed out, most religious people really did yearn for a "picturable" God.

Several of the scientists who wrote essays for Frances Mason's book, perhaps influenced by the *Evolution* movie, suggested a useful heuristic device: imagine snapshots of the earth at intervals in geological history. Then imagine stringing those snapshots together so that they could move: voila! A film version of evolution, evolution made visible. Such a film did exist, of course, and appears to have had some effect. As sophisticated a critic as Edmund Wilson made note of it, describing it quite lyrically, and musing, "When I saw this film, I was much impressed by its possible educational value. If the picture could only, I thought, be shown in Dayton, Tennessee, the inhabitants of that backward region might be shaken in their literal faith in the account of creation in Genesis." After that, however, he attended "a Biblical picture," *The Wanderer*, a 1925 movie based on the story of the return of the Prodigal Son, "and there found a representation, just as plausible, and just as impressive."[63]

Despite the insistence of the scientists defending evolution in the press that evolution and religion were compatible, the religion they had in mind was not that of most anti-evolutionists. Many of these scientists were not so far off the mark when they denied being materialists, but they were certainly not believers in the Christianity of the fundamentalists. Furthermore, in making claims about the religious messages implicit in evolution, they wandered beyond what many people had begun to see as the boundaries of science.

Anti-evolutionists were aware of this and accused scientists of undermining science by pontificating about religion. Le Buffe astutely used Osborn's own words against him. In an interview with science reporter Alva Johnson, Osborn attributed public mistrust of scientists to "the vaporings of scientific men . . . who carry over a reputation gained in the scientific field and write on religious and spiritual subjects of which they know little." Such people "invite a strong reaction against themselves and against science."[64] Le Buffe did not fail to discern an irony in such statements.

Le Buffe also objected to the visual images of evolution on display at the American Museum of Natural History. The venerable images of Eden contrasted in the most visceral way with images of human ancestors resembling brutes. These images—the pictures in people's heads—contained power that words could not dispel. Le Buffe quoted Conklin, who had written: "The theory of evolution has given men sublimer conceptions of the world and its creator than has any rival doctrine."[65] A glance at Neanderthal anatomy made nonsense of any such claim, Le Buffe maintained. Conklin's "sweet words" were "absolutely and unmitigatedly false. For Adam and Eve [evolution] has given us the muzzle-faced, crooked-kneed, sloping-thighed sub-man imagined by Professors Osborn and Conklin, and one H. G. Wells."[66]

Even the stalwart Frances Mason could be horrified by newspaper images of apelike human ancestors. As was—it turned out—Henry Fairfield Osborn. In 1926 the *New York Times* published an account of an address to the Catholic Library Association titled "Ignorance and Evolution." The *Times* reported Monsignor Joseph H. McMahon's observation that "scientists do not assert dogmatically that man is descended from the ape, yet the average man is led so to believe by the exhibitions in our museums."[67] Scientists including Osborn attempted to reassure the public that evolution presented no threat to Christian values or to Christian notions of the human place in nature. And, sympathetic to those who found the specter of brutish ancestors repugnant, Osborn tried to offer an alternative vision.

CHAPTER SIX

Stooping to Conquer, and a Hall Full of Elephants

Early in 1926, Osborn received a note from O. Farneur, a man from Brooklyn, who had read in the *New York World* of attacks by Boston's Cardinal William Henry O'Connell on the human evolution exhibit at the American Museum of Natural History. Farneur suggested that Osborn reply to such attacks in a series of radio broadcasts, beginning with a program about dinosaurs—"most people are familiar with them due to the movie 'Lost World'"—and, having thus prepared the way, Farneur thought, Osborn could "then work down to your many proofs of early man's existence." This sensible plan even included advice about which New York radio stations would be most appropriate.[1]

When anti-evolutionists challenged the human evolution exhibits at the American Museum, however, Osborn chose not to respond in the manner suggested by Farneur's letter. The rhetorical strategies he did adopt highlight the differences between scientists and science popularizers and between scientists and nonscientist intellectuals, and they suggest some of the complications of communicating scientific ideas in the midst of public controversy.

Farneur's suggestions had much to recommend them. Dinosaurs were in vogue. The movie version of Arthur Conan Doyle's *The Lost World*, featuring Willis O'Brien's special-effects dinosaurs, had indeed been one of the cinematic highlights of 1925. Probably no one, however, had done more than Osborn to increase the fame of dinosaurs. He had sent numerous fossil-hunting field crews out to the American West to bring back spectacular dinosaur bones for the museum, and the dramatic exhibits of dinosaur skeletons in his museum—exhibits he had helped to design—taught several generations of visitors to visualize dinosaurs and other extinct animals as once-living, active creatures. Osborn himself had named *Tyrannosaurus rex*, the "tyrant lizard king," a christening that amounted to something of a public relations coup for the dinosaurs. And his book, *The Age of Mammals*, had told the story of the history of life in accessible language illustrated with vivid images.[2]

Osborn did not take Farneur's advice, however. The contrast between his vivid descriptions and exhibits of large extinct animals and his discussions of human evolution illuminates an underlying tension in his work. The debates of the 1920s exacerbated this tension, revealing the complexity of the relationship between scientists, science, and the public.

In his defenses of evolution during the decade, Osborn increasingly emphasized human evolution. Although his book about human ancestors, *Men of the Old Stone Age,* had been well received, the prolific Osborn had not begun his professional life as primarily a physical anthropologist; he became interested in human evolution when new discoveries of fossil hominids and especially of cave art early in the twentieth century piqued his imagination, and that of the public. His greatest passions as a paleontologist to that point had been for two families of large mammals, the Proboscidea, or elephant family, and the extinct family of large ungulate mammals called Titanotheres, and he remained committed to studies of those animals. Throughout the decade of the twenties he worked to complete large monographs on these two groups—his private diaries show that even during the Scopes trial he saved time each day for work on them—and he considered the 1929 publication of the Titanothere monograph the crowning achievement of his career. The Titanothere and Proboscidea monographs were lavishly illustrated volumes and comprehensive studies. *The Titanotheres* included some of his—and his students'—most interesting and insightful theoretical work. But in responding to anti-evolutionists, Osborn focused increasingly on the much more fragmentary evidence for human evolution.

As the evolution debate of the decade progressed, science popularizers tended to be circumspect about human evolution. Henshaw Ward explicitly excluded human ancestry from his widely advertised and well-reviewed *Evolution for John Doe.* Complaining that "the average man . . . thinks evolution is 'the doctrine that man is descended from monkeys,'" Ward eliminated from his book any "attention to the 'monkey doctrine' [or] any reference to any ape-like creature." Ward's solution to the touchy issue of human evolution was extreme, but other authors also deemphasized human evolution. They had good reasons for doing so. As Ward said, the "average man" who associates evolution with monkeys "is so amused or offended at this theory that his whole mind is occupied with it."[3] Ward recognized that the cultural resonance of monkeys and apes was ancient and complex. Long visual traditions, as well as recent popular culture, associated them with racial and ethnic concerns as well as class tensions, and by the 1920s they had also come to express ambivalence about modernity. Complicating the matter, the fossil evidence for human evolution was far less complete in the 1920s than it is today.

Osborn recognized these things, too. Yet in his responses to anti-evolutionists, he focused on human ancestry, and he did so in a way that became increasingly controversial. He argued that although humans were related to the other primates, our connection with the apes was actually quite distant. He distanced humans from any ape ancestry by maintaining that humans and apes had diverged—in different directions—from a common ancestor in a remote past. Neither humans nor apes now resembled the original progenitor—any more than they now resembled one another. The monkeys and apes may have been our distant cousins, but they were not—and did not look like—our grandparents. This claim would annoy some evolutionists, who objected that this amounted to splitting hairs. After all, what the public wanted to know was whether we had evolved from some kind of primate. Although scientists disagreed on the details of the human past, none of them suggested that we were not primates. Osborn's fastidiousness about the precise nature of these relationships confused some members of the public, yet it became something of a leitmotif of popular accounts of evolution.

In Osborn's case, anti-evolutionists who targeted the Hall of the Age of Man at the American Museum forced attention on human evolution. But scientists also focused on human evolution in order to make arguments not just about the fact of evolution but about its metaphysical implications, and their notions of the implications of evolution had been shaped by their thinking about human evolution. All of these things made the issue of human ancestry unavoidable. They also made the debate bewildering. When Osborn and his colleagues argued that humans had evolved not from apes but from very distant ancestors common to both groups, they were widely quoted by journalists and popularizers. And they muddied the waters.

Osborn's shift in emphasis to human evolution was part of a larger rhetorical shift, something that must be understood in the context of the tradition of evolutionary thought to which he belonged. Instead of emphasizing the story told by fossils and trusting that the public would draw the desired conclusions, he increasingly emphasized the reconcilers' claim that evolution was not only perfectly consistent with Christian ideals; it actually confirmed Christian moral lessons. This interpretation of Christian messages inherent in evolution, however, had little in common with the Christianity of most anti-evolutionists because Osborn's evolutionary Christianity was so thoroughly suffused with notions of progress.

His depiction of science also shifted during the decade as he responded to assaults on the museum by insisting on the transparency of the facts on exhibit.

This adamant defense of the truth of the exhibits contrasted with his more customary willingness to discuss the dynamic, contingent nature of scientific hypotheses. Originally, exhibits in the Hall of the Age of Man were intended to include examples of the kind of scientific inference that went into the displays of evolution—for example, reconstructions of hominids showing the skull exposed on one side and fleshed out on the other or "evolutionary tree" diagrams that included the ecological and geological context of organisms. But as critics, following Bryan's lead, targeted attempts at scientific caution, labeling theories as "mere guesswork," Osborn backpedaled.

As more anti-evolutionists criticized American Museum exhibits depicting evolution, Osborn reiterated his protest that there was not a single unproven theory on display at the museum. In designing exhibits for the Hall of the Age of Man, he wrote, "we are scrupulously careful not to present theories or hypotheses, but to present facts . . . put together conscientiously by experts who have been trained to clearly distinguish between fact and opinion."[4] Yet the exhibits in that very hall did include both "facts" and theories that were in dispute among scientists as well as anti-evolutionists.

Osborn was aware of this, of course. Indeed he, along with other scientists writing about evolution for popular audiences, had often emphasized the dynamic, contingent nature of scientific hypotheses. In explanations of museum exhibits and illustrations, for example, he commonly pointed out that the images had changed, and would continue to change, with increasing knowledge. He included such cautionary notes in the exhibits in the famous Hall of the Age of Man. In labels explaining specimens in the hall he acknowledged the fragmentary nature of many of the fossils, and he attempted to incorporate in the exhibits demonstrations of the kinds of scientific reasoning that went into the construction of images of human evolution.

Yet, in response to challenges from anti-evolutionists, who questioned both the scientific veracity of the evidence and the method of its interpretation, he became increasingly intransigent. In response to fundamentalist challenges, he dug in his heels and made dogmatic assertions. And he conflated the evidence for evolution with arguments interpreting its religious significance.

The Hall of the Age of Man

Anti-evolutionists' attacks were directed especially at the Hall of the Age of Man.[5] There is an interesting silence in these attacks, though—and in the later historiography of the museum. Osborn's colleague and protégé William King

Gregory, who had shepherded John Roach Straton through the museum, noted that Straton had just glanced at the display of human ancestors—he had not examined the entire hall. Neither, apparently, had the hall's other critics. Or, at least, they had focused only on one small section. The floor plan and photographs of the hall demonstrate that this was not the Hall of Man, or the Hall of the Evolution of Man. It was the Hall of the Age of Man. Had Straton been less focused on the implications of human evolution, he would have acknowledged that human evolution itself occupied much less of the hall than did the context of that evolution. Furthermore, the section devoted to human evolution was arranged far more didactically than the exhibits surrounding it. Almost half of the hall was devoted to large Ice Age mammals—from the period just before the recent, the period of the most modern human ancestors. The hall included a large case, for example, full of dramatic mounted skeletons of giant ground sloths and their relatives. Murals by Charles R. Knight brought these animals to life.

A narrow band of flat cases along the center of the hall housed the exhibit on human evolution, including hominid bones and casts of bones and the reconstructions of the heads of fossil hominids by J. H. McGregor. Simply in terms of drama and aesthetics, these flat cases would surely have been the least arresting part of the exhibit. The floor plan and descriptions of the hall also show clearly that fully half of the exhibit was devoted to elephants, including complete mounted skeletons of mastodons and mammoths. The notorious Hall of the Age of Man included the widely attacked hominid exhibit, but in the context of an extensive display of the far more physically imposing large mammal contemporaries of early human ancestors. And it included the whole history of elephants, from beginning to end—the panorama of elephant history in mounted skeletons, sculpture, and paintings by Knight. These were not just the elephants that evolved in concert with humans; this was the whole of elephant life.

Why had Osborn so prominently emphasized elephants? In 1923 he wrote: "An insatiable Wanderlust has always possessed the souls of elephants as it has the tribes and races of man. . . . The romances of elephant migration and conquest are second only to the romances of human migration and conquest."[6] In 1911, just three years after being appointed president of the museum, he had proposed an ambitious elephant hall, writing to the famous taxidermist Carl Akeley that he had been interested in elephants for as long as he could remember and assuring the museum's trustees that an elephant hall "would not fail to interest a wide public." The trustees vetoed the idea because, as the most influential trustee, J. P. Morgan—Osborn's "Uncle Pierpont"—wrote to him, "A hall full of elephants was too many elephants."[7]

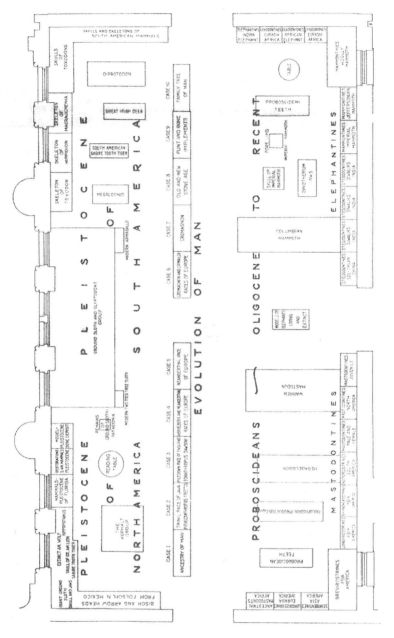

Floor plan of the Hall of the Age of Man at the American Museum of Natural History. A great deal of the floor space in the hall was devoted to displays of the large mammal fauna of the Pleistocene, including giant ground sloths, saber-tooth cats, the Irish elk, and, especially, elephants. American Museum of Natural History Library

The elephants in the room: this is a photograph of the Hall of the Age of Man, taken in 1925. Some of the murals by Charles R. Knight are also visible in this photograph. Image #388984. Photo by Julius Kirshner, 1925. © American Museum of Natural History

Hominid display case, Hall of the Age of Man, American Museum of Natural History. This is one of the displays attacked by anti-evolutionists in the 1920s. Image #38095. © American Museum of Natural History

Osborn really loved elephants. But he was no eccentric. The fossil evidence for the evolution of elephants was abundant, and it was magnificent. In a 1920 article in *Natural History* describing the new Hall of the Age of Man, he explained that "it has been deemed wise" to include in the hall "the entire history of the evolution of the proboscideans, which taken altogether is the most majestic line of evolution that has thus far been discovered."[8] Elephants had evolved rapidly and had traveled across the globe. They had tremendous heuristic value. And they appealed to the public.

Furthermore, the museum's collection provided an extremely complete record of the history of elephant evolution, far more convincing than the fossil record of human ancestors known in 1925. And, as Osborn often noted, elephants—and the rest of the Pleistocene megafauna—were part of the ecological and zoological context of human evolution, the zoology of the human past. So it makes sense that he would have included them in the Hall of the Age of Man—an exhibit to which he devoted a great deal of money and attention and of which he was very proud. Exhibits of big dramatic animals piqued public interest and drew attention to environmental context.

In the context of the evolution debates of the 1920s, however—in the rhetorical strategies he adopted for trying to ease the fears of a jumpy public—Osborn did not emphasize the history of elephants. In the tension between his lively visualizations of the large animals of the past and the didacticism of much of his verbal pedagogy, imagination lost ground. Faced with challenges mounted by anti-evolutionists of the 1920s, Osborn departed from many of his customary practices as a scientist and as a popularizer of science, and he did so in ways that were characteristic of many of the rhetorical strategies of the scientists defending evolution in the press at the time.

First, he focused on the human past in defending evolution, even though the evidence for human evolution in 1925 was full of gaps.[9] In addition, some of the known specimens were controversial among scientists, who had not reached a consensus on their interpretation. But in responding to anti-evolutionists, Osborn chose not to focus nearly as much on the fossil evidence for the evolution of life in general as on the much more ambiguous evidence for human evolution.

Osborn had been presenting the evidence of the history of life to the public for his whole adult life—he had built a museum dedicated to making that story visible. And he had written elegant, abundantly illustrated books and articles detailing the chronological narrative of this history. But in the evolution debates of the 1920s he shifted his rhetorical emphasis. And his emphasis set the tone for the dominant newspaper coverage of the debate.

When anti-evolutionists challenged the truthfulness of the Hall of the Age of Man, Osborn replied that the view of evolution on display in his museum not only was absolutely factual but also, if studied without prejudice, affirmed Christian values. The exhibits in the Hall of the Age of Man, he insisted repeatedly, "demonstrate the slow upward ascent and struggle of man from the lower to the higher stages, physically, morally, intellectually, and spiritually."[10] This notion of struggle upward, of ascent and progress, took a central place in newspaper accounts of evolution in the 1920s.

The argument was threefold. Point one: Evolution was simply true, not a theory. Point two: Evolution was no threat to Christian values. On the contrary, it affirmed those values. Point three—and this became a major emphasis for Osborn—calling evolution the "monkey theory," as newspaper editors liked to do, was a mistake. We had not really evolved directly from monkeys or even from apes. They were our distant cousins, but certainly not our grandparents.

The rejection of an apelike ancestry was an odd shift, correctly identified both by fundamentalists and by a number of other scientists as sophistry or special pleading. Osborn personally paid a real price for making this argument. Ultimately, it isolated him among his colleagues. More and more tenaciously during the course of the debate he insisted that the term *ape-man* was misleading. The "missing link" should be called the Dawn Man. Furthermore, he claimed that the evolutionary link between humans and apes receded in time to the vastly distant past. This argument increasingly stranded him out on a scientific limb, provoking even his own most loyal student, William King Gregory, to disagree with him publicly. And it confused the public.

Newspapers frequently reported that Henry Fairfield Osborn, eminent defender of John Thomas Scopes, had forsaken evolution, saying that we did not descend from monkeys after all—and he was inundated with letters from the public asking whether this meant that he had abandoned evolution altogether.[11] This drove Osborn crazy. He generally had his very capable secretaries send patient and polite notes in response to such queries from the public, explaining that Professor Osborn had been misquoted. His own letters to friends and colleagues reveal his irritation, however. He blamed the press for sensationalizing his message. And he had a point. But other scientists criticized his model of human evolution and were deeply divided as to the wisdom of including the sublime religious message of evolution in their arguments before the public.

Why did he make this rhetorical shift?

There were three reasons. First: distaste for the image of a simian ancestry. Second: race. Third: the pressure of the public controversy and his notion of who belonged to the public he was addressing in this context and what concerned them.

To begin with, Osborn increasingly suffered from a malady William King Gregory wryly dubbed "pithecophobia"—a fear of apes and monkeys. Osborn understood, and to some extent shared, the public distaste for simian ancestors, apemen, and cavemen. He understood that much of the public saw them as defective, as caricatures or parodies of humans. He expressed concern about the long visual tradition associating monkeys and apes with unsavory characters and brutality, including ferocious- or menacing-looking apes and apes kidnapping human women for prurient purposes, as in, for example, an early nineteenth-century sculpture by Emmanuel Frémiet, "Gorilla Abducting a Negress," and the famous World War I propaganda poster of a German ape carrying a symbolic woman and wielding a club labeled "Kultur." Osborn actually got letters from people asking if the rumors that apes kidnapped human women were true; indeed, he mentioned these queries and the Frémiet gorilla in a published article attempting to dispel such misconceptions.[12] This visual tradition had been intensified during the controversy surrounding Darwin's publication of the *Origin of Species* in 1859 and again during the war, when anti-German propaganda associated Darwinism with German militarism.

Osborn was acutely aware of this tradition and wrote that the history of illustrations of primates as brutes caused understandable trepidation about evolution. He believed that if only the public could apprehend the immense time that separated humans from the simian branch of the primate family tree, their fears would be calmed. Most of all, he blamed newspapers and movies for misrepresenting our noble ancestors, portraying them as brutes and "ape-men."[13] In response, he worked to create an alternate, nobler view of human evolution—for example, in his collaboration with Knight on a magnificent mural for the Hall of the Age of Man depicting a Cro-Magnon cave artist at work.

During the 1920s, Osborn divided his evolutionary tree diagrams into increasingly distinct human and simian branches, removing their common ancestry to the remote past and sometimes indicating the significance of the rift by a vertical line. But the diagrams he designed to illustrate the deep bifurcation in the family tree separating humans from the rest of the primates reveal a second reason for his shift in emphasis when it came to human evolution: race.

Osborn was hardly unusual in his racism; racial preoccupations became obsessions in the 1920s. Many of his scientific colleagues were, like him, active members of eugenics organizations. But his racial theories, combined with his pithecophobia, significantly compromised his scientific judgment. His theory of the geographical origins of human evolution demonstrates the problem. In arguing that human evolution did not threaten Christian values, Osborn maintained that human evolution was a story of progress—the ascent of man, he liked to say. Human ascent was the result, somehow, of hard work. This was a kind of natural selection by means of a Protestant work ethic. "The moral principle inherent in evolution is that nothing can be gained in this world without an effort," he asserted, adding, "The ethical principle inherent in evolution is that only the best has the right to survive."[14] As a corollary, he postulated that human evolution must have taken shape in challenging environments. So, he theorized, humans must have evolved on the high plains of northern Asia, probably Mongolia, where the climate was dry and harsh and ingenuity would have been at a premium. Surmising that "the evolution of man is arrested or retrogressive in every region where the natural food supply is abundant and accessible without effort," Osborn concluded that "while the anthropoid apes were luxuriating in the forested lowlands . . . the Dawn Men were evolving in the invigorating atmosphere of the relatively dry uplands."[15]

Why did he think human evolution was retrogressive where the food supply was abundant? It was obvious—look at the people of tropical countries compared, for example, with those mythical heroes of eugenicists like Osborn, the "Nordics." Osborn had provoked much criticism by writing the introduction to Madison Grant's 1916 racist tract *The Passing of the Great Race,* and despite the hostility the book met with both from critics of its racial ideology and from scientists who challenged its scientific pretensions, Osborn continued to support Grant, a close personal friend, and to champion his publications.[16] Osborn also hosted eugenics conferences at the museum—he had pushed to have the Hall of the Age of Man finished in time to be used as a showcase at a 1921 eugenics conference—and proselytized for immigration restriction, or, as Osborn called it, "immigration selection."

Among the problems with Osborn's theory of the Asian plains origins of humans, of course, was the fact that much of the evidence contradicted it. Fantastically expensive expeditions to Mongolia failed for years to find evidence of early human evolution. Although two skulls of *Sinanthropus* ("Peking Man") were discovered in China in 1929 and 1930, Osborn continued to hope for discoveries of much earlier human progenitors than the relatively recent *Sinanthropus.* The evidence until this time was scanty at best and was no match for evidence supporting alternate theories: the much earlier *Australopithecus,* a real missing link, was found in 1924—and a description published in 1925, the year of the Scopes trial—in South Africa, where Darwin had predicted it would be.[17]

Did this affect Osborn's loyalty to his model?

Not at all. He tenaciously ignored *Australopithecus,* even when *Natural History,* the popular magazine published by the American Museum, featured the discovery prominently. He had been a great friend and supporter of the paleontologist Robert Broom, who worked with Raymond Dart on the *Australopithecus* specimens from Taung. Osborn supported Broom's work on mammal-like reptiles, helping him acquire funding for both research and publication and purchasing specimens from him.[18] Broom had been a guest at his home at Garrison, a privilege extended largely to the inner circle and the elite few. And Osborn knew Dart. Yet he never helped when Broom and Dart sought funds for further exploration in search of more *Australopithecus* specimens. He never sent curators to study the famous fossils, even though William King Gregory made a trip to Africa in the late twenties to study gorillas. Nor did he request that J. H. McGregor make a sculptural restoration of *Australopithecus* for display along with the series he had done of the Java, Piltdown, Neanderthal, and Cro-Magnon people. The museum had sent McGregor to study the European and Java skulls, at significant expense, and Osborn had visited them himself in 1921. He made no effort to see *Australopithecus,* or to send his staff to see it; by the end of the decade, he acknowledged and cited it but referred to it as an advanced ape rather than an important step on the path to humanity. Instead he continued to insist that the high plains of Asia had to have been the "cradle" of human evolution, even when expeditions to Mongolia failed to supply confirmation of the theory.

Osborn was not alone in rejecting evidence of an African origin for humanity. The British anatomist Sir Arthur Keith explicitly linked racial prejudice and the search for an Asian hominid ancestor in a feature article in the *New York Times Sunday Magazine* in 1930. Conceding that the evidence for African ancestors for *Homo sapiens* was strong, Keith insisted that the most advanced known human ancestor, Cro-Magnon, must have come from somewhere other than Africa. Why? Setting aside all archaeological evidence, he wrote frankly, "My preference for Asia is founded on a belief in the virtues of race. . . . Everyone has to admit that in all the great developments of human civilization the people of Africa have never been in the van." What mattered was locating the origin of the white race in Asia. Unapologetically he explained: "My racial prejudice leads me to seek for the Cro-Magnon cradle—the evolutionary center of the white man—in . . . Asia . . . partly because the native peoples of Africa lack the progressive genius of the Asiatic."[19]

The evidence for an African origin was compelling, however, while the looked-for abundance of fossils of the earliest hominids from Asia failed to materialize, despite grand efforts to find them. Even at the American Museum, some colleagues dissented from the Asia theory, especially as evidence to the contrary seemed to mount. The anthropologist Nels C. Nelson wrote in 1928 that he was not enthusiastic about the theory of Central Asian origins for humans: "It may be a good paleontological theory," but the evidence for Africa seemed stronger: "I would as soon side with Darwin and bet on Africa."[20] To try to explain the lack of evidence supporting his theory of Asian origin, Osborn repeatedly suggested that the hominids in Mongolia had been so alert, so adept at avoiding danger, that they were seldom caught in natural catastrophes and were therefore rarely preserved as fossils. Gregory warned him privately that it seemed unwise to publish this idea, but Osborn had already published it and continued to do so.[21]

Ultimately, Osborn believed in different rules for elephants and humans. He insisted that human evolution was a story of ascent from lowly beginnings to a noble and glorious result, that our evolution had been dignified. But for elephants there had been no real up or down, high or low. Certainly, elephants had gotten bigger and smarter. He did perceive teleology in the evolution of elephants and other mammals—indeed, he often used the evolution of titanothere horns as an example explaining his theory of orthogenesis, or directional, progressive evolution. But there were no lowly elephants. Elephants had not begun inauspiciously in order to culminate in the modern variety. Elephants, in addition, had begun in Africa—that was acceptable. Osborn described Africa as the cradle of elephant evolution but as a backwater for humans.

It was in his focus on human evolution, especially in the context of the 1920s debates with fundamentalists, that he claimed increasingly adamantly that natural selection could not explain human evolution. Ours was a moral fable, and when he wrote about human evolution, Osborn became, although he denied it, a kind of theologian. Not a careful or sophisticated theologian, but a teller of spiritual tales.

Osborn's increasing tendency to draw moral lessons about evolution raises a third reason for his odd rhetorical shift: the public nature of the debate. In *The Earth Speaks to Bryan* he had insisted that the exhibits in the Hall of the Age of Man included absolutely no speculation, no theories, nothing but established facts. The truth.

A statement by one of his colleagues at the museum, the paleontologist William Diller Matthew, helps to illuminate this perspective. For the zoologist evolution was a theory, Matthew wrote—a good theory, and the only theory that explained all the facts. But for paleontologists, who spent their lives among fossils, mapping them with associated rock layers, it was not a theory but an observation. Paleontologists, who had immersed themselves in years of study of the details of earth history, "have come to love their science beyond all else. It is their home which they have helped to build." For them, evolution "is not a matter of deduction but of observation." It was a fact, not a theory: "Evolution is no more a theory to the man who has collected and studied fossils than the city of New York is a theory to the man who lives in it."[22]

But in making his case before the public, and under pressure from concentrated assaults by fundamentalists, Osborn veered away from the narrative of the fossils and into the realm of the preacher. In writing *The Earth Speaks to Bryan,* he wrote to his publisher, "Imagine yourself not a New Yorker, but a nation-wide American"—someone from a place like Ohio or Tennessee, for example—"homebred and pious, who wants his natural history with a very large dose of religion and spirituality."[23] He had reconciled his own religious beliefs with evolution, and he imagined an uneducated and unscientific public looking to him for religious reassurances. He held a sympathetic image of this poorly educated public, people in "geographically isolated regions," people who cherished, he explained to an Italian colleague, "very deep and sincere religious sentiments" and who were "the direct descendants of the original American settlers."[24] Still, admire them in the abstract as he occasionally might do, the public of his imagination was not made up of people at all like anyone he knew personally. In reminding Scribner, "If I write this book I stoop to conquer," he also insisted—several times—that the book should be printed in "very large type" and that the cover should be designed with the purpose of "arresting attention by conspicuousness—even if it violates the best taste."[25]

He had long been treated with deference and respect by newspaper and magazine editors. They called upon him more and more often as an authority during the evolution debate, especially when anti-evolutionists targeted the museum. And he had used publicity and his access to publicity adeptly, even brilliantly, in building the museum. But his ambivalence about publicity would intensify with the crescendo of the evolution debates, and with good reason. In attempting to balance his ardent defense of theistic evolution, his promotion of the museum, his instinct for publicity, and his commitment to evolutionary theories of racial hierarchy with the dictates of scientific protocols, he ran into trouble.

Hesperopithecus, the Nebraska Primate

In the midst of his 1922 exchange with Bryan in the *New York Times,* Osborn received a small fossil tooth from Harold Cook, a Nebraska paleontologist. Cook thought that the tooth, although quite worn, both during the animal's life and after fossilization, looked very much like a primate molar. Osborn concurred and consulted with colleagues at the museum, who expressed cautious optimism. Osborn's own optimism quickly became much less careful. If this was, as it seemed to be, evidence of the existence of a higher primate—perhaps even an early hominid—in North America, it would be the first such find. And from Nebraska! The irony would be delicious. Abandoning scientific reserve, Osborn soon announced that museum scientists had discovered the first higher primate ever to be found in North America and that it had come from Bryan's own state. In a further departure from his normal practices, he even alluded to the Bryan connection in his formal scientific announcement of the genus. Naming the new animal *Hesperopithecus cookii*, he suggested in his announcement before the American Academy of Sciences that perhaps he ought to have called it Bryopithecus, "after the most distinguished primate the State of Nebraska has thus far produced."[26] The excellence of the story for publicity purposes soon overtook him. Edwin Slosson wrote to Osborn that in publicizing the find through the Science Service, it was "impossible to discuss the subject in the newspapers without using the word 'link,' and the temptation to connect the find with our eloquent ex-Secretary of State was irresistible."[27] Osborn's British colleague Grafton Elliott Smith published a popular article in the *Illustrated London News* referring grandly to *Hesperopithecus* as "the ape-man of the Western world," complete with an imaginative reconstruction by the painter Amedee Forestier of the family life of the now very human-looking animal.[28] Even in his enthusiasm for the contest with Bryan, Osborn's scientific caution did not entirely abandon him, and Forestier's graphic imagination brought him up short. He attempted to inject a note of caution, suggesting that although the tooth appeared to be that of a primate, it did not supply enough information to warrant such a fanciful imaginative picture. He was careful to emphasize that he was not suggesting that the animal was an early human; he claimed only that it seemed to be a primate in the early part of the human lineage. Yet he did attempt, without success, to persuade both Gregory and McGregor to work with Knight to create a painting of the *Hesperopithecus* as it would have looked in life.[29]

He assigned the reliable William King Gregory, a comparative anatomist of formidable skill and scrupulous scientific integrity, to investigate further. After continuing analysis and the addition of new specimens from the same quarry, Gregory ultimately determined that the tooth did not, after all, belong to a primate of any kind. It was not even a molar. It was a premolar of the extinct peccary *Prosthennops*. When, in 1927, Gregory finally published a paper correcting the original diagnosis, fundamentalists had their turn to gloat.[30] John Roach Straton

The fanciful version of *Hesperopithecus* as described by Grafton Elliott Smith
and published in "*Hesperopithecus:* The Ape-Man of the Western World,"
Illustrated London News, June 24, 1922, 244.

sent Osborn an invitation to debate him, mentioning in the same breath Osborn's exalted status as president that year of the American Association for the
Advancement of Science (AAAS) and the "miraculous pig's tooth."[31]

The paleontologist Stephen Jay Gould has suggested that, although anti-
evolutionists have continued to take visible pleasure in citing the *Hesperopithecus*
story as evidence of the gullibility or even duplicity of scientists, the episode really demonstrated science working as it should. A hypothesis was made public,
was tested, and when proved false, withdrawn.[32] This argument has much merit.
But it is also true that in response to his very public debate with Bryan, Osborn

departed from his customary scientific practices. He had, to be sure, occasionally named a species of extinct elephant on the basis of a single tooth, but more commonly his species definitions rested on more complete evidence; elephant teeth, in any case, were far more complicated than primate or pig teeth—even a single elephant tooth contained much more information. Primate teeth were not nearly as complex; primate molars did resemble pigs' molars; and the tooth in question was especially worn. Furthermore, Osborn named *Hesperopithecus* not just as a new species but, necessarily, as a new genus. When he named new elephant species, his formal diagnoses were well documented and well illustrated, following the protocols and conventions of the very precisely defined rules of scientific nomenclature.[33]

Osborn normally took time to consider the evidence. In the case of *Hesperopithecus,* he leaped in precipitously. He yielded to the temptation to make fun of Bryan even in his formal scientific announcement. More to the point, he could not resist the temptation to tell the press that the oldest North American human ancestor came from the home state of William Jennings Bryan. He made it a centerpiece in his rhetorical strategy. In *The Earth Speaks to Bryan* he wrote that "the earth spoke to Bryan and from his own native state of Nebraska, in the message of a diminutive tooth, the herald of our knowledge of anthropoid apes in America. This Hesperopithecus is like the 'still small voice.'"[34]

Hesperopithecus turned out not to have such a small voice after all, however. As the "Ape Man of the Western World," the discovery was featured prominently in the *New York Times* in September 1922. The writer for the *Times* mentioned Bryan repeatedly throughout the very long article.[35] Although Osborn thought of the *Times* as more circumspect and more responsible in covering science than the tabloids were, the editor was prepared, in this case, to exploit fully the ironic potential of the Nebraska ape. R. E. Turpin, the assistant Sunday editor, requesting an extended article from Osborn, asked also for "a drawing of the reconstructed man."[36] He settled instead for a long feature article quoting Osborn at length. The article was illustrated not with a reconstruction of *Hesperopithecus* but with a familiar photograph of the famous captive gorilla John Daniel, whose stay in a New York hotel the previous year had been the subject of a good deal of newspaper attention, especially after he died.

Although Osborn did not attempt to supply the *Times* with a picture of the "reconstructed man," he did speculate, on the basis of the shape of the tooth, that the animal would likely have been much handsomer than a chimpanzee. The *Times* included Osborn's references to the associated faunas by their scientific names, as well as his interpretation of the ecological and biogeographical context of the find, but it brushed aside doubts expressed by other scientists who suggested that the two teeth—the 1922 find and a similar one found in 1909 but put aside because at that time it did not seem to be identifiable—did not warrant the bold conclusions Osborn had drawn from them. And newspaper articles about the find mentioned Bryan so often that there was no mistaking the reason for the publicity.

Naturally, then, *Hesperopithecus* was vulnerable to attack, even before Gregory withdrew it as a named species in 1927. It became a source of some embarrassment. Bryan himself singled it out relentlessly in speeches ridiculing Osborn and his scientific colleagues for their gullibility.

Bryan understood the rhetorical value of *Hesperopithecus* at least as well as Osborn did. Edwin Grant Conklin wrote to Osborn that Bryan had given a lecture at Princeton, attacking both of them by name and making fun of the Nebraska ape, and an acquaintance informed him from Dayton that soon after his arrival for the Scopes trial Bryan began citing *Hesperopithecus* as a perfect example of the flimsiness of scientists' evidence for evolution and of scientific "gullibility."[37] The *New York Herald Tribune* reported that "Mr. Bryan made a triumphal entry into Dayton. . . . He ridiculed scientists, particularly Henry Fairfield Osborn. 'They found a tooth in Nebraska, in a sand pit, took it to Osborn and he and some others made it into the missing link.'" Turning the tables on scientists who attempted to invoke their unique mastery of scientific method, Bryan emphasized the tenuousness of the evidence supporting *Hesperopithecus:* "The scientists would throw out the Bible on evidence that wouldn't convict a habitual criminal of a misdemeanor in any court in Christendom."[38]

Osborn was sensitive enough about the Nebraska ape that Gregory, after publishing his disclaimer in *Science* in 1927, also felt obliged to compose a long letter of apology and explanation to Mrs. Osborn, who took even obliquely implied aspersions on her husband's character extremely seriously.[39] The *Hesperopithecus* episode suggests that the public debate altered Osborn's scientific judgment and behavior, and it was not the only instance.

Piltdown

Like many other biologists, Osborn had initially been skeptical of the Piltdown specimen—and for good reason, since it was ultimately shown to be a hoax. But once he accepted it, he very publicly promoted it as proof that the fundamentalists were wrong, and in the spirit of the argument from authority, he made much of the fact that his change of heart was the result of visiting the site and the specimen in person. In 1916, he related, he had been skeptical. During a trip to England in 1921 he changed his mind and, characteristically, did so with energy and élan. He wrote to E. Ray Lankester that he planned to contribute articles about it to *Natural History* "so as to nail my colors to the mast." In addition, he announced that he would devote an entire display case in the Hall of the Age of Man to Piltdown.[40] When skeptics criticized the exhibit for including the Piltdown material, he answered, "There is no discovery which has been more fully confirmed than that of Piltdown man. I reached this conclusion only after visiting the ground in person and giving the matter my personal study."[41]

Nonetheless, some of his colleagues at the museum remained cautious about Piltdown. In 1925, J. H. McGregor declined to help with a proposed painting by Charles R. Knight, intended to depict the life of a Piltdown group—he was sure Knight would produce a lovely painting but feared there was simply not enough evidence to justify such a restoration, although he had himself created a sculptural restoration of a Piltdown individual. Fundamentalists cited McGregor's Piltdown as evidence of the museum's lack of credibility.[42] Piltdown figured prominently in Alfred Watterson McCann's attack on the museum, *God—or Gorilla,* and other anti-evolutionists took up the theme. The editor of the *Sunday School Times* ran a series featuring Piltdown especially and the Hall of the Age of Man in general as demonstrations of the "amazing credulity of many who pride themselves to-day on being scientists."[43] A series of newspaper articles carried word of harsh criticisms by Boston's Cardinal O'Connell, who claimed that the Piltdown specimen "was a hodge-podge of miscellaneous bones, originally strangers to each other," proving "the grotesque gullibility of so-called scientists."[44]

Ape-Men and Dawn Men

Probably the most significant ramification of the debate, for Osborn, was the increasingly public controversy with Gregory over the term *ape-man.* Osborn had special affinities for his protégés, and for Gregory more than most. Gregory appears to have reciprocated Osborn's affection, but by the end of the decade even he accused Osborn of allowing his scientific judgment to be compromised by metaphysics and by participation in the public debate with fundamentalists.[45] More and more insistently, Osborn pressed his claim that humans had never descended from any apelike animal, that apes had evolved from our common ancestor in an entirely different and less dignified direction, and that science should forcefully renounce the popular term *ape-man* in favor of the more savory "Dawn Man."[46] The Dawn Man became something of a joke even among friendly colleagues—Conklin wrote to Thomas Hunt Morgan that as president of the AAAS and host of the organization's annual meeting, "Osborn said that it was the largest scientific meeting ever held in the history of the world, and I suppose that goes back to the period of the 'Dawn Man.'"[47]

The debate with Gregory over the human relationship to apes became quite public, culminating in an exchange in the *New York Times*. Most of their colleagues agreed with Gregory, eventually more or less isolating Osborn. At a 1927 meeting of the New York Academy of Medicine, Osborn delivered a paper arguing for a separation between humans and apes, including the claim that the public found the idea of a close relationship so repulsive that they turned away from evolution itself. During the discussion, the Columbia University biologist J. H. McGregor, who worked closely with Osborn and the other scientists at the museum, emphatically aligned himself with Gregory in the debate: "The great line of zoological cleavage is not the separation of apes from man, but the cleft between anthropoid apes and man as one single group and the tailed monkeys as another group."[48] Furthermore, as McGregor and Gregory argued in a joint paper, Osborn's pronouncements would not allay the fears of an intelligent lay public; in seeking to mollify, he would only confuse them. Even the "man in the street" recognized the resemblances of the modern anthropoids to humans, and "the general reader may very well mistakenly gain the impression that it is Professor Osborn's deliberate intention to disclaim for the human race all kinship whatever with the anthropoid apes" and to believe that science held that the human family "had kept itself aloof from other vertebrates."[49]

Although newspapers and popular magazines tended to reflect Osborn's statements about humans' separation from ape ancestry, some scientists dissented, especially privately and in their own journal, *Science*. A letter to the editor of *Science* in 1926 juxtaposed a statement from Osborn with another by Darwin, commenting, "Let us be honest. Darwin stated distinctly, rightly or wrongly, that man proceeded from the Simiidae. . . . Hence Bryan was not attacking a pure fiction, nor setting up scarecrows." Furthermore, "Most men of science still believe that man proceeded from a pre-ape or pre-monkey stock," and although the precise scientific classification of that stock was not really crucial to the argument, it was important that scientists make the truth, as they understood it, clear and unambiguous.[50] Osborn's attempts to reassure the public by distancing humans from apes in the family tree seemed to many scientists to be disingenuous.

More exasperating, the newspapers reported on the controversy in terms that Osborn perceived as thoroughly misleading. The president of the American Museum of Natural History, some headlines noted, now denied that humans had evolved from apes. A cartoon actually pictured Osborn being thanked by a chimpanzee for relieving it of the embarrassment of a relationship to humans. He received numerous letters and queries from the public, asking whether he had, after all, abandoned his support of evolution. A speech he gave in New Haven in 1926 especially provoked newspaper responses that he thought misrepresented his remarks as well as queries from confused members of the public. In response to one lengthy query, he complained that he was "extremely disconcerted and annoyed at Associated Press reports all over the country" which had suggested that he had declared evolution and theology irreconcilable. He firmly denied it: "Of course I said nothing of the kind in this or any other address which I have ever given or written."[51]

In the heat of the evolution debates, Osborn's ability to shape his own publicity seemed to desert him, and he blamed journalistic commercialism. In the same letter, he wrote, "I told the reporter who interviewed me regarding my address not to introduce religion; for his own purpose, apparently to sell his news item to the paper, he did not keep his promise."[52] The reporter in New Haven was not alone, however. Other scientists began more and more often to protest the whole "not-from-the apes" line of argument.

Among the public, the idea that the original human ancestor did not really resemble an ape may have comforted some people; many newspapers and magazines during the Scopes trial offhandedly noted that all literate people understood that humans had not really descended directly from monkeys—it was only, as a *New York Post* cartoon illustrated, a scarecrow, meant to alarm the public.[53] Over time, however, reports of Osborn's increasingly adamant denial of an ape ancestry did, as Gregory and McGregor predicted, provoke irreverence and confusion.

Anti-evolutionists pounced on it, proclaiming that this was proof of scientists' inconsistency, confusion, and duplicity. In late June 1925, the *Tampa Daily Times* reported at great length the challenge issued by a retired Presbyterian minister, Dr. J. G. Anderson. Anderson reported on a visit he had made to the museum, especially to the Hall of the Age of Man, which he proclaimed ought to be called the "Hall of the Hoax." Although confused about the name of the museum, identifying the exhibit as "the Hall of the Age of Man in the Metropolitan Museum," he repeated the now familiar litany of criticisms of Osborn and the exhibit. Quoting Osborn's guide leaflet to the hall, "Man is not descended from any known ape, either living or fossil," the article demonstrated the disjunction between Osborn's disclaimer and the visual impression made by the exhibit. Anderson interpreted the hall as "an ocular demonstration of how man was developed from an ape," asserting that the teachers and schoolchildren visiting the exhibit "believe they are looking upon a visible demonstration of the evolution of man." The article went on: "How then, asks Dr. Anderson, does he know, that the ape ever lived, since there is no living or dead evidence that the thing ever existed at all." Anderson joked, "Thus far it has only an imaginary existence, and that imaginary being maintains its habitat in the brain cells of the evolutionist." Osborn, Anderson claimed, was "strangely inconsistent." Yet according to scientists, "this logic is considered so inexorable that any man who demurs to it brands himself as unintelligent." This amounted to "intellectual tyranny."[54]

Other critics outside the scientific community began to challenge Osborn's credibility as a scientist. Wilfred Parson, a Catholic priest, wrote, "We are being bombarded by pseudo-scientific propaganda these days." Citing Osborn especially, Parson noted, "Even Henry Fairfield Osborn, one of the most indefatigable evolutionary propagandists . . . admits he can show no direct non-human ancestor from which man evolved. All he has is analogy, inference, probability. From a chain like that no bridge of evidence was ever erected." Generalizing, he continued, "The quarrel that Catholics—and not Catholics alone but all lovers of truth—have with much popular writing on Evolution, is not that it is science, but that it is bad science. It presents as fact what is only speculation."[55]

Another Catholic critic, Father Francis Le Buffe of Fordham University, acknowledged, "No one in America speaks with greater authority than Professor Osborn on his own specialty." Osborn had strayed beyond his realm of expertise, however, and was now speaking and writing "more as an orator than as an outstanding scientist of international fame." Parsing Osborn's language, Le Buffe claimed that "strong affirmations abound just where facts are scarcest and interpretations and inferences are most plentiful. It is precisely such grandiloquent dust-throwing into the eyes of the public, which has no time to check up assertions with facts, that makes temperate writing on evolution so difficult."[56]

In part, Osborn fell victim to his own imagination and a gift for storytelling. Colleagues at the museum often tactfully restrained his enthusiasms. When in 1924 he suggested it might be possible to create a phylogeny of flint implements for his proposed revision of *Men of the Old Stone Age*, the museum anthropologist, Nels C. Nelson, warned him that while "one may properly compliment you on your courage," the evidence was lacking. Gently but firmly Nelson added, "It seems to me unwise to deal with such a subject except in the very broadest of terms, especially in a general treatise which ought to keep clear of controversial topics."[57] But Osborn had controversial topics in mind; he proposed this revision in part to address the attacks of fundamentalists in clear, nontechnical prose. Nelson reminded him of the perilous consequences for any scientist who "sought to force an interpretation beyond what his facts warrant," advising that "in the main it seems to me that such advanced interpretation as you propose should first be published in some technical journal and thoroughly thrashed over before being incorporated in a general treatise."[58] This was, of course, exactly the procedure Osborn had failed to follow in the *Hesperopithecus* case.

His museum colleagues offered such advice gently. Other scientists could sometimes be more pointed. One article in *Science* criticized him implicitly by mentioning the "great publicity" recently given to the Gobi expeditions, and without mentioning Osborn by name, it warned, "to meet the popular demand for sensations . . . is no easy accomplishment in these days of Sunday sheets and sensational scientific exaggeration. Research workers are in great danger of giving this popular demand for something striking, something unusual, too much heed." It was especially important for scientists to avoid this sort of thing: "If we would dignify our profession . . . we can not afford to fan the sensational flame. The craze for advertising is making us forget professional ethics and personal justice."[59]

Osborn also allowed his confidence in his own authority to blur his vision. If the public understood that he had been to the Piltdown site himself, they would be reassured; and if scientists of standing, both as scientific men and as pious Christians, vouched for the spiritual credentials of evolutionary theory, then public fears should be calmed. He had spent many years studying and living with the evidence of evolution and had acquired skills that could not be mastered overnight. This sense of the years of scientific experience and observation, blended with the conviction of the scientist's personal integrity, was, for Osborn and many of his colleagues, part of the very method and definition of science. It was also part of a persuasive rhetorical strategy. In a letter to Charles Davenport, he wrote that the published record of a eugenics conference held at the museum should include a photograph of the participants "because the personality of the men interested in this movement will go a long way towards popularizing the movement." The status of the leaders, he was confident, would carry a lot of weight with the public.[60]

Science outsiders did not necessarily agree. A review of *The Earth Speaks to Bryan* warned, "If Mr. Bryan is not to be ignored, he should be met fairly and answered fairly. The scientists, of whom Prof. Osborn seems to be typical, have not met his attack satisfactorily. . . . Yet the general public craves exactly what Mr. Bryan demands: a comprehensive statement of the arguments for evolution, reduced in scale so as to make discussion possible and shorn of abstruse scientific terminology. If such a statement had been forthcoming several years ago the Scopes trial might never have been held."[61]

Some intellectuals from outside the community of science saw the evolution controversy as a lost opportunity. As Walter Lippmann pointed out, however, educating the public about the method of science was no easy thing. In an address titled "The Duty of Biology" published in *Science* in 1926, the Colorado botanist T. D. A. Cockerell lamented the sorry state of scientific knowledge on the part of the public. This was "not altogether the fault of the public." Another problem lay with people who "still cling to the old idea that science is only for the elect." Cockerell reserved his most eloquent criticisms, though, for teachers who oversimplified science, "those who abhor the technical or difficult side of science, and wish it made simple and easy. They think the aroma of science is enough, without the substance." This sort of pedagogy, he insisted, actually left students and the public in ignorance of the methods of science. He called on his colleagues to resist "the great modern illusion . . . that we may have pictures without knowing how to draw, music without ability to play, science with only the simple conceptions of a child. This is the voice of barbarism, calling from the darkness of the past." Part of the problem, though, resulted from the constraints of publishing: scientists were not given enough space for complete exposition of their work, so that even colleagues in different disciplines might find the work difficult to understand. "It is almost as if we used signs for language."[62] And of course they did.

When he insisted that the museum's Hall of the Age of Man included nothing but fact, no theories or hypotheses, Osborn strayed from Cockerell's—and ultimately his own—ideas about science education. He did so under pressure. He also did so out of ambivalence about the proper scientific response to a vehement public controversy—and he was no fan of vehemence, often citing the effectiveness of "the still small voice." And, in trying to forge a philosophically and religiously congenial version of evolutionary theory, he wandered beyond the boundaries of science as they were increasingly coming to be understood.

Many people commenting on the debate would object to this sort of thing. One review of *The Earth Speaks to Bryan*, remarking, "This forthcoming trial . . . is getting all cluttered up with experts," complained that "the liberals and naturalists [were] just as bad" as the fundamentalists. "For instance, Prof. Osborn, from the first word he writes, becomes a preacher," the reviewer complained, when "what he means to imply is that evolution is the truth about nature, and that because it develops a complicated structure from a single cell, and an ostensibly perfect man from an inferior barbarian, it has certain aesthetic and religious values." In conflating science with his interpretation of its implications in this way, "he proves that both sides are hopelessly muddling the issue."[63]

From an avowedly secular perspective, an editorialist for the *Truth Seeker* lamented the prominence of the theistic interpretation of evolution: "The Dayton trial laid bare the fact that the press, our scientists, lawyers, and of course, the liberal clergy, are unwilling to accept the challenge to choose between evolution and Christianity, but have preferred to hedge, evade, equivocate, and compromise. . . . Fundamentalism must go. But it will never be abolished by half-baked scientists handicapped by theistic conceptions."[64]

An article in *Life* magazine suggested that the discussion provoked by the Scopes trial was "doing good, abating the tedium of the silly season and diffusing knowledge." Mentioning a feature article by Osborn in the *New York Times*, the article, like so many others, quoted Osborn's familiar statement that "no existing form of anthropoid ape is even remotely related to the stock which has given rise to man." The magazine went on: "So the real authorities do not claim at all that man derives from monkeys. All that theory Dr. Osborn stigmatizes as 'pure fiction, put up as a scarecrow.'" The *Life* magazine article concluded, though, that the images of Neanderthal "cave men" on exhibit at the museum and reprinted in the article were also fictions, "well adapted to scare the Fundamentalists blue . . . fakes and not good evidence at all."[65]

The images in the American Museum of Natural History did a lot to make the ancient past imaginable. They were added to the protean store of ideas Walter Lippmann so provocatively referred to as "the pictures in our heads."[66] The Hall of the Age of Man—if one looked at the whole thing—made a visual argument about the human place in nature. When, in the heat of the debates of the 1920s, Osborn neglected that visual argument and claimed a place for humans beyond nature, he ventured into the territory of religion, and in doing so he created confusion. He spoke most eloquently when he wrote about the fossils themselves, without trying to make them carry theological weight, and through his contributions to the exhibits at the museum. Even at his most eloquent, however, his message would be mixed.

CHAPTER SEVEN

The Pictures in Our Heads

According to H. L. Mencken, "the climax" of William Jennings Bryan's address to the court at the Scopes trial was "a furious denunciation of the doctrine that man is a mammal." Mencken marveled: "It seemed a sheer impossibility that any literate man should stand up in public and discharge any such nonsense. Yet the poor old fellow did it. Darrow stared incredulous. Malone sat with his mouth open." Reminiscing many years later, Joseph Wood Krutch, who had covered the trial for the *Nation,* remembered the moment as one of the highlights of the event. Mencken, Krutch recalled, had made a point of falling noisily from a table, as if to punctuate the absurdity of Bryan's charge.[1]

The transcript of the trial reveals no explicit denial of the human place within the zoological class Mammalia, although a number of witnesses remembered Bryan making such a declaration. Arthur Garfield Hays recalled that Bryan "arose and absolutely and unequivocally refused to be a mammal."[2] Whether his alleged refusal to be a mammal occurred in reality or only in historical memory, Bryan did, according to the transcript, object vehemently to a diagram assigning humanity to the zoological class Mammalia, published in George William Hunter's *Civic Biology,* a textbook routinely assigned to Tennessee high school biology classes. Bryan responded viscerally to this image.

Bryan had a point. He had an eye for ambiguity in evolutionary metaphors. Like many diagrams published by scientists and science popularizers of the time, Hunter's balloons could be interpreted as undermining common written and spoken defenses of evolution, defenses made vulnerable by the claims scientists made, the disarray of evolutionary theory in the 1920s, and a disjunction between public and scientific understanding of scientific illustration. Visual images played an important part in the public discourse associated with the Scopes trial, but they did not necessarily convey the messages their authors intended.

Bryan complained that the *Civic Biology* illustration effaced humans, imprisoned them anonymously among the mammals, in "a little ring . . . with lions and tigers and everything that is bad!"[3] For scientists, the *Civic Biology* diagram fit an established set of visual conventions. Humans held no special place; they resided within the rather small circle allotted to the mammals. That circle was small be-

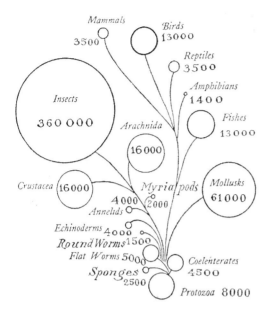

William Jennings Bryan objected to this diagram in *A Civic Biology,* the 1914 textbook assigned to John Thomas Scopes's students. Both at the trial and in his *Memoirs,* Bryan complained that the diagram implied that humans were lost among the mammals in a small insignificant circle. Humans, Bryan maintained, should be afforded the dignity of their own separate circle. Reprinted from George William Hunter, *A Civic Biology Presented in Problems* (New York: American Book Co., 1914), 194.

cause the number of species in the class of mammals *is* small relative to the number in other zoological classes. Scientists familiar with such diagrams understood the chart to describe taxonomic relationships, those of the scientific system for classifying living things; they also understood it to maintain silence on questions of religious or political significance.

Bryan knew better. He recognized the intent of the diagram, but he also understood that, from a lay perspective, it implied something more. And he realized that illustrations of scientific ideas could have a profound impact. In his *Memoirs* he returned to this theme: "*No circle is reserved for man alone.* . . . What shall we say of the intelligence, not to say religion, of those who . . . put a man with an immortal soul in the same circle with the wolf, the hyena, and the skunk?"[4] For scientists this illustration was a version of a familiar branching diagram depicting natural relationships. From Bryan's point of view it seemed to mock traditional verities about human significance. It was the human place in nature that was at stake.

Ironically, many of the defenders of evolution shared that concern, but they believed that the images of evolutionary ideas published in books, magazines, and newspapers, hung on museum walls, circulated in movie theaters, and sent to Dayton for exhibit at the trial would allay the fears of any observer willing to see the truth. Nevertheless, scientific illustrations drew fire from anti-evolutionists who understood visual images to hold a place at the center of the debate. Even before the trial, a newspaper article featured that diagram from Hunter's *Civic*

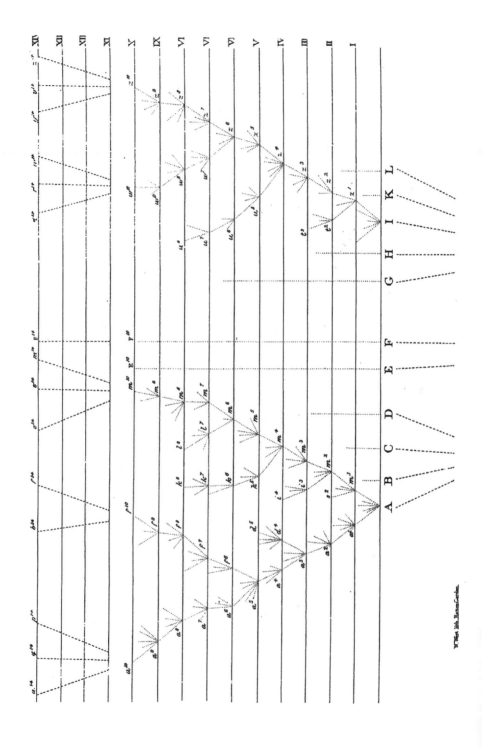

Biology under a heading that announced: "This Started the 'Monkey War.'"[5] Opponents of evolution targeted illustrations because visual images of scientific ideas were eloquent and because they said both more and less than their authors intended.

Chains, Trees, and Branches

As the ancient concept of a *scala naturae*, or a linear and hierarchical chain of being, began to give way in the late eighteenth and early nineteenth centuries, it was increasingly supplanted in zoology by an image of the natural world as organized in a complex branching pattern. When Darwin published *On the Origin of Species* in 1859, the evolutionary tree diagram he included was the only illustration in all 495 pages of the book. Darwin included this diagram because it played a crucial role in his thinking about evolution, beginning long before 1859. Darwin actually formulated his theory by devising and contemplating this mental picture of evolutionary patterns.[6] Darwin's famous tree illustrated his understanding of the complexity, contingency, and fecundity of the evolutionary process. It was an abstract tree: it did not describe the fate of specific organisms or lineages; it described a process. It indicated increasing diversity, without suggesting growth in a single direction. Darwin explicitly intended this tree to be nondirectional, not a literal tree with a main trunk and side branches, but a branching diagram.

The durable notion of a linear chain of being would not give way so easily, though. Trees depicted literally as trees rather than as branching diagrams grew out of a late nineteenth-century tradition. The evolutionary tree motif became familiar to the American public through Ernst Haeckel's family tree diagrams. Haeckel's influential "Pedigree of Man," widely published in the 1870s, was a prototype for later trees. Haeckel's tree was unusual yet curiously representative of several underlying assumptions and conflicts in late nineteenth-century biol-

---

(*Opposite*) Charles Darwin developed this branching diagram, the only illustration in *On the Origin of Species* (1859), to help him think through his concept of evolution. It shows ancestral forms (capital letters) and their descendants (lowercase italic letters) through successive generations (superscript Arabic numerals). Both forms that persist unchanged (F) and those that become extinct (B, C, D) appear. The diagram depicts evolution as a complex process, without a single direction, in vivid contrast to the tradition of presenting nature as a linear chain of being. Reprinted from Charles Darwin, *On the Origin of Species: A Facsimile of the First Edition* (1859; Cambridge: Harvard University Press, 1964), foldout following p. 116.

ogy. Most evolutionary diagrams of the time were simple line drawings; Haeckel's "pedigree" was drawn to look like a real (if somewhat misshapen) tree. But
though it had branches, it revealed an essentially linear concept of evolution, and
an undeniably hierarchical one. Its most obvious feature was that it culminated
in "man," who resided not only at the summit of the tree but at the top of the
main trunk, surrounded by the next "higher" animals, the other primates. Other
kinds of animals occupied outlying branches, which were atrophied and unim-
pressive. They looked like evolutionary dead ends, not like growing branches.
The categories were inconsistent: closely related groups of mammals, such as rodents, and ecological types that did not constitute related groups, such as "beasts
of prey" and "beaked animals," were both given branches. The top quarter of the
tree was devoted to the mammals. Taxonomic categories were amplified toward
the top of the tree, by implication magnifying the significance of the "higher
orders." Lower orders, such as rodents, were confined to single branches—although they might include large numbers of species, including many still living—while a subset of the primate order, the family of the apes, adorned the
crown of the tree; the four genera of apes surrounded the single species "man" at
the pinnacle. Though the diagram in some sense presented a branching concept
of evolution—echinoderms, such as starfish, were not positioned as steps in the
progression toward humans but as a side branch—in essence it retained the old
concept of the chain of being, a main trunk progressing from monerans (organisms of the kingdom Monera) to "man." Groups of organisms that continue to be
abundant and diverse on earth today were represented as "side branches," apart
from the main trunk leading to humans. Furthermore, Haeckel's tree, unlike
Darwin's branching diagram, conveyed no sense of time. It was static, apparently
complete, including no labels or other conventions to indicate time's passage and
no extinct animals. The appearance of the "lower" taxonomic groups on the tree's
lower branches implied that those groups appeared on earth earlier. Starfish and
monerans continue to exist and to evolve, but Haeckel's diagram offered no way
to show this. A position near the bottom of the tree seemed to mean a low position in the hierarchy of nature more than an early appearance on earth.[7]

This tree contrasted markedly with the majority of late nineteenth-century
tree diagrams published by scientists in journals intended primarily for other scientists. These diagrams tended to be simple line drawings. Haeckel also published several very different diagrams, described by reviewers as "fernlike" and
branching.[8] The anatomist Sir Arthur Keith, calling Haeckel the "pioneer and
prince of pedigree-makers," saw in the fernlike trees a "storm-blown appearance,"
perceiving that the branches of such trees, "under the stress of evolutionary

PEDIGREE OF MAN.

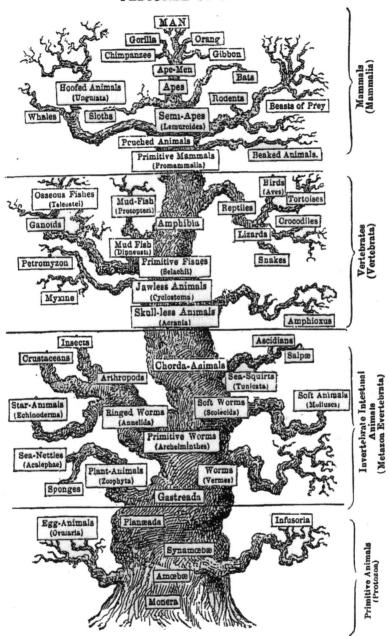

Widely published in popular books about evolution, Ernst Haeckel's "Pedigree of Man" (1866), unlike Darwin's branching diagram, depicts a literal tree and conveys an essentially linear vision of evolutionary relationships. It includes no convention for revealing change over time or for distinguishing between extant and extinct taxonomic groups. Reprinted from Ernst Haeckel, *The Evolution of Man* (1866; New York, 1896), 189.

Amniota { Paired nostrilled or Amphirrhina with Amnion, without gills.											
Reptiles, Reptilia.								Birds, Aves	Suckling animals, Mammalia.		
Primeval Reptiles, Tocosauria.	Lizards, Lacertilia.	Snakes, Ophidia.	Crocodiles, Crocodilia.	Tortoises, Chelonia.	Flying Reptiles, Pterosauria.	Dragons, Dinosauria.	Billed Reptiles, Anomodontia.		Billed Animals, Monotremata.	Pouched Animals, Marsupialia.	Placental Animals, Placentalia.

This fernlike diagram by Haeckel contrasts starkly with his much more widely reproduced "Pedigree of Man." Scientists who were critical of the "Pedigree" cited this diagram as much more scientifically correct. Ernst Haeckel, *The History of Creation*, trans. E. Ray Lankester (London: John Murray, 1875–76), vol. 2, opposite p. 223.

winds, tended to grow in one direction; the most progressive branch of all—the human branch—occupying, as it should do, the extreme right."[9] The fern-shaped trees remained relatively obscure, however, while the "Pedigree of Man" became an archetype for popular images of evolution. The story of evolution "from monera to man" carried a singular resonance, and trees of this essential type would remain prominent.

Trees like Haeckel's "pedigree" caused a backlash among scientists. William King Gregory included a copy of Haeckel's pedigree in a discussion of human family trees but was careful to note that "man is for convenience placed at the center of the tip of the tree, although Haeckel was anything but anthropocentric in his teachings." In the most current diagrams at the American Museum, Gregory added, "mankind is shown as a curious side-shoot from the anthropoid stock."[10] Looking back from 1934, Arthur Keith wrote that trees like this "reflect human vanity rather than zoological justice." They suggested "that the tree of life grew and groaned all through these past geological aeons just to produce humanity at its topmost shoots, and that all other forms of life are but abortive branches of the great human stem."[11]

Despite the qualms of some scientists, the evolutionary diagrams published in textbooks and books for the lay public in the 1920s often repeated the literal tree motif of Haeckel's famous "pedigree" and implied a similarly linear view. A characteristic example is the tree in Benjamin Gruenberg's *The Story of Evolution*, a highly stylized, conventionalized rendering, including no information about time or extinction, conveying the impression that there is a "main line" of evolution culminating in "man" (in a suit!). Gruenberg's use of this kind of tree is especially intriguing because the text of his book reveals a far more nuanced understanding of evolutionary patterns. Gruenberg seems to have been much quicker to challenge received wisdom expressed in words than to question the conventions of illustration. A similar tree, used as the frontispiece for a 1921 book, Adam Whyte's *The Wonder World We Live In*, outfitted the human atop the tree in the garb of the stereotyped "caveman."[12] Cavemen in trees proliferated.

Another diagram frequently reproduced in the popular science literature of the 1920s reinforced the impression of a linear ascent to humans. Originally published by Thomas Henry Huxley in *Man's Place in Nature* in 1863, the sequence of skeletons of the four great apes—a gibbon, an orangutan, a chimpanzee, and a gorilla—and a human appeared amid the debates following publication of *On the Origin of Species*. In that context, Huxley's primary concern was to establish relationship and thus the fact of evolution. In his debate with the anatomist Richard Owen over human origins, Huxley's strategy was to argue that anatomical similarities among the brains of the primates implied a close relationship. Although he sought to establish the physical relationships of humans and the great apes, Huxley conceded "the vastness of the gulf between civilized man and the brutes . . . whether *from* them or not, he is assuredly not *of* them."[13] Even so, his ape-to-man series implied that the distances separating the four great apes from one another were no greater than the distance between the gorilla and the human, and the left-to-right sequence suggested progress. Those would be the obvious inferences when the sequence was removed from the anatomy lab and reproduced in popular books. And it was widely reproduced, beginning in the 1870s in successful books by Haeckel and continuing in science popularization in the 1920s.

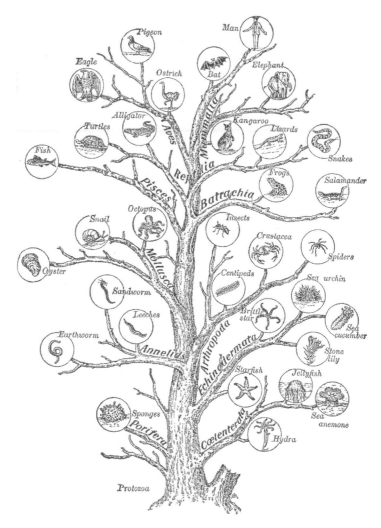

Benjamin C. Gruenberg's 1919 family tree diagram, depicting living animals perched on branches of a realistic tree, is typical of evolutionary diagrams in textbooks of the period, as is the whimsical use of clothing in the figures representing humans. The cut-off branch near the base of the tree represents a trunk for families of plants, depicted on another page. Reprinted from Benjamin C. Gruenberg, *The Story of Evolution: Facts and Theories on the Development of Life* (Garden City, NY: Garden City Publishing Co., 1919), 71.

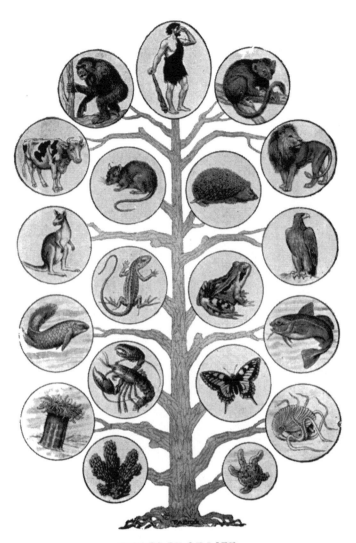

THE TREE OF LIFE.

It was not unusual for artists to design family tree diagrams in which the humans—almost always occupying the top of the trees—were outfitted as stereotypical cavemen. The cavemen at the tops of the trees frequently wore one-shouldered fur garments and carried clubs; they seem almost never to have been cave women. Reprinted from Adam Gowans Whyte, *The Wonder World We Live In* (New York: Knopf, 1921), frontispiece.

Originally published as the frontispiece to Thomas Henry Huxley's 1863 book *Man's Place in Nature,* this sequence appeared often in the 1920s. Though Huxley conceded "the vastness of the gulf between civilized man and the brutes," the human figure is positioned no farther from the gorilla than the gorilla is from the chimpanzee. Reprinted from Thomas Henry Huxley, *Man's Place in Nature* (1863; Ann Arbor: University of Michigan Press, 1959), frontispiece.

Scientists often cited the cultural distance between humans and apes; some of them, like Osborn, prized that distance as much as Bryan did. But Huxley's diagram implied something different. Based on anatomical studies of extant species, the series included no reference to evolutionary time or ecological context, and the proximity of the human figure to the apes, along with the left-to-right direction from apes to human, would evoke familiar echoes of the old notion of a linear chain of being. From the perspective of people unfamiliar with its context, Huxley's ape-to-man series could imply that the distance separating the human from the gorilla was no greater than that between the gorilla and the chimpanzee. This visual series survived as a recurrent motif for the 1920s literature.

Huxley's diagram provided a template for a visual cliché—easily used in a standard evolution joke. This cliché appeared in press coverage of the Scopes trial in 1925. A cartoon in the *New Yorker* invoking it illustrated the cultural backdrop of the evolution debate. Called "The Rise and Fall of Man," the series progressed from chimpanzee to Neanderthal to Socrates, then—by implication, descended—to William Jennings Bryan. Using the same motif, the "Evolution Number" of *Judge* magazine in the summer of the Scopes trial showed an evolutionary progression to a stereotype of the decade, a "cake-eater."[14] For the editors of such magazines as the *New Yorker* and *Judge,* the evolution debates were more about cultural issues than about the substance of science, and those magazines—and cartoons like these—played an important part in keeping cultural conflict at the center of the debate. Such cartoons caricatured evolution in order to make jokes about human culture. The joke might consist in the exposure of some human pretense in the face of our animal origin, or it might highlight the peculiar status of humans as evolutionary anomalies. In general, though, the final figure supplied the punch line, diverting attention from the assumptions underlying the sequence, including the very notion that it *is* a sequence, marking progress from left to right, as in a language of words. The cliché became a kind of common knowledge. The common knowledge in this case included the linear, goal-directed version of evolution and human descent from apes, views that were far more controversial among scientists in 1925 than published diagrams revealed.

Early in the twentieth century paleontologists at the American Museum of Natural History pioneered complex tree diagrams. Osborn, who had described and named the evolutionary pattern known as adaptive radiation, in which a lineage of organisms undergoes rapid evolutionary diversification, published innovative diagrams of adaptive radiation in mammal evolution, especially the evolution of elephants, animals he dearly loved and the subject of his last great work, the two-volume *Proboscidea*. These diagrams could include a variety of data—especially ecological, stratigraphic, and anatomical. By implication they associated anatomical change with ecological context. A diagram of the evolution of elephants, for example, separated browsing elephants—those adapted to feed on leaves and branches—from grazing elephants, or grass eaters. Adaptations for grazing struc-

The Rise and Fall of Man

Primate   **Neanderthal Man**   **Socrates**   **W. J. Bryan**

"The Rise and Fall of Man." The cartoon shows the steady ascent of humans from primate to Neanderthal to Socrates, followed by—this being the joke—a descent to William Jennings Bryan. The cartoon is from the *New Yorker*, June 6, 1925, 3. Used with permission. © The New Yorker Collection 1925 Rea Irvin from cartoonbank .com. All Rights Reserved.

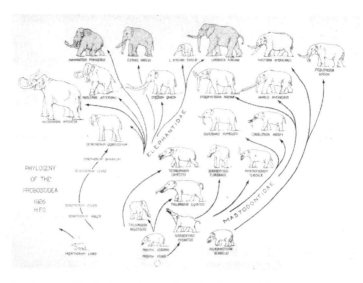

"Phylogeny of the Proboscidea, 1926." A family tree of elephants: this is
one of many elephant family tree diagrams designed by Osborn. In general,
Osborn's elephant tree diagrams are bushy, some of them illustrating the
principle he named "adaptive radiation" and many of them including ecological context. Image #311495. © American Museum of Natural History

tured one of the key narratives used to persuade the public of evolution, the story
of horses. Curator William Diller Matthew designed diagrams of the evolution of
horses that incorporated geological data in the form of a labeled stratigraphic column: the layers of rocks and their names were juxtaposed with the evolutionary
changes in horse anatomy associated with rock layers—in other words, evolution.

Illustrations of the evolutionary history of horses, especially those Matthew
produced for exhibit at the American Museum, held a prominent place among
the most frequently reprinted tree diagrams of the 1920s. Representations of the
fossil record of horses had been favorite heuristic illustrations of evolution ever
since Thomas Henry Huxley, touring the United States as Darwin's most famous
public advocate, visited the fossil horse collection of the paleontologist Othniel
Charles Marsh at Yale College in 1876. The Yale horses made such a convincing
case for evolution that Huxley immediately included them in lectures during his
American tour. Huxley and Marsh collaborated on a diagram that became an archetype—or, as Stephen Jay Gould has suggested, canonical—and which exemplified the linearity of the view of evolution that horses were used to demonstrate.

Writers of popular science in the 1920s continued to cite horse evolution as
the most complete possible fossil record and the most convincing case of evolu-

tion. The horse diagrams most widely published in the 1920s were two of those designed by Matthew. Creatively incorporating geological strata and associating them with changes in parts of horse anatomy over time, these diagrams were exhibited at the museum and published in successive museum leaflets. By 1925 Matthew realized that horse evolution was much more complex than his own most famous illustrations of it might indicate. He had highlighted the net direction from *Eohippus* to *Equus* partly to demonstrate evolutionary sequence. In the face of challenges to evolution in the 1920s, science popularizers used the horse diagrams heuristically—to prove the fact of evolution even though they simplified the pattern by emphasizing large trends and net direction and neglecting complexity and side branches. And museum display and publication fixed these images in a canon, obscuring their dynamic function as "working hypotheses."[15]

The construction of museum exhibits—which was expensive—extended the life span of Matthew's most linear horse images. The prominence of the American Museum exhibits meant that those diagrams were widely reproduced in

THE EVOLUTION OF THE HORSE.

Horse evolution has long been used to illustrate and to prove evolution. This diagram, executed in 1902 and on display at the American Museum of Natural History in the 1920s, was widely used in science popularizations and textbooks of the decade. It incorporated geological and anatomical observations to depict the pattern of change over time. It also conveyed a linearity that could mislead anyone unfamiliar with the complexity of horse evolution. Image #35522. © American Museum of Natural History

books and magazines for popular audiences. By 1930, they were also included in books for children such as *The Earth for Sam*.[16] Matthew's familiar horse diagrams implied a linear, teleological evolutionary pattern that could readily be extrapolated to human evolution. And it was. Suggesting that the model of horse evolution had influenced the way scientists envisioned evolutionary patterns in general, Gregory argued, in his debates with Osborn over human evolution, that the pattern of horse evolution should not be used as a general template. The rich fossil record of horses, however, along with the tradition of illustrating ancient horse sequences as the best demonstration of evolution, ensured the continuing prominence of these images. Furthermore, from the perspective of anyone trying to convince the uninitiated of the fact—as opposed to the pattern—of evolution, the example of horses was very persuasive. And in 1925, evolution meant human evolution. If horse evolution followed a linear trajectory, the obvious conclusion was to extrapolate and assume a linear ascent to the "highest" form of animal—humans.

This view of evolution as progress appeared everywhere in the popular culture of the middle 1920s. It was evident in the 1923 movie called *Evolution*, made with help and advice from curators at the American Museum, which again made the rounds of New York theaters in the summer of the Scopes trial. This movie was probably among the versions of evolution most familiar to the public in the 1920s: a New Yorker who attended John Roach Straton's famous 1924 sermon attacking the American Museum sent an amused account of it to Osborn, noting that Straton had also denounced the "faked film" with its images of life arising from "green scum." But, the writer perceived, such indignation "showed that even to him the film was very effective." And it clearly told a story of a linear ascent "from monad to man." As a reviewer noted, the purpose of the movie was to "[trace] by fact and scientific deduction the derivation of animal and plant life and from thence via the apes into mankind."[17]

Linear views of the ascent of humans appeared in more solemn form in American popular culture as well and continued to do so throughout the decade. The minister of the Fourth Unitarian Church in Brooklyn, S. R. Mayer-Oakes, wrote to Osborn in 1930: "We have recently done a daring thing for a Church. Departing from the conventional ecclesiastical figures of saints and angels, we have gone to Natural History for our symbols." A new memorial window, "symbolizing 'Human Aspiration,'" had been installed, and surrounding it, "we have the whole chancel frescoed to depict the Evolution of Life forms rising from protozoa to Man." The project had been suggested, Mayer-Oakes implied, by inspirational interpretations of "the upward surge of life," including Osborn's. "As you might expect, it shocks some people until it is interpreted to them. Then the significance of Evolution grows on them."[18]

Osborn's emphasis on evolutionary ascent had unquestionably made an impression. The complex branching view of evolution and the linear notion of progress and ascent could be intertwined in complicated ways, but intertwined they were. Gregory, graced with a wry sense of humor, produced a whimsical version of the linear evolution of a human "from fish to man" set on the limb of a tree. Although the linearity of the sequence resembles Huxley's shorter linear chain, the tree branch suggests that Gregory was keeping the larger context in mind. Anyone familiar with many of his diagrams would be aware of this context, but it could easily be lost to viewers of the single image.

One of Gregory's tree diagrams made explicit the dimension of geological time left out of the popular cliché. Most evolutionary biologists understood this dimension to be implied in linear diagrams. The familiar sequence appearing at the top of the tree represented the living remnants of a long history, products of a more complex branching pattern. Gregory reversed the conventional left-to-right order by placing humans on the far left, perhaps intending to undermine the usual connotation of progress toward humans. His inclusion of "undiscovered ancestors" at the base of the diagram highlighted its function as "working hypothesis." Significantly, diagrams this implicitly methodological were not common in popularizations.[19]

In his writings Gregory suggested that earlier taxonomists, under the influence of the old chain-of-being model of linear evolution, had simplified evolutionary trends by assuming that animals evolve from simple to complex or from generalized to specialized forms. In reality, he suggested, each species of animal is a palimpsest of primitive and advanced characters, its history recorded in its structure. Some features have greater diagnostic value than others, and he designed a number of trees illustrating the evolution of diagnostic features, such as the primate hand. Trained as a comparative anatomist, Gregory began with skeletons and reasoned backward through all the forces that contribute to the forms that skeletons take: the constraints of physiology, development, genetics, and evolutionary history. His scientific work had a strong ecological bent. The animal we see in the modern world is the complex result of a subtle combination of factors, of history and ecology, or as Gregory wrote, of "heritage and *habitus*." An evolutionary branching occurs under the influence of a particular set of ecological circumstances, and once taken, that branching is irrevocable. So the ecological history of an animal is a part of both its past and its future. Animals evolve in

MAN AMONG THE VERTEBRATES

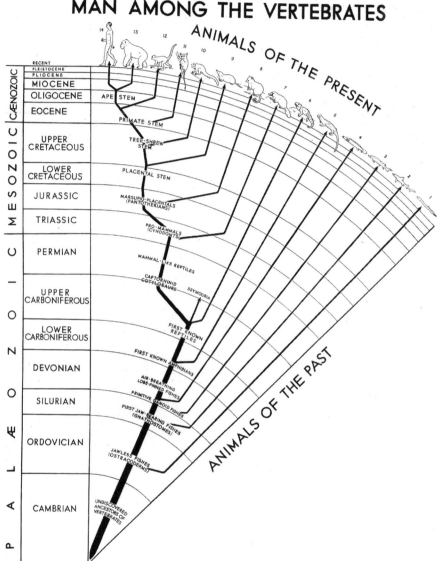

William King Gregory's diagram makes explicit a dimension—that of time—understood by scientists to be implied in linear illustrations like Huxley's but often neglected in depictions of evolution in popular texts. By including "undiscovered ancestors," this diagram represents scientific process—and the role of diagrams in scientific hypothesis testing—more realistically than most popularizations did. Image #313712. © American Museum of Natural History

response to ecological circumstances, but they have to use the raw material their evolutionary history has given them.[20]

Gregory's diagrams reflected these ideas in creative and witty fashion and give us a sense of the resonance and the linguistic complexity of this visual lexicon among scientists. He designed a mural featuring an evolutionary tree diagram called "Man among the Primates" which represented the races of humans in stereotyped clothing and included Cro-Magnon and Neanderthal figures taken from paintings by Knight. This especially treelike tree, with its symbolically attired figures, contrasts interestingly with a 1916 diagram of the human evolutionary tree drawn by Gregory for a more strictly scientific publication.[21] The 1916 diagram resembles a tree in only the most abstract way. In its abstraction, the earlier diagram emphasizes the branching pattern and avoids the appearance of a linear chain or hierarchy with humans on top. Both trees actually branch in a Darwinian fashion, although neither one includes a label of the horizontal axis. The later tree, however, designed as a museum exhibit, does convey, in its "treeness," a hierarchy, since the primates on the "lower" branches appear to be positioned there simply because they are "lower" primates—they have not, after all, ceased to evolve. Intriguingly, Gregory's diagrams appear to have grown more imaginative and symbolically richer as he intensified his efforts to communicate with the nonscientific public. But to appreciate their wit and subtlety, one would need to know more of the scientific literature than most nonscientists would have. Anyone familiar with Gregory's scientific papers as well as his many imaginative drawings and diagrams for the public would have realized that he understood evolution to be a complex branching process, but the diagrams—like the diagram in the mural "Man among the Primates"—that appeared most often in museum exhibits and popular books commonly tended, even though branching, to suggest human ascent. Furthermore, many of the diagrams expressed convictions and biases that came from outside the realm of science proper.

In claiming that tree diagrams "express most clearly the author's meaning," Osborn was both right and wrong. The function of the diagrams as working hypotheses of science was belied by their authority as truth and their longevity on display and in print. The effort to use them as persuasive devices obscured the syntax of scientific convention necessary to comprehend the full intentions of their authors. Extrascientific messages, intended or not, would often be more evident to audiences.

The reader need not have been a creationist, an anti-intellectual, or a fool to react with discomfort to the hierarchical diagrams used by defenders of evolution. Several of the "fish-to-man" diagrams suggest an evolutionary hierarchy

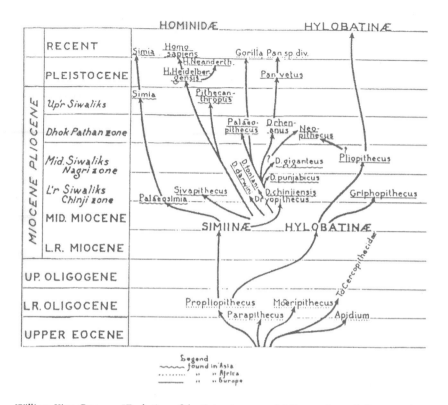

William King Gregory: "Evolution of the Primates to 1916." Gregory's 1916 diagram of the phylogeny of primates, illustrating a technical scientific paper, contrasts strikingly with his later, more fanciful trees, designed for exhibit at the museum. From William King Gregory, "Studies on the Evolution of Primates. I. The Cope-Osborn 'Theory of Trituberculy,' and Ancestral Molar Patterns of the Primates. II. Phylogeny of Recent and Extinct Anthropoids, with Special Reference to the Origin of Man," *Bulletin of the American Museum of Natural History* 35 (1916): 337.

redolent of social and political preoccupations of the time. According to the caption for the frontispiece of Gregory's 1929 book *Our Face from Fish to Man*, the series culminates in a "Roman athlete." The figure beneath the Roman athlete—but superior to the chimpanzee—represents, the caption tells us, a Tasmanian. This face contrasts oddly with its antecedents. Half smiles seem to play across the faces of the "lower" animals, giving them benign, almost friendly appearances. The chimpanzee's is the most anthropomorphic of the animal faces, suggesting a progression of sentiment from fish to humans. The Tasmanian, unlike the lower animals, seems to scowl, betraying a distinctly less friendly attitude—or, perhaps a great sadness.

Gregory does not tell us that the Tasmanian in the picture was a real person, with a name. The face of this Tasmanian unmistakably came from a frequently reproduced late nineteenth-century photograph of a middle-aged woman, Trucanini,

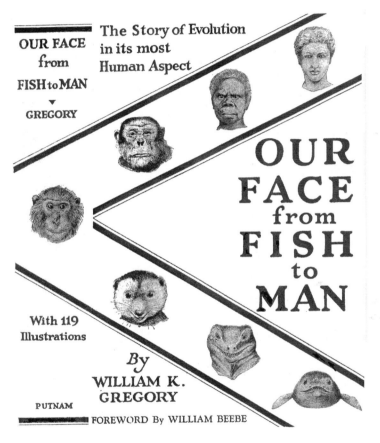

The dust cover of William King Gregory's 1929 book *Our Face from Fish to Man* displays a treelike diagram. Within the book the topmost figure is identified as "a Roman athlete." The penultimate figure, described only as a Tasmanian, was based on a real historical figure, a woman who became famous in the late nineteenth century as "the last Tasmanian." Her name was Trucanini. Reprinted from the dust jacket of William King Gregory, *Our Face from Fish to Man: A Portrait Gallery of Our Ancient Ancestors and Kinsfolk Together with a Concise History of Our Best Features* (New York: G. P. Putnam's Sons, 1929).

believed to have been the "last Tasmanian." When the British settled in Tasmania in 1803, the indigenous population was estimated to have been about four thousand. Introduced diseases rapidly reduced that number, and the British government relocated the few remaining Tasmanians several times, confining them on small reservations or in places that might today be called "internment camps."

The photographer Charles Woolley was commissioned to photograph the five remaining Tasmanians for exhibit in the International Exhibition in Melbourne in 1866; Trucanini was one of them. She was also the last of this small remnant population to die, in 1876, and as anthropologist Vivienne Rae-Ellis puts it, "Woolley's melancholic image . . . is the portrait of an elderly woman in despair, recorded for posterity as her four companions faced imminent death, leaving her in fear of surviving alone, the last of her people."[22] Trucanini's image had been displayed at the exhibition and reproduced in photography books of the late nineteenth century. It was a famous image, evocative of Victorian understandings of race and culture.

By the early twentieth century, it would have also echoed the popular sentimentalizing, in the United States, of the romanticized figure of the ostensibly "vanishing" American Indian.[23] For evolutionists like Osborn and Gregory, especially those committed to a view of human evolution through progressively improving stages from "primitive" to Nordic, the remnant population of Tasmanians had extraordinary interest as—they preferred to think—a species of "living fossil."

To some members of the public, the diagrams conceived at the American Museum might seem to include entirely too many ideas. Gregory, like Osborn, his mentor at the museum, was active in the eugenics movement. As has often been pointed out, racial and eugenic themes were prominent in the famous Hall of the Age of Man exhibit, whose construction began in 1915; indeed, Osborn had made special efforts to complete the exhibit in time for the International Eugenics Congress held at the museum in 1921. Since many of the biologists of the time were active in the eugenics movement, it would be hard to say that eugenicist views were disproportionately represented among the scientists defending Scopes in the press. But it is certainly true that many defenders of evolution held such views and voiced them relentlessly.[24] Textbooks—including Hunter's *Civic Biology*—routinely incorporated messages about racial hierarchy and eugenics. It is reasonable to ask whether those ideas limited their audience—Osborn, for example, referred to "our" Nordic heritage in his articles in the American Museum's popular magazine *Natural History*.[25]

Americans of non-Nordic ethnic heritage might have found the linear view of evolution implied in many diagrams less than compelling, since the linear model was implicitly—and sometimes explicitly—hierarchical. Diagrams presenting humans as the apex of the evolutionary process often positioned particular humans at the pinnacle and others closer to the animals or to more "primitive" branches of the family tree. Even diagrams intended to convey the complexity of evolutionary patterns, as many of Gregory's were, could reinforce the message of racial hierarchy. Osborn had photographs of one of Gregory's tree diagrams from the Hall of the Age of Man sent to Dayton, Tennessee, for the edification of the Scopes trial jurors and the public. The same tree appeared in newspapers during the summer of the trial, in some cases with a photograph of Osborn as an inset. This diagram resembled Gregory's "Animals of the Past, Animals of the Present" turned on its side, suggesting some of the themes in that diagram but with several significant differences. Focused on a single, bifurcated branch of a larger tree whose existence was indicated by a trunk at the left side of the picture, it implied that human evolution occurred somewhere other than at the pinnacle of the tree. But the progression carried the label "Line of ascent to man." The lineage of the primates was decisively separated from that of humans and their ancestors. And the living members of the primate family aligned themselves in the familiar chain-of-being pattern, this time vertically, at the right-hand side of the diagram. At least one newspaper that published a photograph of this diagram included an exceptionally long caption, explaining to the public what the image meant. The verbal explanation emphasized that the white race of humans—represented by an "American"—belonged on the "topmost twig" of this vertical hierarchy, and that "on the same stalk, in lower order, are placed the Australian native, the negro [sic] and the Chinese."[26]

Another diagram, this one designed by Osborn, incorporated the same sort of racial hierarchy, increasing the depth of the bifurcation of the main trunk separating the simian stem from the humans.[27] Osborn intended a hierarchy here, too, and he ordered it along what he argued was an increase in brain size, this time from right to left. In this case, though, the "white" race occupied only the penultimate position in the brain-size hierarchy, the honor of ultimate position being awarded to the extinct Cro-Magnon. Only the white race on this diagram appears on the same branch with the Cro-Magnon, a species of early humans Osborn romanticized and admired. Cro-Magnon people were blessed with extremely large brains, and they produced sophisticated cave art. Osborn wrote popular articles extolling their virtues and took a good deal of pride in having rehabilitated their reputation with the public. Indeed, he often suggested that they were in many respects our own superiors, attributing our decline to a decadent civilization. So it meant a lot when he placed them on a branch of the human family tree that included the living "white" race but no other living human races.

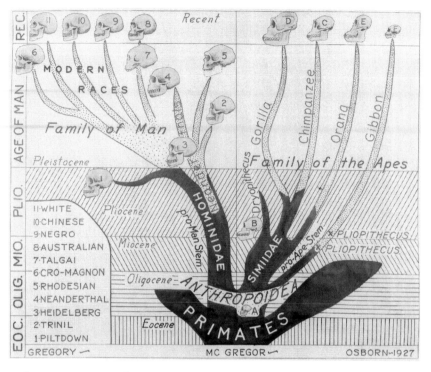

In the 1920s, Henry Fairfield Osborn designed tree diagrams that featured strong bifurcations between the anthropoid and human lineages. They also often made explicit Osborn's racial theories. Image #312268. © American Museum of Natural History

During the course of the 1920s, Osborn divided his tree diagrams into increasingly distinct human and simian branches, removing their common ancestry to the remote past and sometimes, as in this case, indicating the significance of the rift by a vertical line.

Osborn insisted more and more forcefully—and publicly—during the debates of the 1920s that the lineage that led to humans had split off from the rest of the primates as long ago as the Eocene—at the base of the human family tree—a theory with which very few evolutionary scientists agreed. Osborn modified Gregory's bas-relief of the Family Tree of Man from the Hall of the Age of Man for his collection of essays on the evolution debates, *Evolution and Religion in Education,* published in 1925. The drawing of Gregory's original included the same hierarchy of relationships, but Osborn expanded the time allotted for the most recent period, the Quaternary, or the "Age of Man," relative to earlier geological periods. In her biography of Gregory, Sheila Ann Dean accuses Osborn of having "distorted the drawing so that the human skulls graphically predominated the entire Cenozoic."[28]

This pattern could be justified as implying a logarithmic scale—many graphs are designed in this way. In a logarithmic plot, the numbers on one axis may follow a logarithmic scale, the same kind of scale used for computing decibel levels or the seismic numbers assigned to measure the force of earthquakes. In such a scale the numbers increase by powers of ten instead of simply increasing arithmetically, and the reason for doing this in the case of a graph of change over time

EXISTING FACTS OF HUMAN ASCENT

Osborn published this diagram in his collection of essays about the evolution controversy, *Evolution and Religion in Education*. It is a variation on the diagram sent to the Scopes trial as evidence. It is also the only illustration in the book, an uncharacteristic thing for Osborn, whose books were more often lavishly illustrated. Henry Fairfield Osborn, *Evolution and Religion in Education: Polemics of the Fundamentalist Controversy of 1922 to 1926* (New York: Charles Scribner's Sons, 1926), 206.

would be to allow more room for detail in the recent period, which is, of course, the period of interest when it comes to human evolution. Does this explain Osborn's arrangement of the time scale on his human family tree?

An idiosyncratic graph of evolution published in *Scientific American* in 1925, "Evolution from the Nebula to 1925," was accompanied by an explanation of the use of this kind of scale. The left-hand column, labeled "Years Ago," the author explained, was divided into equal spaces that represent unequal time intervals: "Each space represents ten times as long a duration as the space next below it." This allowed a closer focus on the more recent, and relatively shorter—but most interesting!—time periods. In characteristic *Scientific American* style, the author noted, "If we should attempt to represent the entire chart on the same scale as the lowest space, it would be more than 300 miles long!"[29] Significantly, the author, who wrote regularly for *Scientific American,* a periodical boasting a scientifically literate audience, felt it necessary to explain how this sort of scale worked. The editors of *Scientific American,* in other words, assumed that the average reader would not be familiar with the scientific convention of plotting variables against a logarithmic scale. Osborn's tree diagrams lengthening the time of recent geological periods relative to earlier times did not indicate that they had been plotted in this way. Though Osborn might have imagined that such a convention was understood, and to scientists something like it might have been intuitively obvious, it seems unlikely that it would have been familiar to the general public. In the absence of any explanation of the time scale, the uninitiated could easily have gotten the impression of an extended duration for human evolution.

Osborn, perhaps ironically, was ultimately as concerned with creating distance between humans and the other animals—especially the other primates—as was William Jennings Bryan. As Gregory so wryly noted, Osborn had become a "pithecophobiac." Emphasizing the length of time separating humans from the other primates was one way to establish this distance. Another way was by dwelling on the notion of progress. Osborn's view of evolution had always been progressive—he remained among the staunchest adherents of the idea of orthogenesis. The extrascientific implications of this position became much more explicit in the course of the debates with fundamentalists in the 1920s.

In this context it is clear that linearity was indeed linked inextricably with notions of progress. Lucretia Perry Osborn's 1925 book, so tellingly called *The Chain of Life,* included a starkly linear time line that can perhaps be taken as the unvarnished version of evolution discussed at the Osborn dinner table on Fifth Avenue or at the family estate on the Hudson River. It was no accident that Henry Fairfield Osborn insisted so resolutely on substituting the word *ascent* for Darwin's term *descent* of humans through evolutionary processes. His diagrams, and his wife's diagram, consistently expressed his belief in human evolution as an ascent, an idea he elaborated most fully in his 1927 book *Man Rises to Parnassus.* And an illustration of increasing brain size in *The Chain of Life,* like the "line of ascending intelligence" in some of Osborn's own tree diagrams, demonstrated both the linear progression of the process and the essence of the human ascent.[30] Affirming this streamlined view of evolutionary progress, Osborn added the *Chain of Life* chart to the 1929 edition of the museum's guide to the Hall of the Age of Man.[31]

Osborn's trees often combined branching patterns with linear hierarchies. And though his branching diagrams reveal an interpretation of the past that was not strictly or simply linear—for example, putting Piltdown and Cro-Magnon on separate, dead-end branches rather than on a main trunk leading to modern humans—he also arranged J. H. McGregor's reconstructions of human ancestors in the linear series that so offended Alfred Watterson McCann and John Roach Straton. And both journalists and other scientists occasionally objected to the implications of some of his linear diagrams as well as his statements about the evolutionary distance between humans and apes (though less often, unfortunately, to his ideas about race). In 1920, J. H. McGregor had written in a letter to Edwin Grant Conklin that although he was proud of the scientific credibility of his sculptural restorations of fossil hominids, he disapproved of photographs of them

American Museum of Natural History exhibits emphasized the scientific reasoning that went into the reconstruction of these human ancestors by the biologist and artist J. H. McGregor. Nevertheless, anti-evolutionists seized on them as examples of scientific imagination—or duplicity—run amok. Image #313682. © American Museum of Natural History

lined up in a row: "My only objection to this sort of photo is that most people jump to the conclusion that the heads represent . . . stages in a monolinear series which I don't believe for a moment."[32]

Another colleague, William McDougall, wrote to Osborn in the heat of the Scopes trial publicity, "Surely the question the public is interested in is not whether man is ascended from some existing species of ape, but whether from an ape-like form." McDougall wondered whether Osborn's argument was not disingenuous: Osborn illustrated his statements discarding the "ape-monkey theory," McDougall noticed, with diagrams supporting that very theory.[33] McDougall objected to the claim that humans did not share a common lineage with other primates. Osborn was not being disingenuous; he was quite sincere. But the diagrams he designed revealed a source of some confusion. Images of evolution as a linear progression, intended to demonstrate a comforting notion of progress, betrayed the verbal argument. Despite the insistence of Osborn and many of his colleagues that humans were not directly descended from any living forms of apes or monkeys, the linear chain-of-being images of evolution so prevalent in the popular literature probably did more to confirm such a connection than words could have done.

Letters to Osborn from the public and letters to editors of newspapers and magazines suggest that at least some people were indeed confused. An editorial in the *Lexington (KY) Herald* in 1922 objected to a tree diagram published in the paper and taken from a popular textbook: "An outline of classification of animals . . . can scarcely be designated a 'family tree.' Merely to mention the gorilla as a living type of animal of the Anthropoids (an undeniable truth) certainly does not commit the author to the advocacy of man's descent through this animal." Then, in a statement echoing confusion expressed frequently in letters to scientists and letters to editors from readers, the editor asserted, "One cannot be descended from a contemporary."[34]

The controversial sculpture *Chrysalis*, by the famous explorer Carl Akeley, embodied the idea of descent from a contemporary. Akeley had devised innovative exhibit techniques for the American Museum and was therefore associated with it in newspapers. When he entered *Chrysalis* in a prominent New York art show and it was rejected, he accused the judges of censorship, claiming it had been excluded for its evolutionary theme: it portrayed a man (or Man) emerging from the skin of a gorilla. In response Charles Francis Potter, Unitarian minister and vocal advocate for evolution, exhibited it in his Upper West Side church. But Osborn had the museum issue a statement to the papers saying that even though Akeley was associated with the museum, *Chrysalis* was not. Osborn's disclaimer may have implied a judgment of the quality of the sculpture as art, but he undoubtedly wished to distance the museum from this view of evolution as well. *Chrysalis* did express an understanding of evolution that could lead lay audiences to ask: If we evolved from apes, why are there still apes? The museum received many such queries, and Gregory attempted to answer them in lectures and essays. He was vexed by Osborn's increasingly adamant—and public—insistence on the distance between humans and apes, fearing that it exacerbated public confusion.

Albert Ingalls, a writer for *Scientific American,* voiced concern with the line of reasoning that denied human descent from apes. Such arguments were misleading and amounted to a form of "diluted Darwinism." Whatever our precise relationship to the living species of apes, we certainly belonged to the same family, as two tree diagrams accompanying his article made clear. Exonerating scientists themselves from any such lack of candor, Ingalls blamed popular writers, imploring them: "Let us then be frank and not designate our ancestor by some misleading phrase if we mean thereby to give the impression that the 'common parent stem' was a much more noble creature than the living apes and quite different from them."[35]

Ingalls exempted scientists from his criticism entirely too generously. No one was more dedicated than Osborn to demonstrating the nobility of human ancestry. He wrote repeatedly that the history of illustrations of primates as brutes caused people understandable trepidation about evolution and that if only the public could apprehend the immense time that separated us from the simian branch of the primate family tree, their fears would be calmed. Most of all, he blamed newspapers and movies for misrepresenting our noble ancestors, portraying them as brutes and "ape-men." In an article in the *New York Times,* for example, he complained, "The myth of ape-ancestry lingers on the stage, in the movies, in certain anti-naturalistic literature, in caricatures of our pedigree, even in certain scientific parlance."[36]

Despite the best efforts of evolutionary biologists, the average man or woman had good reason for associating evolution with ape-men. Newspaper accounts of the Scopes trial referred endlessly to the Dayton event as the "monkey trial." Moviegoers could see Dr. Challenger confront and vanquish ferocious ape-men in *The Lost World.* Songs making fun of monkeys had proliferated in the wake of Darwin, and in the twenties songs like "You Can't Make a Monkey Out of Me" were available in recorded form as well as in sheet music. Children in Kentucky schools learned anti-evolution songs featuring monkeys.

Visual images colored the way scientists themselves thought about evolution. In the 1928 collection of essays by scientists edited by Frances Mason, *Creation by Evolution,* several authors of chapters described evolution in words by suggesting that the reader conjure up a picture of a stratigraphic sequence of fossils, or of still images of slices of past time, being connected and run like a movie, allowing us to see evolution happening. "If we were to take a set of photographs of [a sequence of] fossils from the base of the series to the top, and copy them on a cinematograph film," wrote the geologist Francis A. Bather, for example, "we could see evolution taking place before our eyes." In the same volume the paleontologist Richard Swann Lull suggested that a "succession of drawings comparable to a moving picture film" might allow us to envision the evolution of birds from dinosaurs. And the biologist Herbert Spencer Jennings, in an article called "Can We See Evolution Occurring?" tied the idea to the metaphor of the tree: "We cannot directly see the growth of a tree, but by taking photographs at intervals and running them through a moving picture machine we can see the tree grow. . . . Ought we not to be able to get some sort of a moving picture of evolutionary change?"[37] It is intriguing that this notion occurred to three of these scientists. Mason's correspondence with them suggests that they had not read one another's chapters in advance.[38] They would all, undoubtedly, have seen or at least been aware of the movie *Evolution,* circulating in theaters. Visualizations like this movie—or the existence of movies in general, and the acute awareness of moving pictures as a way of seeing—affected the way people, including scientists, conceived of evolution. And when scientists debated evolutionary patterns and mechanisms among themselves, slides, photographs, diagrams, and drawings were part of the argument, both evidence and language.

If Osborn's arguments about ape ancestry and his linear hierarchies of human evolution had been exceptions to a general rule, they would still have been interesting, and even influential, because he was among the most visible of scientists in the public forum and because the pictures made in his museum were disseminated so widely. But the scientists who defended evolution in the press disproportionately shared his view of evolution as progress and adopted an emphasis on progress as part of their rhetorical strategy in writing for the public. And they often borrowed American Museum lantern slides for lectures and published copies of its illustrations in books and pamphlets. Osborn enjoyed extraordinary access to the press and to the public, and the visual images he designed were widely reproduced. His views of evolution were the views most conspicuously on display.

The continued use of these diagrams illustrates an important principle. Diagrams were used to formulate theory and to test theory—they were in an important sense analogous to experiments—but they were also used as didactic and persuasive devices. In this context they could convey a very different set of ideas not only about how evolution worked but also about how science worked. So they demonstrated two things: one, that scientists assumed a set of visual conventions, a lexicon that might not be visible to the public; and two, that the longevity of diagrams on display in museums and in successive editions of books implied a static notion of science that created—and still creates—misunderstandings about how science works.[39]

Not that Osborn—or Gregory—did not intend to convey a progressive view of evolution. They intended to convey exactly that. But they also intended more. Both Osborn and Gregory believed that humans were the highest and best form of life that had yet appeared on earth. And Osborn's view of evolution was an orthogenetic view of linear progress. Nevertheless, Gregory and Matthew disagreed with Osborn's notions of evolutionary progress in important ways, especially with his views of human evolution. Furthermore, the supremely confident Osborn welcomed debate with them publicly, even if he had private misgivings. And all of them expressed the contingent, dynamic nature of scientific theory and the idea that drawings, diagrams, and other kinds of illustrations could be used to test theories and to generate them. Yet the popular diagrams would lose this sense of scientific synthesis while emphasizing the single idea of progress, of evolution in the form of a linear, improving chain of being.

Osborn and the curators at the American Museum tried hard to prove that evolution allowed humans a noble place in the scheme of things. As the museum's *Guide to the Hall of the Age of Mammals* declared, "Like a book [the hall] is to be read from left to right, beginning with the Characters of Mammals and the Family Tree of Mammals at the left of the entrance, and ending with Man, who is a species of Mammal and considered by himself to be the head of his class."[40] Bryan was not mollified, however. And Osborn's attempts to ennoble the Cro-Magnon caveman and to "banish the myth and bogie of the ape-man ancestry" would be eclipsed by the irreverence of monkey images that dominated media coverage of the 1925 Scopes trial.[41]

CHAPTER EIGHT

Scientists and the Monkey Trial

It was no coincidence that William Jennings Bryan's denunciation of the family tree diagram at the Scopes trial came up during a dispute over the relevance of scientific testimony. Bryan's was one of three passionate speeches adding drama to the discussion of expert testimony. The other two were by Dudley Field Malone for the defense and by the chief prosecutor, Tom Stewart, who until that moment had been widely lauded for his reserve, focus, and dignity. That these addresses were three of the four most impassioned speeches at the trial was also no coincidence. The defense aimed to build a case around the arguments of "reconciler" scientists that evolution was not incompatible with the Bible. This argument had been prominent in newspaper coverage of the evolution controversy, and the eloquent speeches on this issue were widely reported in the press, even excerpted at length in some major papers. The statements of scientists arguing for the harmony of evolution and Christianity also received a good deal of newspaper attention. Yet other aspects of the trial ultimately eclipsed most serious discussion. Among other things, the momentum of the story got out of hand. And the casting of the event as the "monkey trial" ultimately precluded clear vision of the issues.

The Scopes Trial: Chronology

John Thomas Scopes was the perfect defendant. He was probably the only person at his trial *not* accused of being a "publicity seeker," and for this simple thing he was widely admired. One of the reasons Scopes made such a satisfactory defendant in the case was that he resolutely refused offers for movie contracts and book deals capitalizing on his notoriety. Headlines praised his modesty and his reluctance to claim celebrity, and scientists, in rallying to Scopes's support, vouched for his sincerity, as evidenced by the fact that he turned away from easy fame.

The entertainment newspaper *Variety*, among many others, marveled at considerable length on Scopes's refusal to profit by his celebrity. According to the paper, he had received numerous offers: The paper was especially impressed that he had "politely declined a brilliant offer from William Morris for a world lecture tour under most dignified auspices and conditions." William Jennings Bryan, who had joined the prosecution in the case, had arranged for a lucrative post-trial lecture tour, the paper claimed. Scopes, in contrast, expressed interest only in returning to his studies, even though he had become, as *Variety's* sub-headline put it, "the Biggest Current 'Name' for Any Kind of Show Business."[1] Newspaper editors found him remarkable because so many of the people involved in the trial seemed to be publicity hungry. Many of the ubiquitous Scopes trial cartoons in the summer of 1925 singled out publicity seekers—including the town of Dayton itself—as objects of jokes and derision. Skepticism about publicity was a common refrain in a society preoccupied with, and ambivalent about, fame and money making.

There was a growing awareness—some said an obsession—in the twenties about the new professions of public relations expert and publicity agent, about the work of people like Edward Bernays and "Poison" Ivy Lee who did so much to publicize publicity. The word *propaganda* came up often, but it was used in an interestingly different way than it would be later in the century—in many cases, as if it were a synonym for *publicity.* The word *propaganda* sometimes, but not always, carried vaguely disapproving connotations, but so did the word *publicity.*[2]

Complaining that "Greenwich Village is on its way to Rhea County," an editorial in the June 11, 1925, *New York Evening Post* predicted: "There will shortly descend upon Dayton, Tenn., the greatest aggregation of assorted cranks, including agnostics, atheists, Communists, Syndicalists and New Dawners, ever known in a single procession." The serious issues raised by the Tennessee trial of John Thomas Scopes for teaching evolution were in danger of being obscured by such "limelight lizards." According to the *Post,* "Teachers, research workers, biologists and other men of science are being smothered in the rush of long-haired men, short-haired women, feminists, neurotics, free-thinkers and free-lovers who are determined to shine in reflected glory."[3] Like so many writers commenting on the Scopes trial in 1925, the editor at the *Post* offered in capsule form a litany of the cultural preoccupations of the decade. The trial symbolized cultural rifts of various kinds, not only science versus religion but also the specter of the city versus stereotypes of small-town America, modernity and cultural radicalism against tradition, and serious questions about the problem of democracy in a modern technical society. The temptation to cast the trial in terms of such dichotomies seemed at times to be irresistible. One of the few things about which most people agreed was that publicity amplified everything and probably altered things, too. Both the volume and the tenor of publicity certainly worried scientists.

The events leading up to the Scopes trial began almost haphazardly, yet for some people, at least, the issues at stake were quite serious—as the *New York*

Evening Post noted wryly, "serious enough to laugh at."[4] But even those who agreed that the trial raised important issues often disagreed about exactly what those issues were. The efforts of anti-evolutionists to persuade state legislatures to outlaw the teaching of evolution in public schools culminated in Tennessee on March 21, 1925, when Governor Austin Peay of Tennessee signed the Butler Act into law. Under the new law, written by John Washington Butler, public school teachers were forbidden to teach "any theory that denies the story of the Divine Creation of man as taught in the Bible, and to teach instead that man has descended from a lower order of animals."[5] Governor Peay, who was hoping to run for the U.S. Senate, issued a statement declaring the law a symbolic act. No one, he affirmed, expected it to be enforced. It was meant, he wrote, as "a distinct protest against an irreligious tendency to exalt so-called science." In the same statement, however, he predicted, "Nobody believes that it is going to be an active statute."[6] Indeed, the state had approved high school textbooks that included evolution.

Governor Peay was mistaken. Beyond Tennessee, some people did expect the law to be enforced. And the debate over evolution had become far too prominent in the news for the Tennessee law to lie dormant or be discreetly ignored. The directors of the American Civil Liberties Union (ACLU), established to defend people caught in the repression of civil rights associated with the world war and the Red Scare that followed, decided to use the Tennessee law as a test case, to try to bring the issue before the Supreme Court. From its headquarters in New York the ACLU issued a statement in April 1925 disclosing the results of a recent survey that found that "freedom of teaching . . . has been more interfered with by law in the past six months than at any time in history."[7] Since there was no ACLU office in Tennessee, they advertised in Tennessee newspapers for a teacher willing to participate in a test case.

George Rappleyea, a mining engineer originally from New York who was living in Dayton in 1925, spotted the ACLU ad in the *Chattanooga Daily Times*. In a soon-to-be famous scene at Fred Robinson's drugstore and soda fountain (also the town textbook outlet), Rappleyea, Robinson, and several other town movers and shakers—motivated by a hope that the publicity would help restore a flagging local economy—decided to stage a test case in Dayton, recruiting John Thomas Scopes to volunteer as the defendant.[8]

William Jennings Bryan soon announced that he would join the prosecution, representing the World's Christian Fundamentals Association (WCFA). Though he had not actually tried a case in some thirty years, his enormous following among fundamentalists and his fame acted as catalysts shaping the events that followed. Getting wind of Bryan's participation, H. L. Mencken, the journalist and editor of the *American Mercury,* persuaded Clarence Darrow to join Scopes's defense team. Darrow may well have been the most famous lawyer in the country at the time; certainly he was among the most controversial. At the time of the Scopes trial he was still widely vilified for defending, the previous year, young thrill-killers Nathan Leopold and Richard Loeb. Joining him on the defense team was Dudley Field Malone, who had served under Bryan as third assistant secretary of state in the Wilson administration and who was renowned as an eloquent (and elegant) lawyer specializing in divorce cases. Malone was also known as a feminist—his wife, Doris Stevens, a member of the feminist "Lucy Stone League," apparently caused some confusion (and gossip) in Dayton by appearing with him under her own name. The defense team also included ACLU lawyer Arthur Garfield Hays (named after three presidents), who, a reporter for the *New York Times* noted admiringly, had in the past "gone to jail for what he believed to be right."[9] In the media, and in subsequent historical memory, however, the conflict would often be cast as a contest between the views of the two most famous adversaries in the Dayton courtroom, Clarence Darrow and William Jennings Bryan.

Clarence Darrow was not the defender the ACLU had in mind; indeed, many leaders of the organization tried to persuade him to step down as Scopes's legal advocate. His reputation for animosity toward organized religion—even, some people were convinced, toward any religion at all—threatened to obscure the civil liberties issues that were the original intent of the test case. Darrow noted in his autobiography that the Scopes trial was the only time he donated his services— and he was normally very well paid indeed. Unlike the leaders of the ACLU, he hoped to use the trial to highlight what he saw as religious bigotry and the antidemocratic potential of organized religion.

Many members of the ACLU's governing board would have preferred to hire more conservative and less controversial defense lawyers than Darrow and Malone, arguing that Darrow's participation, in particular, would cloud the issues. In fact, Scopes later recalled that the ACLU was rather bitterly divided over the issue. Some of its members, Scopes remembered, thought of Darrow as a "headline chaser."[10] In addition, they intended to focus on the constitutional issue of separation of church and state, and they feared that Darrow's flamboyance and his reputation as an agnostic would steer the trial toward a debate about religion itself. They predicted that a confrontation between Bryan and Darrow would inevitably turn into a "circus," deflecting attention from the serious issues at stake. Darrow, they calculated, would put the question of science versus religion at the top of headlines, and because he and Bryan were both so well known, the trial would inevitably be sensationalized. Scopes had long admired Clarence Darrow, however. He pointed out that, Darrow or no Darrow, the prospect of Bryan's arrival had already made a circus atmosphere inevitable. Indeed, he said, back in Dayton the carnival had already started. Hays, who managed the defense, concurred, and the ACLU's governing board narrowly voted to include Darrow and Malone. Scopes was, after all, the defendant, and it was his decision.

Scopes admitted to having broken the law, although he may not have done so, strictly speaking: he was a math teacher and a coach, and he taught biology only occasionally as a substitute. He did not actually remember having taught evolution but agreed that in principle he would have if the occasion had arisen. The regular biology teacher was a less convenient person to use in the test case because he had a wife and children and therefore had more to lose. Scopes affirmed, in any case, that he had assigned the passages on evolution included in the state-mandated textbook, George William Hunter's *Civic Biology*. He was sure, he said, that it was impossible to teach biology adequately without evolution. No one, including the defense, expected anything but a conviction. The field of battle compelled interest, not the outcome.

The trial and the activity surrounding it have been variously described as comedy, farce, and tragedy, but nearly everyone agreed that it all made for extraordinary drama. Journalists, photographers, moviemakers, vendors, evangelists, and eccentrics with trained chimps converged on Dayton. New telegraph lines were installed to handle reporters' dispatches on the trial. Movie newsreel makers and a Chicago radio station arrived. Reporters expected that the anticipated confrontation between Bryan and Darrow would be magnificent and that the oratory, as H. L. Mencken put it, would be "gorgeous."[11] Bryan promised that it would be a "duel to the death" between Christianity and evolution. Mencken advised Darrow: "Nobody gives a damn about that yap schoolteacher. The thing to do is make a fool out of Bryan."[12]

The lawyers for the defense never claimed, or tried to claim, that Scopes had not broken the law.[13] Instead, they recruited a team of expert witnesses—scientists and religious scholars—to demonstrate that the law was inconsistent and should not stand. The prosecution challenged the relevance of the expert testimony, and the judge, John Raulston, a conservative Methodist, concurred. The witnesses were allowed to enter their statements on the record for the sake of the appeal, which was the focus of the defense attorneys' attention all along, since they hoped to take the case to the Supreme Court. And the statements of the witnesses were distributed to the press and published in newspapers and magazines, another of Darrow's objectives.

On the penultimate day of the trial, the crowd was so large that the proceedings were moved outdoors, maximizing the number of spectators for the remarkable culmination. Left without expert witnesses, and therefore without a case, but also acutely aware of the riveting dramatic effect it would have, Darrow, in an unprecedented move, called William Jennings Bryan to the witness stand to testify as an expert on the Bible. The result became legendary. Historians continue to discuss why Bryan would have agreed to be cross-examined by one of the great courtroom lawyers in the country. One reason seems to be that he expected to be allowed in turn to put Darrow on the witness stand. He may have expected to be asked about evolution, rather than about the Bible: he had been challenging evolutionists in lectures on evolution and in newspaper exchanges and was confident in his ability to answer questions about his objections to evolution. Darrow did not query him about evolution, however: He hammered Bryan with question after question about the Bible. The questions were all calculated to make Bryan's claim to believe in a literal interpretation of the Bible appear ridiculous—as Mencken had said, to "make a fool out of Bryan." Darrow asked about the passage in the Bible in which Joshua made the sun stand still; where Cain got his wife; and how it was possible to know the length of the six days of creation if day had not been divided from night until the fourth day. Bryan insisted he had never thought about Cain's wife but conceded that he did not believe that the sun revolved around the earth or that it had in the past. He admitted that he did not think it necessary to postulate that the six days of creation had lasted exactly twenty-four hours each—he had in fact never objected to the interpretation of Genesis that allowed for an ancient earth, nor did most other fundamentalists of the time.[14] Attorney General Tom Stewart tried repeatedly to put a stop to this testimony, and eventually Judge Raulston did call a halt to it, but not before, in the eyes of most journalists, at least, Bryan's reputation had been badly damaged.

The next day, Scopes was found guilty (to make sure, Darrow requested that the jury return a guilty verdict so that the defense could appeal the decision in a higher court), and the judge levied a fine of one hundred dollars, which was paid by the *Baltimore Sun*. A few days later, William Jennings Bryan died in his sleep. The ACLU did appeal the case in the Tennessee Supreme Court, where it was thrown out on a technicality—according to Tennessee law the judge should not have imposed a fine larger than fifty dollars without consulting the jury—thwarting the defense strategy. Sandbagged on its way to the U.S. Supreme Court, the case of Tennessee versus John Thomas Scopes ended, and the Butler Law remained on the books for another forty-two years.

A Drama That Had Everything

These are, in brief outline, some of the facts of the case. It has been discussed, taught, dramatized, and analyzed ever since. The Scopes trial has become part of American mythology for many reasons: it provided the takeoff point for a Cold War allegory in the play and the movie *Inherit the Wind* and was immortalized in Frederic Lewis Allen's 1931 popular history *Only Yesterday;* it embodied important issues; and it made a lively story. The first trial ever broadcast on radio, the Scopes trial was an early media event. More than a hundred journalists came to the small town of Dayton, Tennessee, to cover the trial, along with photographers, movie-makers, radio technicians, telegraph operators, and hangers-on. Virtually every newspaper in the country carried daily stories on the trial, and commentary came from around the world. Newspapers sent their star reporters to Dayton and ran stories of the trial on their front pages. Like later such events, the trial gave the daily news a feeling of continuity and excitement. Today's news added a piece to the puzzle of yesterday's news. It all seemed to form itself irresistibly into a good story. As H. L. Mencken enthused, the Scopes trial, from the point of view of journalists, "had everything."[15]

And it did. It had drama, and it had meaning. Journalists crafted stories of the trial in the shape of allegory and the language of melodrama. In the contemporary literature it was billed as the clash of titanic forces and ideas. Nominally about the teaching of evolution in public schools, it was also widely viewed as the confrontation of small-town America versus the city, tradition versus modernity, science versus religion, academic freedom versus repression, even reason versus revelation. Above all, it featured famous and colorful figures, especially Clarence Darrow confronting William Jennings Bryan, with H. L. Mencken cheering in the wings. Mencken, who covered the story for the *Baltimore Sun,* also did much to shape it, at the time and later. For any number of reasons people found the story irresistible. A story carrying the inherent drama of the Scopes trial may have a certain momentum of its own. During the trial reporters were as impressed as anyone else by the journalistic spectacle the trial had created—and the spectacle they had themselves created, reporting often on their own activities.

Even the show-business newspaper *Variety* covered the trial. Under the headline "Ballyhooed Hullabaloo," the paper described the scene in Dayton: "Ballyhooed by the broadest and blaringest hullabaloo in American history, Dayton, Tenn., the seat of the great serio-comic evolution monkey business is today a teeming, seething bedlam." The article caricatured the town, claiming that every resident was Protestant and anti-evolution. It dwelled on the various attractions:

vendors selling monkey souvenirs—no one could avoid mentioning those, of course—but also lectures including slides on the life of Christ and acting troupes, Tom's Comedians and the Paramount Players, performing under large tents to small audiences. All in all, the paper noted, Daytonians were disappointed at the small attendance of people from out of town. Of an estimated six hundred visitors, probably two hundred were reporters—"the flower of the land's feature writers and star reporters." The rest, according to *Variety*, beyond the small number of attorneys and trial witnesses, could be dismissed. They were "as vicarious and gregarious a collection of nuts, bigots, jumping and howling missionaries, and carnival riffraff as ever gathered within the precincts of one small burg."[16] This was fairly typical language—most newspapers sounded like entertainment papers, at least some of the time.

Not all newspapers were blatantly anti-fundamentalist, of course. Nonetheless, the reporters from the powerful wire services generally were, and their stories reached all corners of the country.[17] Although Mencken's columns for the *Baltimore Sun* were not generally syndicated, his Scopes trial dispatches were purchased for publication by newspapers nationwide.[18] Reporters from nonfundamentalist periodicals outside the South tended to treat the event, and especially Bryan and the local people, sardonically, but Mencken was the master of irony. In reminiscences published after an interval of nearly forty years, Joseph Wood Krutch still recalled Mencken's jubilant participation as one of the highlights of the event: "He enjoyed in the courtroom what were perhaps the happiest moments of his life contemplating, and in a sense presiding over, a spectacle which seemed arranged for his delight. . . . Had he invented the Monkey Trial no one would have believed it."[19] In fact, he may have done as much to invent it as anyone did.

Mencken persistently referred to the rather gentle and likable defendant as "the infidel Scopes." He related with glee the improbable friendship he struck up with the evangelist T. T. Martin and Martin's attempts to save his (Mencken's) soul. His descriptions of the townspeople's reactions to Darrow were deliberately humorous and certainly fanciful. The faithful, Mencken claimed, were convinced that the Almighty was sure to strike Darrow down at any moment, and "they kept away from him, for they didn't want to be present when the lightnings from Heaven began to fall . . . whenever a thunderstorm blew up . . . everyone save a few atheists began to edge away from Darrow in the courtroom." Hays and Malone, Mencken added, "were also regarded as doomed, but it was generally believed that Darrow would be knocked off first, so the bolder spirits sometimes approached them quite closely, especially when no thunderstorm was in prospect."[20]

Mencken's humor was often cruel, even brutal, and unfair. He regularly referred to the local people as "yokels," "rustics," "simians," "primates," and "morons." He referred to Bryan's admiring followers as his "customers." He described a decision by Judge Raulston as "a masterpiece of unconscious humor. The press stand, in fact, thought he was trying to be jocose deliberately and let off a guffaw."[21] But he reserved his most derisive sneers for Bryan, calling him a "buzzard," a "buffoon," and a "poor clown." He labeled Bryan's major speech at the trial "touching in its imbecility" and pronounced Bryan himself "a tinpot pope in the coca-cola belt and a brother to the forlorn pastors who belabor halfwits in galvanized iron tabernacles behind the railroad yards."[22] After Bryan died, Mencken published a stunningly vicious attack in a parody of the standard eulogy.[23]

Mencken was undoubtedly the most unfair, and possibly the most cynical, journalist, but much of the reporting from Dayton was written in similar voice. Reporters often turned the principal actors into symbols, even while focusing on personalities. Even the normally staid *New York Times* indulged in florid language, emphasizing personalities and charisma. Dramatic language was not altogether unwarranted: the performances were an important part of the message. Darrow's most famous speech at the trial reiterated familiar themes invoking medievalism and the Inquisition, portraying anti-evolutionists as "bigots" and as threats to all the advances in civilization, enlightenment, and tolerance achieved with the advance of science. He was eloquent—even Henry Fairfield Osborn sent him a warm note of praise and congratulations, and newspapers called him brilliant. But, as Mencken wrote, people who merely read the text without seeing the performance could have only an incomplete appreciation of the address: "You have but a dim notion of it who have only read it. It was not designed for reading, but for hearing. The clangtint of it was as important as the logic."[24] Although the *New York Times* reproduced long sections of Darrow's address verbatim, the paper also reported in colorful detail on his delivery, making it sound very much like great theater. The *Times* writer pictured Darrow as a vivid performer: he "prowled around inside the big arena of the court room, his voice sinking to a whisper at times, again rising in a burst of rage. . . . He knew he was defying the lightning."[25]

Reporters from cities in the North seemed to relish descriptions not only of Darrow's brilliance but of the folksiness they assumed he had adopted for Dayton's benefit. In counterpoint to this, they relentlessly portrayed local people as rustics—condescendingly referring to men in overalls and women dressed in calico and nursing infants. Even reporters who objected to one-dimensional portraits of the local people seldom managed to transcend such stereotypes. Russell

Owen, who covered the trial for the *New York Times,* wrote of the "simple people" of Dayton that they had been unfairly maligned as "bigots." They were ignorant, yes, but no more so than people in rural parts of New York, and though "their minds are simple," they were eager to learn. His description was headlined: "Dayton's Remote Mountaineers Fear Science."[26]

William Jennings Bryan was without question the most maligned character at the trial. Cartoons portrayed him as an Inquisitor or as a Don Quixote, and cartoon monkeys made fun of him endlessly.[27] Reporters scoffed at him, ridiculed him, and made a symbol of him. Shortly after the trial ended, Owen published an article called "The Significance of the Scopes Trial," defining it as "a symptom of the age-old struggle between rationalism and faith." It was, Owen thought, "perhaps the most exciting of all such meetings since they were settled by the sword." Bryan, of course, represented all that was medieval—a favorite word among supporters of evolution. Like many others, Owen seemed to relish calling Bryan by the nickname "the Commoner" or "the Great Commoner." Also like most reporters, he emphasized personality and indulged in stereotypes. Darrow was "brilliant." Judge John Raulston was "a raw-boned mountaineer . . . a typical product of the primitive hill life of Tennessee . . . born in a little place known as Fiery Gizzard."[28] Details had seductive power. Reporters noted with apparent relish that Raulston was born in Fiery Gizzard. Had he been born in Knoxville or Dayton, it would not have been funny. But naming Fiery Gizzard as his birthplace created an indelible impression. Newspapers often observed that Judge Raulston seemed to relish having his photograph taken and to enjoy being the object of publicity in general—publicity and the craving for it were so much in evidence that anyone who appeared to covet fame immediately invited ridicule.

Russell Owen highlighted Bryan's famous testimony on the last day of the trial by careful framing: "It was the meeting of the great forces of skepticism and faith, and there was only one possible result."[29] This pronouncement had the ring of inevitability of Greek tragedy. It also fed the myth of the cataclysmic confrontation between science and faith, eliminating the myriad other issues involved. In a colorful—and typical—dispatch, the *New York Evening Post* described the confrontation between Darrow and Bryan, with Bryan on the witness stand, as both an extraordinary event in legal history and—more emphatically—as performance. The paper characterized Darrow's interrogation of Bryan as "yesterday's amazing spectacle in the Bryan-Darrow debate—unprecedented in the annals of Anglo-Saxon jurisprudence."[30] Setting the scene, the *Post* noted that people filled the benches surrounding the outdoor platform where the encounter took place and that the spillover crowd included people hanging out of windows, perched on the roofs of automobiles, and watching from the branches of trees. The crowd was central to the story: "Bryan and Darrow spoke as frequently to the throng as they did to the Court. Cheering, hooting, laughter, applause and cat-calls were interspersed freely." The reporter wrote as if producing fiction, drama-tizing the emotional jousting of the two protagonists, Tom Stewart's repeated an-guished entreaties to the judge to put a stop to it all, and the judge's apparent relishing of the spectacle. "I object with all the strength I possess to this thing's going any further," Tom Stewart protested. The judge allowed the drama to con-tinue. "Bryan and Darrow grew acrid in their exchanges. The crowd cheered and yelled applause when Bryan scored." Historians' accounts of this event have often, with good reason, been guided by readings of the transcript of the trial, but for the writer at the *Post*, the content of Darrow's questions and Bryan's re-sponses seemed almost secondary to the emotional dynamic of the encounter. "Bryan kept his temper better than Darrow. He was a better debater. His answers were prompt and spontaneous. He said just what he wished to say. The crowd was with him." The encounter sounded almost like a boxing match: "Darrow was badly nettled at times. . . . Bryan was magnificent in his poise. Darrow was per-sistent. He went at Bryan like a bulldog. He growled when Bryan shook off his at-tacks." But this was all rhythm, building up to excerpts of exchanges making Bryan appear foolish. Bryan claimed to be acquainted with some "worthy scien-tists," for example, but could not, when pressed, name any of them. And so the story built up to a crescendo in which Darrow persisted and Bryan finally lost his composure. "The questions and answers shot back and forth like machine gun fire. Both men were livid. Bryan's fan worked frantically. Ire shot from his eyes." Darrow faced Bryan as if taking lethal aim: "His eye squinted like a man looking along a gun sight." The judge, according to the writer, gave every indication that he enjoyed the show. "There was good reason to believe the Court had welcomed the performance, if not actually staged it."[31] The example from the *Post* was typi-cal of press coverage. The conflict between Bryan and Darrow did take center stage—and reporters treated it as theater. From the very beginning, editorialists took Bryan and Darrow as representatives of ideas, habits of mind, and ways of life. Their roles as symbols made up an important part of the drama.

Journalists often portrayed Bryan as bumbling but also as mean and even vi-cious. A story in the *New York Evening Post* described him as something of a pred-ator: "He is alert, grim and conveys the impression of lying in ambush."[32] In his autobiography, Darrow contrasted Bryan in his old age with the younger Bryan. He characterized the anti-evolution movement as a "campaign against educa-tion" and recalled that at the trial he "used to sit in the courthouse and note how exactly Mr. Bryan represented the spirit of intolerance that he sat fanning into flame."[33] Other defenders of evolution characterized Bryan in similar ways. In her account of the Dayton trial, Marcet Haldeman-Julius read her assessment of Bryan's character into his physical features, noting the "fanatical light in those hard, glittering black eyes" and the "cruel lines of his thin, tight-pressed mouth." His very physiognomy, she decided, "proclaim[ed] . . . that he would stop at nothing to attain his own ends."[34]

Narratives of this kind emphasized personality in a way that slid imperceptibly into interpretation. It was easy, reading most newspaper accounts of the trial, despite their attention to the larger issues involved, to perceive the event primarily in terms of personalities. In the case of the Scopes trial, interpretations shaped by the focus on personalities were especially predictable because most of the characters were colorful, some of them quite charismatic. But they were all undoubtedly more complicated than the legend allowed them to appear.

And the casting of Bryan and Darrow as binary opposites and symbols of what the trial meant fed a mythology that portrayed the issues as simple dichotomies. Interpretations of the trial as a clash between Bryan and Darrow, and the theatrical tone of reporting on the encounter, magnified a tendency to see the trial in simple binary terms. These two men were easily cast as symbols of pairs of opposing forces and ideas that had been offered as keys to the cultural tensions of the decade—even of modern life in general.

The emphasis on Bryan and Darrow as colorful individuals left aside the many complex issues, including the question of exactly what ought to have been at stake in the trial, as well as the roles of fame, of famous people in the unfolding of the event, and of the complex relations of science and modern life. It eclipsed significant questions about the nature of the relationship of science not only to religion but to a larger culture more and more influenced by media, publicity, and celebrity at a moment when these influences had reached something of an inflection point. It suggested that the North was arrayed against the South, the city against the country, the sophisticate against the rustic, tolerance against bigotry, rationalism against superstition, and science against religion. But nothing was really this simple. The issues were not really clean dichotomies—none of them were—though they seemed on the surface to arrange themselves that way.[35]

More recent work has corrected much of the old mythology,[36] demonstrating convincingly, for example, that the dominant view of Bryan in the earlier Scopes literature was highly misleading, unfair, and grossly oversimplified. Bryan, for all his faults, sincerely believed in majority rule and strenuously opposed elitism. Bryan's anti-evolution campaign long puzzled historians, who saw it as out of character, an aberration that made no sense in light of his long career as a populist and a progressive. This view, recent historical interpretations have shown, was a misapprehension. Bryan's opposition to evolution did not grow out of biblical literalism but out of a conviction that the teaching of evolution was inextricable from the pernicious spread of social Darwinism. Having read Vernon Kellogg's *Headquarters Nights,* about life in Germany in the aftermath of the Great War, and Benjamin Kidd's book *The Science of Power,* relating Darwinism and Nietzsche, Bryan came to associate evolution with social Darwinism and with German militarism, an especially evocative connotation in the wake of the war. In addition, his themes at the trial, as throughout the debate, always prominently included the principle that the majority in the community should be allowed to control the curriculum in a tax-supported school.

Bryan's fears of social Darwinism were not entirely fantastic. Many of the scientists who defended evolution and academic freedom were also active proponents of eugenics. So were some of the most successful popularizers of science, such as Albert Edward Wiggam, whose books about evolution and eugenics were best sellers. The biology textbook assigned by the state of Tennessee, George William Hunter's *Civic Biology,* showcased eugenics as one of the accepted advances of modern biological theory, as did many widely used textbooks of the time. And some writers, at least, acknowledged that the claim Bryan raised about the right of the majority to control the curriculum in tax-supported schools was not easy to dismiss out of hand. The issue of how to reconcile the demands of democracy with the logistics of expertise in an increasingly technical society defied simple solutions. Many people—including critics of Bryan—expressed uneasy pessimism about democracy in the twenties. In 1930 Felix Frankfurter wrote in the *Atlantic,* "Epitaphs for democracy are the fashion of the day."[37] Many people explicitly linked their forebodings to the growth of science and scientific or technical expertise.

Making Bryan into a symbol meant not only misrepresenting him but also obscuring much of the quite complicated debate between religious modernism and traditionalism that culminated in the anti-evolutionism of the 1920s. Reinhold Niebuhr, Walter Lippmann, and J. Gresham Machen, among others, published subtle and provocative essays at the time, demonstrating that the issues at stake were neither trivial nor easily dismissed.[38] Evolutionists found it easy to parody and to ridicule much of the fundamentalist rhetoric of the 1920s. But fundamentalism was not a monolithic movement. The early Scopes literature selected examples of those fundamentalists most easily parodied and held them up for magnification, virtually ignoring the more thoughtful and well-educated religious traditionalists of the time.

With the perspective of time, scholars have gained analytical distance, allowing them to see beyond some of the stereotypes embedded in the earlier stories. Recent histories have importantly located the trial in the cultural context of the 1920s in America and in the context of the rifts within Protestantism that were, at least as much as objections to evolution, at the heart of the controversy.[39]

Scientists, Publicity, and the Authority of Experts

The public—and publicity-inflected—nature of the evolution debate and eventually of the trial also had implications for the relationship between science and popular culture. The difficulties for scientists of communicating with the public in the context of a sensationalized public debate had already become evident; the trial only revealed these difficulties more dramatically, perhaps. But that was no small thing: scientists, already concerned about the possibility of miscommunication of their ideas and the compromising of their authority in such a highly charged atmosphere, found many of their fears about publicity and celebrity confirmed.

The roles of scientists at the trial, as well as in the evolution debate that preceded it, unsettle any notion of simple binary pairs of ideas around which the trial can be arranged. The ACLU worked with scientists to recruit experts who were also believers and who could therefore testify that evolution was compatible with religion. The defense strategy took this shape in part because of the wording of the Tennessee law, which was ambiguous and vague and included forbidding anyone "to teach any theory that denies the story of Divine Creation of man as taught in the Bible." The defense argued that evolution did not necessarily deny any such thing. Science reconcilers had been making this argument all along, and these were the scientists who had dominated news coverage and popularization since 1922.

Early during the course of planning for the trial, officers of the ACLU consulted Michael Pupin, president that year of the American Association for the Advancement of Science (AAAS), seeking help to recruit scientists of standing who might testify, or help them arrange for expert testimony, at the trial. Not surprisingly, Pupin suggested that the organization work with Henry Fairfield Osborn and Edwin Grant Conklin, who had served on the AAAS committee responding to anti-evolutionists and whose credentials as "first-string" biological scientists were impeccable.

Both Osborn and Conklin participated in the effort, but, perhaps because he enjoyed the services of a large support staff at the museum, Osborn played a particularly energetic role in soliciting participation by other scientists and in trying to shape the public debate. He wrote to a number of friends that he was devoting several hours every day to work on Scopes's behalf—and Osborn was a man who allocated his time very carefully and efficiently. Scopes visited Osborn on his trip to New York to consult with the ACLU, as did three members of the prospective defense team, Clarence Darrow, Bainbridge Colby, and Dudley Field Malone. Osborn guided them through the museum, posed with them for a photograph, and advised them to focus on finding respected scientists who would emphasize the compatibility of evolution with traditional religious—meaning Christian—beliefs. He devoted a good deal of time to correspondence with colleagues whose credentials as churchgoers and believers could be demonstrated, recommending that they testify at the trial, collecting sworn affidavits from those who could not or would not go, and dictating one himself to be sent to Dayton for use by the defense.[40] He had the museum send a photograph of an evolutionary tree diagram from the Hall of the Age of Man for exhibit at the trial, and he instructed that his publisher, Charles Scribner, send one thousand copies—copies Osborn paid for himself—of his little book *Evolution and Religion*. He rushed *The Earth Speaks to Bryan* into print for the occasion, dedicated it to Scopes, and procured a copy of the Dayton telephone book so that his secretaries could send copies to the most prominent citizens of the town.[41] He sent a letter to John Washington Butler, the author of the anti-evolution bill, claiming somewhat disingenuously that the two of them came from very similar pious backgrounds, and he exchanged numerous urgent telegrams with Scopes, George Rappleyea, and the defense lawyers. Several scientists who had been asked by the defense team to testify sent him telegrams asking if he recommended going. He signed his many letters on behalf of Scopes "Chair of the Scientific Committee to Defend John Thomas Scopes," and Rappleyea and Scopes seemed to see this as his role. Supporters and denouncers of Scopes alike referred to Osborn as the leader of the scientific defense, as did the newspapers. Privately, however, he remained significantly at odds with the defense team. He confided to a number of friends that he found the ACLU role at the trial regrettable, for he believed it to be a politically radical organization, and he disapproved of radicalism. He wrote to Judge John R. Neal, co-counsel for the defense, that he feared that "the American Civil Liberties Union would queer this case and give it a wrong start in public opinion. The Union includes the leading pro-German, anti-war, radical, pro-Bolshevist, anti-religious group in this country."[42]

The ACLU expected him to go to Dayton, however, and released statements announcing that both he and Conklin would testify at the trial; the Associated Press reported that they would.[43] *Popular Science Monthly* reported that "the entire membership of the American Association for the Advancement of Science" stood "ready to offer expert testimony."[44] Many newspapers picked up the story, reporting that a team of the most eminent scientists, headed by Osborn and Conklin, would testify at the trial.

Osborn and Conklin did not travel to Dayton, however. They did not go despite many telegrams imploring them to do so. George Rappleyea wrote to Osborn assuring him that there would be a comfortable place for him and his wife to stay and that they would be well cared for. Darrow himself took time to write Osborn a personal note, urging him to come. Osborn and Conklin steadfastly resisted all entreaties.

Why did they refuse to attend the trial?

Osborn urged Conklin to join the defense team in Dayton, writing that he would be delighted to do so himself were it not for the fragile state of Mrs. Osborn's health. Conklin also declined to go, citing health problems of his own. He suggested that Osborn reconsider. Both of them protested that they were entirely too busy with work. None of these claims were false, precisely, but they were rationalizations. Both men *were* very busy; and it was true that Osborn's wife had weathered several serious illnesses that spring and that Conklin suffered from diabetes. Osborn did find time, however, to take a "much-needed" vacation in the American West that summer, going to a Colorado Springs resort and stopping on the way to visit a fossil field camp in Nebraska—on his doctor's orders, he was careful to say—without his wife, and Conklin traveled extensively that spring and spent the summer working at the Marine Biological Laboratory at Woods Hole, as was his custom.

It has been suggested the real reason Osborn avoided attending the Scopes trial was that enough evidence against *Hesperopithecus* had accumulated by that time for him to fear a confrontation over it.[45] Certainly William Jennings Bryan arrived in Dayton making fun of Osborn's claims about the Nebraska primate. And some biologists, including Osborn's very reliable colleague William Diller Matthew, expressed early skepticism about *Hesperopithecus*. Osborn himself drew back from the excessive claims for the specimen in newspaper headlines. He may indeed have had trepidations about *Hesperopithecus* in June 1925, but as late as October of that year he wrote to a correspondent, "We have recently confirmed the Nebraska tooth." And in April 1927 he gave a talk before the American Philosophical Society which he subsequently published in two important journals, accompanied, in both cases, by an evolutionary diagram that included the label "Nebraska Hesperopithecus Fossil Bone Implements" at the foot of the family tree, with the legend "Most Ancient Evidence of Man"—and a question mark.[46]

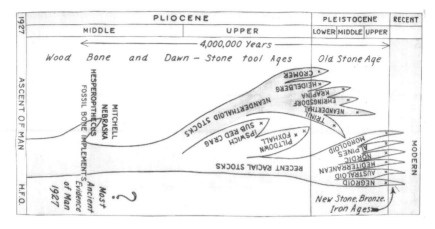

"Ascent of Man." A human family tree diagram designed by Henry Fairfield Osborn, dated 1927 and including the Nebraska specimen called *Hesperopithecus*. Image #121788. Photo by Charles H. Coles. © American Museum of Natural History

It is possible that Bryan's taunts and a growing skepticism about the Nebraska tooth kept Osborn away from Dayton, but perhaps not: he seems not to have abandoned his hopes for *Hesperopithecus* so easily. In any case, there are other, more convincing explanations.

One plausible explanation is that Osborn was wary of public association with Darrow for political reasons. He did express concern about Darrow's reputation: in a memo to the museum's publicity department, he included a cautionary note instructing that Darrow's history of defending "criminal elements" should be played down in press releases about the museum's role at the trial. He also fretted about the ACLU. Despite his energetic activity on behalf of Scopes, Osborn commented frequently to friends and colleagues that he regretted the involvement of the ACLU in the case—he feared that its radicalism would, as he put it repeatedly, "queer" the case in the eyes of the public. And he worried about the kind of notoriety the trial was likely to provoke. He emphasized in private correspondence and in publications that he had not committed himself to the defense cause until he had met Scopes personally, in order to make sure that this was a serious young man, neither subversive nor motivated by a desire for publicity. He repeatedly advised Scopes to avoid being portrayed in the news as a radical. He was not alone in this respect; other scientists observed that Scopes's lack of interest in celebrity was a factor in their support for him. And Conklin and some of the leaders of the AAAS would later balk at lending organizational support to Maynard Shipley's Science League of America out of disapproval of his political commitments.[47] But it is likely that Osborn and Conklin also hesitated to be involved in the circus at Dayton for another reason as well.

In a vivid reminiscence some years later, Winterton Curtis suggested the real reason for the reluctance of the older and better-known scientists to participate in the trial itself: "To the discredit of scientists is the fact that some of our senior zoologists, who were the first ones called upon by the Civil Liberties Union, declined to appear at Dayton, fearing, as I learned later from conversations at Woods Hole, that their dignity might suffer."[48] Looking back on the trial decades later, Curtis revealed that "in two such cases . . . the reason was that the trial would be only a piece of buffoonery and beneath the dignity of a zoologist of national reputation."[49] He may have heard this directly from Conklin; the two had long been friends, and both spent time at Woods Hole that summer.

A group of scientists did go to the trial, but some newspapers noted that the eminent scientists the ACLU had numbered among its defense witnesses appeared to have deserted the cause.[50] Curtis, who went to the trial to testify for the defense, later acknowledged that "those of us who went to Dayton were second string scientists as to age and reputation in 1925."[51] Osborn's private correspondence reveals that he and Conklin were not the only scientists who worried about whether they would lose face among colleagues if they went. Quite a few wrote or wired Osborn seeking his advice in this matter. Some feared retaliation by their institutions if they participated. Conklin's former student H. H. Lane, a professor at the University of Kansas who had advocated publicly for a form of theistic evolution, sent Conklin a telegram, Conklin informed Osborn, "saying that the University authorities at Kansas strongly object to his having any part in the trial and that, in all probability, if he did this, he would sacrifice his position there."[52] A significant number, though, made it clear that they feared that the situation in Dayton would compromise the dignity required of a "scientific man." More and more people were referring to Dayton as a circus.

The members of the Scopes defense found it much more difficult to recruit scientific experts than they had expected it to be. Officers of the Science Service also worked to try to recruit scientists to testify at the trial, and the archives of the organization include numerous letters and telegrams from prominent scientists (and in some cases, from their wives or assistants) explaining why they could not.[53] The ACLU did successfully recruit scientists to testify, at last. Clarence Darrow telephoned the anthropologist Fay-Cooper Cole, of the University of Chicago, and according to Cole's son, Darrow said to him, "Malone, Colby, and I don't know much about evolution. We don't know who to call as witnesses. But we are fighting your battle for academic freedom and we need the help of you fellows at the university." That afternoon, Darrow and Cole met with the University of Chicago zoologist H. H. Newman and with Shailer Mathews, modernist dean of the Divinity School at the same university. All agreed to testify at the trial.[54] Curtis later revealed that he had received a call from Dayton as late as July 1925 asking him to come and that he was pleased to do so, although, he noted, "The bubble of my conceit was punctured later when I learned that several of the then elder statesmen of zoology had been first solicited as witnesses by the American Civil Liberties Union but had made excuses of various sorts."[55]

The young chair of the geology department at Harvard, Kirtley Mather, recalled that when he learned that Bryan would join the prosecution and "denounce evolution as an enemy of Christianity" and that Darrow would "cross swords with the man whose religious beliefs he rejected, I wondered who would be there to defend a religion that was intellectually respectable in the light of modern science." Mather had for some time been arguing for the compatibility of evolution and religion, and he contacted Roger Baldwin of the ACLU "to suggest that, in mobilizing the presumably eminent expert witnesses for the defense, scientists should be selected who were not only held in esteem by their fellow-scientists but were also active members of a church." In addition to his offer to help mobilize others, Mather volunteered his own services at the trial.[56]

That scientists who came to Dayton would be allowed to testify was not a foregone conclusion; to the contrary, it was one of the primary bones of contention. It turned out also to be the focus of some considerable oratory at the trial. Prosecuting attorney Tom Stewart contended that the only issue before the court was whether Scopes had broken the law. Allowing scientific testimony would be tantamount to accepting the defense attempt to make the trial a test of the law.

The judge permitted one of the scientists, the Johns Hopkins University zoologist Maynard Metcalf, to take the witness stand for questioning—though without the jury present. Metcalf's performance would help the judge decide whether expert witnesses should be allowed to testify before the jury—whether, that is, their testimony was relevant to the case. Darrow spent a good deal of time asking Metcalf about his education, experience, and credentials in order to establish his authority. He tried in several different ways to ask whether Metcalf believed that there were any credible biologists who did not believe in evolution, and though the prosecution objected to the question and the judge sustained the objection, Metcalf managed to say that he was acquainted with most of the important biologists in his field and that certainly all of them accepted evolution as an established fact. He went on to explain what scientists understood evolution to mean. H. L. Mencken called Metcalf's testimony "one of the clearest, most succinct and withal most eloquent presentations of the case for the evolutionists that I have ever heard."[57] Other journalists were also favorably impressed and reported the court-room audience to be so as well. Describing Metcalf as "a smiling, good-natured man with a ready wit, and well able to establish his knowledge without asserting his pedagogy," the *New York Evening Post* said: "Opposing counsel and a gaping courtroom crowd hung on the words of men of learning, who traced the character of organisms back through brain-reeling ages to primitive man. . . . It was a fascinating story." However, as compelling as many journalists found Metcalf's statement, it led to the eruption of a "terrific battle of words among counsel."[58]

It was no accident that it was on this question of the admissibility of expert testimony that William Jennings Bryan delivered his first, much-anticipated speech. This was the address in which Bryan complained about the classification of humans with the other mammals pictured in the diagram in Hunter's *Civic Biology*. Bryan began by objecting to the idea that "the minority" could "make the parents of these children pay the expenses of a teacher to tell their children what these people believe is false and dangerous." He asked whether the court had seen the tree diagram in Hunter's textbook, showed it to the judge, and went on to object at length to the diagram for, as he put it, "shutting man up in a little circle like that with all these animals, that have an odor, that extends beyond the circumference of this circle, my friends." The courtroom audience laughed, and Bryan went on to complain that Hunter's text instructed students to copy the diagram in order to learn it. "That is the great game in the public schools to find man among animals, if you can." He went on to introduce Darwin's *Descent of Man* into evidence, and he read from it a passage on the branching of the tree of life into simian stems. He referred again to William Bateson's controversial speech questioning natural selection at the 1921 meeting of the AAAS in Toronto, but, acknowledging the argument that Bateson was contesting only natural selection, he insisted that there was ultimately no difference between rejecting natural selection and rejecting evolution. He insisted that scientists could not prove any example of descent from one species to another, "and yet they demand that we allow them to teach this stuff to our children, that they may come home with their imaginary family tree and scoff at their mother's and father's Bible."[59]

Dudley Field Malone responded to Bryan's rather long speech—it was a speech, addressed to the courtroom audience more often than to the court—with the most-lauded address of the trial, an eloquent statement that by all reports stunned everyone present. Although Malone was not the orator that Bryan was, he understood drama. Scopes later observed that as he began his speech Malone made a gesture that assured him the undivided attention of everyone present—

he removed his coat. In the dreadful heat of that summer, only Malone had declined to appear in shirtsleeves—until this moment. He began his remarks by making it clear where they were directed, saying that "whether Mr. Bryan knows it or not, he is a mammal, he is an animal and he is a man." Claiming Bryan as a friend, Malone said, "This is not a conflict of personages; it is a conflict of ideas." Nevertheless, he intended to use Bryan's rhetoric to challenge prosecutor Tom Stewart's contention that the case should be confined to the question of whether Scopes had broken the law: "I defy anybody, after Mr. Bryan's speech, to believe that this is not a religious question." Against Stewart's insistence that the case should be narrowly interpreted, Malone declared, "Oh no, the issue is as broad as Mr. Bryan himself has made it." He warmed up: "The children of this generation are pretty wise. People, as a matter of fact I feel that the children of this generation are probably wiser than many of their elders. . . . We have just had a war with twenty-million dead. Civilization is not so proud of the work of the adults. . . . For God's sake let the children have their minds kept open—close no doors to their knowledge; shut no door from them. Make the distinction between theology and science. Let them have both." Referring to Bryan's many iterations of his claim that the trial would be a "duel," Malone objected, "There is never a duel with the truth. The truth always wins and we are not afraid of it. . . . We feel we stand with progress. We feel we stand with science. We feel we stand with fundamental freedom in America."[60] The courtroom audience exploded; members of the press gave Malone a standing ovation; people on all sides of the issue applauded. Bryan himself, Scopes later recalled, told Malone that it was the greatest speech he had ever heard.[61] The *New York Times* printed the texts of both addresses on its front page, calling this "the greatest debate on science and religion in recent years."[62]

The judge decided, nevertheless, in favor of the prosecution, that the testimony of scientists was not relevant to the case. The issue to be decided was not whether evolution contradicted the story of creation in Genesis but simply whether Scopes had broken the law. Judge Raulston did allow the defense to enter expert testimony into the record, in the form of written affidavits, for purposes of a possible appeal. The defense could have submitted the testimony of these experts orally, but Bryan had astutely noted that if they were to testify then the prosecution must be allowed to cross examine them. As the historian Edward Larson has pointed out, this would have meant that they could then be asked whether they believed in such doctrines as the Virgin Birth. Although they were all prepared to confirm their belief in the compatibility of evolution and Christianity, their own Christian theology did not include miracles; the defense did not, therefore, want to submit them to cross-examination. Eight scientists, then, and four experts on religion submitted written affidavits, and Hays read excerpts aloud. The statements generally repeated the kinds of explanations of evolution that newspapers and book publishers had by this time made easily available. Some of the statements also included implicit responses to Bryan's address in court.

Maynard Metcalf and Winterton Curtis emphasized the distinction between evolution, in general, and Darwin's mechanism of natural selection, maintaining that the conflation of these two things had caused confusion among the public. Equating evolution and natural selection was, Curtis wrote, a "common misconception" among the public, and it was understandable. Curtis included in his statement an excerpt of a letter to him from William Bateson, assuring him that he had not, in his address at Toronto three years earlier, meant to suggest for so much as a minute that he would deny the fact of evolution. Bateson's only dissent was from natural selection. Curtis suggested that anti-evolutionists had "exploited" the misunderstanding generated by Bateson's address in Toronto. Echoing Dudley Field Malone's address to the court, he stressed the intellectual honesty of science. "Science has nothing to conceal," he said: it "invites you to view the evidence."[63]

Several of the affidavits elaborated on the assertion that it was impossible to teach biology without including evolution. All the scientists reiterated evidence and reasoning that would have been familiar to anyone who had followed popular expositions of evolutionary theory over the previous several years. There were four main categories of evidence, Metcalf said: evidence from comparative anatomy, from comparative embryology, from paleontology and geology, and from geographical distribution.

Several statements cited evidence from comparative anatomy. Charles Hubbard Judd, director of the School of Education at the University of Chicago, said that "the fundamental pattern of the human brain is the same as that of the higher animals."[64] H. H. Newman focused on comparative anatomy, especially the principles of homology and analogy. The variations in homologous bones of vertebrate animals "remind one of variations upon a theme in music."[65] There were, he said, vestigial structures in the human body "sufficient to make of a man a veritable walking museum of antiquities." Perhaps in response to Bryan's objection to the diagram in Hunter's *Civic Biology*, he attempted to elucidate the interpretation of evolutionary tree diagrams. Man, he affirmed, "is obviously a mammal."[66]

Fay-Cooper Cole, the University of Chicago anthropologist, stressed development from simple to complex. "The skeletons tell much of man's history."[67] Repeating the frequently made denial that a belief in evolution required acceptance of apelike grandparents, Cole noted that anthropoid apes "have specialized in their way quite as much as man has in his, so that while they are very similar, yet it is evidence that man's line of descent is not through any of these anthropoids." Humans and apes shared "a common precursor," he acknowledged, but "the anthropoids must have branched off from the common stock in very remote times." The recently discovered Taung specimen, *Australopithecus*, demonstrated that "nature at a very early period was making experiments toward man."[68] Curtis confirmed this idea: "It is not that men came from monkeys, but that men, monkeys and apes all came from a common mammalian ancestry millions of years in the past."[69] Using a familiar visual metaphor, Cole contended that we could read the record of these fossils "as clearly as though we were reading from the pages of a book." Wilber A. Nelson, the state geologist of Tennessee, also said that the rock layers of the state could "be read as clearly as the leaves of a book."[70]

More than any of the others, Kirtley Mather seems to have tried to answer what he took to be the qualms of Christian believers and to respond to Bryan's arguments. Mather emphasized especially his view of the relationship between evolution and Christianity. Clarifying the separate realms of science and religion would demonstrate their compatibility: science, correctly understood, did not aim to trespass on the territory of religion. "Science . . . deals with immediate causes and effects, not at all with ultimate causes and effects." The religious faith of scientists should put the fears of the faithful to rest: "Although it is possible to construct a mechanistic evolutionary hypothesis which rules God out of the world, the theories of theistic evolution held by millions of scientifically trained Christian men and women lead inevitably to a better knowledge of God and a firmer faith in his presence in the world." The tenets of theistic evolution as understood by many scientists "do not deny any reasonable interpretation of the stories of divine creation as recorded in the Bible, rather they affirm that story and give it larger and more profound meaning." Mather had long taught Sunday school, and he had attempted to understand the point of view of religious objectors to evolution. He pointed to the disjuncture between the first and second books of Genesis to show that the Bible did nôt demand faith in a young earth. "A day in the sight of the Lord is as a thousand years, and a thousand years as a day." That he made this point actually suggests that Mather did not know much about his fundamentalist contemporaries, of course; as the historian Ronald Numbers has shown, very few creationists in the 1920s insisted upon a young earth. Nonetheless, Mather evinced empathy for religious conservatives that not all evolutionists did. Many people, he understood, feared that "when the scientist enthrones evolution as the guiding principle in nature he dethrones God," but it was not so. More than a few people were alarmed by evolution because they associated it with the phrase "survival of the fittest." They took that to "imply that the selfish triumph, the most cruel and bloodthirsty are exalted, those who disregard others win. Obviously," Mather conceded, "this is the very antithesis of Christianity." If it were so, opposition to evolution would make all the sense in the world. With a nod to Bryan, Mather maintained that people were not turned away from evolution "by the fear of discovering that their bodies are structurally like those of apes and monkeys; it doesn't bother us to discover that we are mammals, even *odorous* mammals." The problem Christians had with evolution was that "survival of the fittest" sounded so profoundly un-Christian. But there was a remedy: fitness did not most often imply selfishness or cruelty; especially "in the strain that leads to man," it more often meant altruism, cooperation, and unselfishness, "the survival of those who serve others most unselfishly." Believing scientists could therefore be natural allies of the faithful— at least of those who shared Mather's faith. "Here, if nowhere else, do the facts of evolution lead the man of science to stand shoulder to shoulder with the man of religion." Not only was science compatible with religious faith, but it could actually illuminate it.[71]

That science improved our understanding of God had become a commonplace of reconciler literature, and several of the scientists testifying at the trial invoked that idea. Jacob G. Lipman, the dean of the College of Agriculture at the State University of New Jersey and director of the New Jersey Agricultural Experiment Station, a soil scientist, wrote of the perfection of forms resulting from evolution, concluding that science offered "a clearer vision of the great laws of nature and of the methods of the Divine Creator."[72] Even more ambitiously, Maynard Metcalf claimed that "God's growing revelation of Himself to the human soul can not be realized without recognition of the evolutionary method He has chosen."[73]

Taking advantage of the chance to reaffirm in writing the point that Darrow had attempted to elicit in his courtroom testimony, Metcalf noted, "I am somewhat acquainted personally with nearly all the zoologists in America who have contributed extensively to the growth of knowledge in this field . . . of all these hundreds of men not one fails to believe, as a matter of course, in view of the evidence, that evolution has occurred."[74] Inevitably, others also made various forms of the argument from authority. Comparative anatomy, H. H. Newman acknowledged, was a very technical science: "No one but a trained comparative anatomist can reasonably be expected to appreciate the dependence of this subject upon the principle of evolution."[75]

Though understandable, and perhaps necessary, this kind of argument could shade into statements that might be read as a little condescending. Curtis, for example, wrote that "evolution has been generally accepted by the intellectually competent who have taken the trouble to inform themselves with an open mind."[76] After making his argument that "the fact of evolution—of man, of all living things, of the earth, of the sun, of the stars—is as fully established as the fact that the earth revolves around the sun," Metcalf added, "There is no conflict, no least degree of conflict, between the Bible and the fact of evolution, but the literalist interpretation of the words of the Bible is not only puerile, it is insulting, both to God and to human intelligence."[77] Two other scientists also used the term *puerile:* Curtis wrote that "it is not only impossible but puerile to separate man from the general course of events," and Newman used the term to describe denials of the evolutionary implications of homologous structures.[78]

These scientists could assume a rather reverent tone when talking about their devotion to science, and to science as the arbiter of truth. Maynard Metcalf, for example, declared, "For a teacher to fail to bear testimony to essential scientific truth is as unworthy, as cowardly, as essentially sinful as for a man to fail to stand by his religion. . . . Truth is sacred and to hinder men's approach to truth is as evil a thing, as unchristian a thing, as one can do."[79]

Although all the scientists who testified understood that their mandate was to make the case for the compatibility of evolution and religion, and though all of them made that argument honestly, they did not necessarily understand the religious perspectives of anti-evolutionists. All of them were, to one extent or another, religious modernists, and two of them, Fay-Cooper Cole and Winterton Curtis, would, after the trial, develop long, close, and meaningful friendships with the agnostic Clarence Darrow, as would Scopes. But not even the sympathetic and devout Kirtley Mather really spoke the language of fundamentalists.

The journal *Scientific Monthly* reprinted the scientists' statements, as did many newspapers, in full or in part. Some papers reported at considerable length on the testimony of scientists. Editorials praised the scientists' testimony and concluded that the trial, though farcical, had at least given science a platform from which to educate the public. During the trial, the *New York Times,* citing an informative essay by Osborn in the same day's edition of the paper, suggested that the trial had "given to scientists and teachers a splendid chance. They will now have a larger and more alert audience than they have ever known. Such an opportunity for popularizing, in the best sense, scientific truths can rarely have presented itself."[80]

A 1926 article in *Collier's* on publicity and education cited widespread disparagement of the sensationalism in media reports about evolution, but it concluded that the publicity surrounding the Scopes trial had ultimately increased general enlightenment and could be counted as a victory against ignorance, superstition, and repression. Many commentators saw this as one of the consolations of what had been, in most respects, a dismal affair—at least the public had been offered an education in evolution. Scientists and educators had a chance to reach an engaged and enlarged audience.

Some scientists reported that the publicity surrounding the trial had increased interest in evolution among their students. According to the reminiscence of his son, Fay-Cooper Cole reported that after he went back home to Chicago following the trial many people called for his dismissal from the university, but on his return to the classroom to teach a course on evolution he found that the room was packed.[81] Of course, this may in part have reflected Cole's new status as a local celebrity, but bookstores and libraries reported increased interest in evolution as well. A story in the *Newark (NJ) Sunday Call* announced an "unprecedented" demand for books on evolution at the local library. At the height of interest, shelves specially allocated for a display of books on evolution had to be refilled daily; patrons checked out some twenty-five books a day on evolution.[82] In a post-trial article in *Harper's*, W. O. McGeehan observed that people in Dayton during the trial showed a "touching curiosity" about evolution. Robinson's drugstore sold out of copies of Hunter's *Civic Biology*, and "the place was swamped with orders for forbidden literature concerning evolution."[83]

Postmortems

But did all this discussion add to the general fund of knowledge about science or increase the number of people who understood how science worked? Not everyone thought so. There was a good deal of discussion in newspapers and magazines after the Scopes trial about whether all the attention had amounted to mere sensationalism and whether its effects had been destructive or, perhaps, ultimately educational. Not everyone was sanguine about it.

For one thing, it was difficult to banish the memory of the more unfortunate aspects of journalists' reports on the trial. In its aftermath, some journalists expressed regret at the unfair depictions of Dayton and its citizens that had dominated news coverage of the trial. The *New York Evening Post,* for example, lamented: "Dayton, Rhea County and the lovely old state of Tennessee were treated badly and shamefully by this trial. Dayton is no 'hick town,' as it has been painted. Rhea County is no haven of Holy Rollers. Its sons and daughters are as 'progressive,' as intelligent and well dressed and as keen about major league baseball scores and college football as the young hopefuls of 'big towns.' " Dismayed at the conduct of the principal actors at the trial, the *Post* characterized Bryan and Darrow as "two quarrelsome and embittered men, who were not above venomous personalities in a court of law."[84]

An editorial in the *Post* denounced both Bryan and Darrow: "In their shameful debate upon Science and Religion they were allowed to vilify the beliefs of millions of their fellow citizens while the Court, looking on, approved." Their confrontation had done nothing but harm: "Science was not advanced by denouncing Bryan as a 'bigot.' . . . Nor was the Law exalted by turning the court into an open-air circus, where its dignity, prestige and majesty were flouted by the mob." Though the encounter had been expunged from the trial record, it would not soon be eliminated from American memory. "It will be hard to forget the spectacle of a man who three times has offered himself for President of the United States confessing that he neither knows nor cares anything about Science." Darrow had behaved just as badly. "In the guise of trying a lawsuit he permitted himself to sneer at the most sacredly tender beliefs of millions of Christians."[85]

The editor of the *New Republic,* describing the "circus conception of the trial atmosphere," suggested, "Behind the whole performance some Barnum is at work." This was not an unusual observation; many people saw something Barnumesque in the event. The *New Republic* did not exult in this the way some other journals seemed to. "If ever a great issue was cheapened by descent to the ridiculous, the issue of freedom of inquiry and freedom of teaching is cheapened by the clap-trap of the histrionics masquerading as a legal trial in Tennessee. It is hard to decide whether to lament or to rejoice that out of the monkey-play some good may come." Perhaps the newspaper frenzy might have some positive results: "It may be good to get a fundamental issue on the front page of the newspaper even though it gets there as a clown. Farce in the news column may be recompensed by serious discussion on the editorial page."[86] The editor was not really hopeful, though. The memory of the trial would remain sour. In addition, the truth or viability of evolution was beside the point. The point of the whole exercise was this: "The advantage of the trial is not primarily the effort to have the law declared unconstitutional. Neither is it the opportunity to present evidence in support of the theory of evolution, the truth or falsity of which is decidedly a minor issue. The really important thing is the attempt to maintain the general principle of freedom of speech and thought for educators."[87] The issues were serious.

W. E. B. Du Bois also denounced the jokiness that dominated newspaper and magazine coverage of the Scopes trial, writing that "the learned American press is emitting huge guffaws and peals of Brobdingnagian laughter combined with streaming tears. But few are deceived. . . . The truth is and we know it: Dayton, Tennessee, is America." He took the trial very seriously: "This is America and America is what it is because we believe in ignorance."[88]

Walter Lippmann would agree with Du Bois's assertion that "Dayton, Tennessee is no laughing matter."[89] He became convinced, however, that the cure for ignorance, especially about science, would prove elusive. In *American Inquisitors*, a remarkable book about the Scopes trial and the issues it raised, Lippmann imagined a Socratic dialogue in the voices of Socrates, Thomas Jefferson, and William Jennings Bryan and a conversation between an astute fundamentalist and a modernist. Already impressed with the potential power of the media to manipulate public opinion and "manufacture consent," Lippmann argued that very few people would ever really be able to understand science in any profound sense. Most people, Lippmann maintained, wanted to feel that they knew what was true, but for science, by definition, all truths were provisional. People, by and large, did not want truth to be provisional. They wanted certainty. His fictional fundamentalist queried his modernist counterpart: "Are you not in fact blithely proposing to teach the scientific method to a mass of people who haven't the remotest chance of understanding it?" The cost of such knowledge would be unacceptable to most people. Scientific truths could not be guaranteed; they would always be subject to change, and "the mass of men won't tolerate this much uncertainty. . . . They will make incontrovertible dogmas out of scientific hypotheses. You are not teaching science when you teach a child that the earth moves around the sun." A few people might be capable of understanding scientific method and fewer still willing to accept its unsettling implications. "But the rest merely acquire odds and ends of more or less obsolete information which, while it destroys the authority and the majesty of their inherited religion, is in itself morally worthless." Lippmann endowed his fundamentalist with a Lippmannesque eloquence. "It is of no consequence in itself whether the earth is flat or round. But it is of transcendent importance whether man can commune with God." The stakes for the fundamentalist could hardly be higher: "An eternal plan of salvation is at stake." The conflict between the traditional worldview and the modern scientific spirit, Lippmann concluded, was irreconcilable. "They are certified by different systems of thought."[90] Like Maynard Shipley, of the Science League, he diagnosed the malady as a conflict between reason and authority; unlike Shipley, he was not sanguine about the possibility of a cure.

Science was, in Lippmann's version of the story, very much central to the set of tensions expressed in the trial, but not in any simple way. For those people who asked difficult questions about the authority of science and about its boundaries, simply "selling" evolution to the masses did not necessarily add to the general state of education or enlightenment of the public. In particular, many voices denounced the selling of evolution by packaging it as a theologically palatable doctrine.

The ACLU came under fire from people who criticized the strategy of recruiting scientists who would argue for the compatibility of science and religion. The intent of the challenge to the Butler Act was to try to set a legal precedent establishing academic freedom; for many people in the ACLU camp, the evolution issue itself was secondary. Arthur Garfield Hays, explaining the trial strategy of the American Civil Liberties Union in the *Nation,* acknowledged, somewhat defensively, that much criticism of the defense strategy had been leveled at the decision to emphasize the compatibility of science and religion. "Even if the Bible and the theory of evolution were entirely contradictory," critics had objected, "the law is a bad one, so why take the position that they can be reconciled?" In response, Hays tried to suggest that the ACLU "felt there was a real value . . . from a public point of view, in presenting the position of millions of churchgoers who are defenders of science."[91] Not everyone could assent to the form that those arguments took, however. Just as there were people who argued that putting the truth of evolution on center stage was a mistake because the truth of evolution was not the issue, there were also people who objected to the compatibility of evolution with religion as a fair—or necessary—defense of evolution.

The trial had thrown the arguments of reconcilers into vivid relief, and objections to the reconciler argument grew in number and in volume after the trial. Nonscientists often complained that the scientists who testified at the trial and defended evolution in newspaper articles lacked courage. One letter to the editor of the *Nation* remarked on the "absence of intellectual and moral integrity among our well-known men of science." Discounting the possibility that the scientists who chose to involve themselves in the debate were actually religious believers, the writer asked: "What are we to think of the integrity of scientific men today who, in the face of a challenge which demands honesty above all things, are denying the existence of any necessary antagonism between the practically universal teaching of our churches and the plain logic of science?"[92]

Not all scientists believed that there had been a net gain, either. Scientists continued to argue about whether writing for a general audience was a good thing. An editorial in the *New York Times* six months after the Scopes trial noted that Dr. O. W. Caldwell, a scientist who spoke at a dinner of the New York Alumni of Sigma Xi, had charged that "the blame" for the Scopes trial "rested not on an unenlightened section of the populace, but on the men of science themselves" because scientists had publicized their discoveries prematurely. Caldwell did not specify which scientists or which discoveries he meant, but he was not alone among scientists in grumbling that publicizing scientific findings could be seen as a dangerous form of "playing to the galleries" that might backfire. Scientists lodging such complaints may have had such cases as *Hesperopithecus* in mind, and they may have been reluctant to name Osborn explicitly because he remained an important and influential figure. The editorialist at the *Times* disagreed. The insights of science should be published. "And, besides, it does the public good to be shocked. . . . And in the end evolution was 'sold' to a vast body of people who never had heard of it before."[93]

Accusations like Caldwell's could sting, however. As Osborn's attempts to popularize his Dawn Man imagery and his experience with *Hesperopithecus* suggest, popular debate could carry scientists onto unexpectedly treacherous ground. As the debate wore on, scientists found themselves increasingly arguing from difficult and sometimes untenable positions. Many of them began to withdraw from the public controversy, including Osborn, and many of those who had shunned public participation all along expressed chagrin at the shape the discussion had taken. Scientists' attempts to write for the general public, having increased during the early part of the decade, reached a peak in 1926, the year following the Scopes trial, and declined steadily after that.[94] Of course, not all scientists withdrew from the public arena, but many found the experience of the trial, even if they had experienced it only vicariously, sobering indeed. A number of scientists expressed alarm at the tendency of newspapers to misquote them and to simplify and sensationalize complex issues. They also found that communicating scientific ideas to the public required a good deal more time than they were prepared to sacrifice. Most discouraging of all, perhaps, they found their ideas persistently misunderstood, and often for very good reasons.

Thomas F. Gieryn, George M. Bevins, and Stephen C. Zehr have argued, in an important article, that the scientists at the Scopes trial worked to establish the authority of science, in the context of professionalization of science in that decade, by demonstrating that science and religion occupied different realms and could not, therefore, be incompatible.[95] This is an important observation, and it was true of many of the scientific defenders of evolution in the period. It is not, however, the whole truth about scientists who defended evolution during that decade, especially at the trial: even though scientists defending evolution sometimes asserted that science and religion addressed different kinds of questions, the trial strategy emphasized establishing the compatibility of evolution and religion. Many of the "reconcilers" went much further, arguing that evolution offered real insights on matters of spiritual import. The boundaries separating science and religion, observed through the lens of the reconciler literature, could be rather fluid.

The public nature, as well as the timing, of this debate exacerbated the difficulty of communication across a widening gap dividing scientists from the public. Even those scientists most involved in communication with a lay audience— partly under pressure of the media-driven nature of the debate and partly out of defensiveness—probably obscured just those aspects of science that might have helped them to bridge the gap. When anti-evolutionists challenged the meaning of terms like *hypothesis* and *theory* and insisted that scientists confine themselves to unvarnished truth, they raised a thorny issue. For scientists, how to communicate things they understood to be true, but true in a sense now under attack, turned out to be exasperatingly difficult. Commentators often celebrated the new platform the trial offered scientists for educating the public, but for many people this began to look like a mixed blessing at best. The *New Republic* editorialized: "Any Sunday editor would commit suicide if he found he had been careless enough to go to press without an article explaining Darwinism in words of one syllable, accompanied by photographs from the American Museum of Natural History."[96]

The Irreverent Monkey Trial

Newspapers did reproduce images from the American Museum remarkably often. Yet, ironically, Henry Fairfield Osborn found it more and more difficult to control the use of the images so carefully constructed at his museum. Nor did he find publicity as malleable a quantity as he once had. In the year after the trial, frustrated by the coverage of his statements trying to replace notions of "apemen" with the more dignified image of a "Dawn Man," Osborn complained about the pernicious effects that the writers of headlines could have. He also warned on several occasions about the dangerous influence on young people of "the irreverent funny pages." For a scientist who had worked as hard as he had to dispel images of human evolution as a story of descent from brutish-looking apes, the publicity attending the Scopes trial may sometimes have seemed like an unmitigated disaster.

In a letter to Merritt Bond, the editor of the *New York Evening Post*, Osborn objected to the language featured in the *Post*'s coverage of the upcoming trial. His tone was polite: "Although quite unintentional on your part, this is doing irreparable harm all over the United States." Emphasis on the circus atmosphere in Dayton would, he warned, undermine the serious purpose of the trial and would alienate scientists. Signing himself "Chairman, Scientific Advisory Committee in Scopes Case," he declared that the scientists engaged in the controversy were "extremely highminded men who are deeply interested in the large question raised by this trial and who are grossly offended by the term 'monkey trial' that is put upon it by Bryan and which for sensational purposes has been featured in headlines."[97] On the same day, he wrote to former secretary of state Bainbridge Colby, who had been involved with the ACLU in the initial planning stages of the trial: "I especially appreciate the dignified part you took in the conference and I trust you will persuade our friends of the press to cease talking of this as 'the monkey trial,' as by doing this they simply fall into Bryan's hand, whose only weapon is ridicule."[98]

Ridicule was among the most effective weapons in the debate, it might have seemed. Logically, anti-evolutionists mounted often potent arguments challenging proposed "missing links" between humans and other primates. On a more emotional level, visual images of missing links could be even more arresting. No matter how often scientists objected to monkey imagery, monkeys, apes, apemen, and cavemen proliferated during the Scopes trial. One journalist wrote: "Dayton, you know, is the town where monkeys cannot climb the family trees."[99] In the *New York Evening Post*, a cartoon Bryan tugged at the tail of a cartoon monkey, attempting to yank it out of the family tree.[100] Speaking at the Chautauqua in Boulder, Colorado, Billy Sunday said, "My ancestors didn't hang from trees by 'prehensile tails.' "[101] Cartoon monkeys mocked Bryan and congratulated Darrow. As persistently as evolutionists—and many journalists—reiterated the claim that evolution did not mean that humans had descended directly from apes or monkeys, cartoons, metaphors, and images of monkeys and apes dominated the popular debate.

Osborn was not just being fastidious in objecting to the kind of levity that found expression in monkey references. People on all sides of the debate deployed humor as a rhetorical device, a commodity, and a weapon. "Missing links" turned out to have terrific rhetorical power during the Scopes trial. They could be used for humor, for irony, for social commentary of all kinds. And they were.

Newspapers included monkey cartoons commenting on nearly everything. Critics referred to adversaries as "simians" or "Neanderthals."[102] Although some newspapers diligently offered serious articles explaining evolution and reprinted the major speeches and exchanges at the trial verbatim and at great length, it would be difficult to dispel the overwhelming impressions conveyed by humor and impossible to dislodge the iconography of the missing link. The name "monkey trial" would stick, and the name would be an influential part of the legacy of the trial.

CHAPTER NINE

Redeeming the Caveman, and the Irreverent Funny Pages

An advertisement for Alfred Watterson McCann's book *God—or Gorilla*, devoted to excoriating Osborn and challenging the veracity of the Hall of the Age of Man, claimed that the book disproved the "Tadpole and Monkey Theory of Evolution."[1] Both anti-evolutionists and newspaper and advertising copywriters seemed to find the word *tadpole* useful, heuristic, or humorous, perhaps implying a spoof on scientists' attempts to explain the embryological mechanisms at work in evolution, as well as the still nearly ubiquitous development analogy.

Anti-evolutionists and their detractors drew on an arsenal of evocative terms. These terms appeared even when evolution was not the subject. The author of a book review in the *Forum*, for example, used the phrase "since the first Neanderthal man crawled from the slime."[2] *Neanderthal* had become a figure of speech and the term *slime* familiar shorthand for much that was offensive about the idea of evolution. A New York man wrote to Osborn in 1922 that "the statement that our common ancestors were monsters of the slime and chattering apes of the forest [is] the most monstrous scientific blunder the world has ever known." The implications were staggering: "Why if this were possible there would be no order in the organic world. To all intents and purposes it would be a world of chance, of monstrosities of hopeless and irredeemable confusion."[3]

When fundamentalists argued that evolution denied humans the dignity of having been created in the image of God, evolutionists, taking a cue from Edwin Grant Conklin, countered that the biblical notion of God making humans from the clay or from mud hardly seemed more exalted than the idea of evolution. Anti-evolutionists, parrying in an escalating competition of unsavory images, invoked "slime" as a point of origin distinctly more degraded than mud—and funnier. Science advocates adopted the term in a spirit of sarcasm. Phrases like "tadpole theory," "jungle theory," and "the slime" blended humor with evocative images in a modern alchemy of irreverence. The term *jungle*, a favorite of the colorful baseball player–turned–evangelist Billy Sunday, appeared everywhere, as did "jungle" cartoons, often carrying implicit racial connotations.

Probably the most frequently invoked mental image, though, was that of the "caveman." John Roach Straton's famous denunciation of the Hall of the Age of Man included a colorful reference to the "gruesome bones in that Hall of death, and . . . the crawling beasts and degraded cave men whose pictures cover its walls."[4] The intent of the exhibits in the Hall of the Age of Man, of course, was precisely the opposite: Osborn worked closely with museum curators, exhibit technicians, and artists to create a vision of cavemen ennobled, rather than degraded. "I am perhaps more proud of having helped to redeem the character of cave men than of any other single achievement of mine in the field of anthropology," he revealed in a 1925 essay in the *Forum*.[5]

Cavemen needed redeeming, at least from Osborn's point of view. Monkeys, apes, ape-men, missing links, and cavemen had populated the public imagination for a long time. The "men" at the tops of family tree diagrams were often cavemen. Cavemen turned out to be rather protean symbols, adapting easily to Progressive Era preoccupations, and by the time of the revitalized evolution debates of the 1920s, they had been fixed in a common cultural vocabulary. As the Cornell University geologist Harold Whitnall wrote in *Scientific Monthly*, the caveman, "dragged from his prehistoric past to be the scapegoat of the race has, symbolically, received the sins of all his descendants."[6] It was a heavy burden.

The *God—or Gorilla* advertisement, referring to evolution as "barnyard materialism," featured a familiar image of *Pithecanthropus*, better known to the public as "Java Man." This choice of illustration was no coincidence. The reconstruction used was one of the busts executed by the Columbia University biologist J. H. McGregor for the American Museum, part of the series of human ancestors on display in the museum's famous Hall of the Age of Man, widely reprinted in textbooks and popular books about evolution. McCann, a lawyer, devoted much of the book to his attack on this exhibit: accusing museum scientists of duplicity and deception, he charged that reconstructions like McGregor's were deliberately misleading. The fossil evidence, he insisted, was too fragmentary to support such reconstructions. They represented speculation, not science.[7]

But McGregor's series of reconstructions of human ancestors enjoyed wide circulation in popular science books, textbooks, magazines, advertisements, and the press, often used in a spirit of irreverent humor. All this attention drew more attacks on the Hall of the Age of Man. McCann leveled two kinds of objections at McGregor's series of busts: first, that the arrangement of the reconstructions in a linear series was deceptive, suggesting a biological lineage where none existed; and second, that in any case the heads represented fabrications. The evidence was insufficient to support them.

Osborn had McGregor's busts arranged at the American Museum and for photographs in a sequence from "Java Man" to Piltdown to Neanderthal to Cro-Magnon. McCann argued in detail and at great length that the sequence was an illusion. He challenged the assumption that the busts represented a real chronology, claiming that the provenance data did not allow any such inference. He pointed out that the specimens could well have overlapped in time and that they came from bodies that might have coexisted.

Osborn paid a good deal of attention to the orientation of the busts in photographs, wanting to display them from angles that showed off the differences among them. He had long been accustomed, after years of close attention to scientific illustration, to lining up specimens in similar orientation for purposes of comparison. In papers comparing fossil bones, Osborn's illustrations were methodologically exemplary, making sure to compare bones from the same side of an animal, for example, in order to avoid visual confusion. It was easier to "read" and compare bones if they were all oriented in the same direction and if one were not trying to compare, for example, the right side of a jaw of one animal with a left side from another. American Museum curators took some pride in the exhibition of McGregor's reconstructions, displayed in order to reveal the meticulous scientific reasoning underlying such work.

McCann objected to this convention in the case of the hominids, however, claiming that the aesthetic similarity of the styles of the heads perpetrated an illusion. Citing a different illustration of "Java Man," by A. Rutot, appearing in Osborn's *Men of the Old Stone Age*,[8] McCann claimed that Rutot's reconstruction conveyed more humanity and dignity than did McGregor's primitive-looking busts. The McGregor "Java Man" bust resembled a "short-haired hideous creature suggesting a slightly improved gorilla," whereas Rutot had created a "pious creature, looking heavenward with no expression of squat ferocity but rather with a soft sweetness, emphasized by two armsful of gorgeous vegetation, palm leaves, fern and other symbols of docility and peace." The Rutot sculpture proved, McCann implied, that the general impression of ferocity or docility was entirely subjective; it also demonstrated that the impression of chronological sequence was more aesthetic than real. If, in the Hall of the Age of Man, Osborn had substituted Rutot's "Java Man" for McGregor's among the series of McGregor busts, the impression of evolutionary sequence would have been disrupted. Without the appearance of a linear sequence, McCann suggested, the case for evolution would be weakened, and he implied that selective display to create a false impression of linearity was a deliberate strategy on the part of museum scientists.

The anti-evolutionist lawyer Alfred Watterson McCann complained in his book *God—or Gorilla* that the sequence of busts by J. H. McGregor was misleading simply by virtue of being arranged in a series. If this sculpture of the "Java Man" had been substituted for McGregor's, he suggested, the appearance of a sequence would be disrupted. Henry Fairfield Osborn, *Men of the Old Stone Age* (New York: Charles Scribner's Sons, 1916), 73.

McCann insisted that, in any case, the busts represented inventions of McGregor's and Osborn's imaginations. The Java, or Trinil, specimen, for example, was constructed from nothing more than a single skull cap, several teeth, and a femur found in different places and on different occasions. Eugène Dubois, a Dutch doctor, had set out to look for "the missing link" in the late nineteenth century, and in 1891, in Trinil, Java, he had found the specimen he named *Pithecanthropus,* after Ernst Haeckel's hypothetical ape-man. Biologists accepted the Trinil specimen as a genuine member of the human lineage, but Dubois had been frustratingly reluctant to allow other scientists access to the original bones, a fact that McCann enlisted as further evidence of scientific malfeasance.[9]

McGregor himself disliked the linear arrangement of the specimens, believing that it conveyed an idea of nonbranching evolution to which he did not subscribe; however, he defended the scientifically rigorous method he used in his restorations. He was most proud of an exhibit that revealed the skulls on one side of each individual, with the fleshed-out head as it would have looked in life on the

other, meant to display his method of anatomical inference.[10] Osborn asked Mc-
Gregor to contribute an essay to *Natural History* describing "the strictly scientific
basis of [his] mode of restoration."[11] He was very attentive to the development,
display, and reproduction of these reconstructions. For a journal article he suggested that the heads be photographed in profile because "it would bring out their
distinctive evolutionary characters better."[12]

Osborn and his colleagues devoted a good deal of time and attention to the
methods of inference in the Hall of the Age of Man in part because anti-evolutionists targeted the hall so insistently and so vigorously. A dedicated critic, Mc-
Cann scrutinized the evidence assiduously, focusing on the controversial nature
of much of the data. The Piltdown specimens had stimulated debate from the
moment of their discovery, in 1912. The combination of a modern cranium with
an extremely apelike jaw struck many anatomists as wildly improbable, but the
fabrication was executed so cleverly and with such intricate knowledge of the details of fossil preservation that scientists who examined the bones found them
remarkably convincing, and it was not until the 1950s that the specimens were
finally proved to have been faked. Although Osborn himself had originally expressed reservations, he changed his mind on visiting the Piltdown site in England in 1921, and from that time he converted to staunch support of the validity
and significance of the specimen. Accusing Osborn of concealing from the public evidence that Piltdown was a hoax, McCann cited other scientists, including
the American Museum paleontologist William Diller Matthew, who had written
that the Piltdown jaw appeared to be that of a chimpanzee.[13] *Men of the Old Stone
Age*, written before Osborn had resolved his doubts about Piltdown and on which
much of the work in the Hall of the Age of Man was based, included numerous
attempts to explain the contingency of the evidence, which Osborn conceded was
scanty but steadily accumulating. He incorporated such caveats in the labels accompanying the exhibit and in the museum guide to the exhibit, on principles of
scientific objectivity. But efforts to express scientific caution could backfire.

Osborn's attempts to be circumspect carried no weight with McCann, who
claimed they betrayed not scientific scruples but uncertainty, borne of a paucity
of convincing proof. Quoting *Men of the Old Stone Age*, McCann dissected out
and made fun of contingent and qualified statements, counting the number of
qualifiers per page, for example, and accusing Osborn of building an ornate
edifice out of material he knew to be flimsy. McCann noted that some 937,000
people had visited the museum in 1920, and he wondered why "not one" of these
visitors had been "given any hint" of the reservations expressed in Osborn's cautious written language. "If truth, the whole truth, and nothing but the truth is the

chaste objective of science, how are the professors of the American Museum to explain the wholly misleading compound of indirection, innuendo and suppression now posing in the Hall of the Age of Man as a 'scientific fact'?"[14]

Other anti-evolutionists across the country took up McCann's criticisms enthusiastically. Boston's Cardinal William Henry O'Connell echoed them in a series of addresses and interviews that set off a heated newspaper exchange with Osborn. A retired Presbyterian minister from Tampa, Florida, Dr. J. G. Anderson, claimed that scientists assembled bones found in different places, and undoubtedly from different creatures, and reconstructed from them "an ape-man," objecting, "There is not one word of truth in the whole business."[15]

Some speculation necessarily found its way into such efforts as McGregor's restorations, of course. In 1927 Osborn asked McGregor to modify his Trinil restoration "omitting the protruding lips."[16] The hint of a Neanderthal beard aroused a good deal of curiosity. The museum's files include many letters from students and teachers asking how it was that scientists knew that some ancient people had beards and others did not.[17] Osborn acknowledged that "the reason for the present absence of hair on the face is purely conjectural." But he speculated on the basis of his understanding of the biogeography of living human populations: "Few tropical or southern races have any hair on the face and very sparse hair on the body, while northern races are hairy."[18] In 1921, after a trip to England to look at the Piltdown material, he wrote to McGregor that the bust of Piltdown need not be revised in order to remain consistent with new information because "our recent knowledge of the brain makes the pre-human look which you have given the head legitimate, although I am inclined to think the face many have been more hairy than you have represented it."[19]

The facial hair was one of several subtle interpretive refinements that served to separate the more recent hominids, Cro-Magnon and Neanderthal, from the earlier ones. McGregor's earlier hominids, Pithecanthropus, or "Java Man," and Eoanthropus, the fabricated Piltdown specimen, were somewhat open mouthed, if not quite slack jawed, in contrast to the resolute-looking Neanderthal and the Cro-Magnon, who vaguely resembled Thomas Jefferson. In his article in Natural History, McGregor wrote that in his reconstructions he had "tried to be conservative, to follow only the guidance of anatomical fact, minimizing my personal equation in the work as far as possible, and avoiding any inclination to make the result either bestial or brutal." He admitted that "as a concession to popular taste" he had added hairstyles and, in the case of the Neanderthal figure, a beard—obviously not deducible from fossil evidence. "But," he added, "the Neanderthal species was human, not brute."[20]

Osborn made a concerted effort to separate humans from any vestige of an ape. He did this by his design of family tree diagrams. He also did it through his influence on reconstructions of the anthropoid and hominid past. There is an irony in the sequence of McGregor restorations lined up as a series, for although he did intend to convey an idea of progress by so arranging them, Osborn did not mean to claim that the earlier hominids were actual progenitors of more recent species. The Neanderthals might be perfectly respectable relatives, but they never resided on the same branch of the family tree as modern humans. Osborn subscribed emphatically to the school of anthropological thought that assigned Neanderthal and Cro-Magnon to separate species. He believed that if only the public could be helped to visualize them in detail, and to understand the differences between them, much confusion and many misconceptions could be eliminated: "The cave man bore, and still bears, an evil reputation of being a brute, because few people realize that there were two entirely different types of man." The public had undoubtedly been put off by images of cavemen that resembled Neanderthals, but Neanderthals represented "an extremely ancient lower order," one that had been supplanted in Europe by "one of much higher order, known as the Cro-Magnon race of artists."[21]

The Cro-Magnon people were distinctly modern, large-brained hominids, true *Homo sapiens,* and Osborn admired them. He admired them for their large brains, of course; they were, he often commented, "our equal if not our superior in intelligence."[22] In particular, though, he admired them for their art. In 1916, sending a copy of his book *Men of the Old Stone Age* to French colleague Marcellin Boule, he noted that "Cro-Magnon was the art-loving race."[23] The ability to produce art held implications of the utmost importance. His explorations of the notion of "creative evolution," Osborn claimed in 1926, "were first aroused by the realization of the sudden emergence of the moral, intellectual, and spiritual abilities of the Cro-Magnon race." The "intellectual and spiritual evolution of man" had remote origins with the Dawn Man and culminated "with those dramatic transitions in the intellectual life of man in which, entirely without antecedent experience, he suddenly emerges from Nature, like Minerva from the brain of Jove, fully equipped for supreme intellectual and spiritual tests."[24] Osborn was not alone in perceiving the appearance of art as uniquely important in the evolutionary history of humanity. Recent discoveries of Cro-Magnon cave art in Europe attracted a great deal of attention, adding a new dimension to popular notions of Stone Age people and suggesting a satisfying accommodation between the ideas of animal evolution and human uniqueness.

In 1925 G. K. Chesterton published a book that dwelled at length on the public image of cavemen. Popular culture was littered with nonsense about cavemen, Chesterton wrote, none of it anything but unfounded speculation. It was pure fantasy to derive any message from the existence of cave people, with one exception. The one thing known without a doubt about cavemen was that Cro-Magnon people produced a subtle and sophisticated art. We could therefore infer that they were human. "Art is the signature of man," Chesterton declared, arguing that the one thing that really mattered—the human soul—was not something that science could contemplate. Art, and therefore the human soul, had appeared suddenly and complete: all at once: "Monkeys did not begin pictures and men finish them; Pithecanthropus did not draw a reindeer badly and Homo Sapiens draw it well."[25]

Although his vision of the evolutionary context of the development of the human soul took a more complicated form than Chesterton's, Osborn expressed a similar appreciation of the profound significance of art in human development. The series of murals by Charles R. Knight for the Hall of the Age of Man, designed in close consultation with Osborn and with the advice of other scientists most familiar with fossil hominids, represented a concerted attempt to revise clichéd notions of the human past. Osborn worked closely with Knight on his paintings reconstructing scenes from prehistoric life, intending them to convey his understanding of the meaning of evolution. In one typical letter, he wrote of his determination that the series of Ice Age paintings for the Hall of the Age of Man should "express the great broad truths of Pleistocene life," which included especially "the severe conditions of life characteristic of the Pleistocene."[26] Although he and Knight often had difficulty negotiating the subtleties of a very complicated personal relationship, Osborn deeply respected Knight's abilities as an artist. In particular, Knight shared his own reverence for the animal world and his sense of animal life as a romantic thing. Osborn praised a Knight painting of a tiger as "the most beautiful work of the kind in existence. Contrasting with the charging, or rampant tiger, it represents the reflective or scholastic tiger looking quietly out on the River of Time."[27]

In 1919, during the composition of the mural of a Neanderthal family, called *Flint Workers*, Osborn instructed Knight: "Each man must be in a pose natural to wild men, without chairs, who are accustomed to damp or stony ground, and therefore squat, or kneel on a rough piece of skin."[28] Part of the mandate he gave Knight included recording the wildness of these people and the primitiveness of their culture. Knight showed several of the men in the picture in profile, highlighting their sloping foreheads and receding chins. They were short; Knight referred to them as "hardy little people."[29] They were also alert, watching for dangers in their environment. They worked, appearing to have organized a division of labor of sorts; the mural depicted a coordinated social group and, since the group included a woman and child standing in the shelter of a cave, a family. One of the men prepared flints, an activity that Osborn informed readers of *Men of the Old Stone Age* required formidable skill. His descriptions of Neanderthals revealed both distaste and a grudging respect: "This fossilized hunting race of the Neanderthals, low-browed, small-statured, ungainly, hideous of aspect, with retreating chin, broad nostrils, beetling eyebrows, is nevertheless human, beyond challenge."[30] In what did their humanity consist? In 1924 Osborn wrote to his friend and publisher Charles Scribner that he was thinking of adding to *Evolution and Religion* something on "Evolution of the Soul . . . in which I am intensely interested."[31] In one of the essays in that book he made it clear that ungainly as they may have been, the Neanderthals' comprehension of death proved that they were beings with souls—humans. "They had tender sentiments, they revered their dead, they believed in the future existence of the hunter in 'happy hunting grounds,' as evidenced in their inclusion of the finest flint implements in the burial of their dead."[32]

Osborn was at considerable pains to correct the public image of the caveman, as were many of the science popularizers. Aspiring writers of prehistoric tales wrote to him, expressing disapproval of the romanticizing of Stone Age fiction, and vowed to raise the standard.[33] One of the popular writers of "Stone Age

Charles R. Knight's mural *The Flint Workers*, painted for the Hall of the Age of Man, was intended to portray Neanderthal people as active, alert, and technologically innovative, if primitive, in contrast to popular images of Neanderthals. Image #39441A. © American Museum of Natural History

fiction" inspired by Osborn's work, George Langford, who had sent him caveman poems from time to time, wrote in 1920 asking Osborn to write an introduction or a recommendation for a novel he had written called *Pic, the Weapon Maker,* about a Stone Age boy of the Mousterian (Neanderthal) period. Pic, the protagonist, lived in harmony with his animal friends, for the book represented "an effort that aims to picture the men and animals of pre-history as pioneers worthy [of] our deepest respect and to close the hostile gap that exists in many minds, between man and beast." Langford's concept harmonized well with Osborn's ideas about human evolution, for, as Langford wrote, "the human character is idealized. I do not hint at his Simian affinities, and although primitive, he is a true human being."[34] On the recommendation of curator William Diller Matthew, who read the manuscript and assured him that it committed no scientific gaffes, Osborn agreed to contribute an introduction.

Langford wrote to Osborn that most people—average readers, as he put it— imagined prehistoric life as "a sort of hell on earth with half human devils running around in it." Scientifically informed fiction, he suggested, could help put the average reader straight: "Those who shudder at the mere mention of 'missing links' and 'simian ancestors' might view their shady past with equanimity if the matter were presented to them in palatable form." If only Stone Age life were presented attractively, readers "might come to realize that the word 'cave-man' does not necessarily stand for dulness [*sic*] and brutality; that the best in us sprang from the minds and hearts of our much-despised ape-like ancestors."[35]

According to one reviewer of Langford's book, "The acceptance of Darwinism implied to the crude mind that somewhere and sometime there was a glorified ape who was the real first ancestor of man." The public had been subjected to "a considerable flight of books on the life, love and labors of the original manmonkey, and their collateral idiocies of the Tarzan type." Langford offered a rational corrective to this literature, "a tale of the real primitive man . . . adventures in that crude morning of man's world and its personal romance" tempered by "a reverence for scientific verisimilitude that earn for the book and its author" the approval of "so high an authority as Henry Fairfield Osborn." In sum, the book merited praise as "an attempt to reconstruct the adolescence of our race, which, however romantic, is professedly and actually serious."[36]

Stereotypes often overtook lessons like Langford's about the sophistication of Stone Age technology, however. One journalistic advocate of evolution and education wrote, in a review of *The Earth Speaks to Bryan,* that "if Mr. Bryan looked around the American Museum of Natural History a bit, particularly at the clubs kept by the Neanderthal men, . . . and at the diaries and needlework left by the

Cro-Magnons, the artists . . . he would see that life has evolved with something like progress to the present high type of politician and land agent." Although he took Osborn to task for adopting the mantle of the preacher, the reviewer still thought that the book included "a lot of first-class dope on the caveman."[37]

The reviewer must not have read *The Earth Speaks to Bryan* very carefully or have been especially familiar with the museum. The description of the book as including "first-class dope on the caveman" must have been calculated to sound flippant rather than accurate, and the idea that the museum offered exhibits of Cro-Magnon diaries and "clubs kept by the Neanderthal men" was pure nonsense, testifying to the tenacity—and probably to the popular appeal—of the caveman cliché.

Of course, Osborn could not banish the popular caveman. Popular stereotypes of Neanderthals continued to include clubs. Visual traditions like the caveman with the club revealed their tenacity in the illustrations created or sanctioned by scientists as well. Among the most familiar cavemen for newspaper readers would have been a caveman by Charles R. Knight for the frontispiece of Osborn's 1915 book *Men of the Old Stone Age*. This caveman contrasted strikingly with the family in the mural *The Flint Workers*. The caveman in the frontispiece slumped passively before a cave, in the company of another individual who knelt, examining a pair of rocks or flints but appearing singularly unresponsive to his companion, inactive, stoop shouldered, and bearing a club, which hung limply from his hand. The image of the stoop-shouldered caveman—without his somewhat more engaged companion—was reproduced in advertisements for H. G. Wells's phenomenally successful *Outline of History* and in many newspaper articles about the evolution controversy. When Straton attacked the Hall of the Age of Man, newspapers carried pictures featuring cavemen, especially this caveman with the club from the frontispiece of *Men of the Old Stone Age*.[38]

In a projected second edition of *Men of the Old Stone Age*, Osborn planned to jettison this image, telling Knight that it had been superseded by recent research, but the reason was apparently that William King Gregory convinced him that the picture was "radically wrong in the character of the feet." The feet in the picture were too modern—Marcellin Boule, who had described Neanderthal anatomy at length, claimed that the Neanderthal ought to be depicted with a divergent hallux (big toe), like that of an ape. Gregory objected to Knight's *Men of the Old Stone Age* frontispiece not because it was too primitive but because it was not primitive enough. Knight had given his figure "a perfectly modernized foot." Neanderthals, Gregory suggested, should be portrayed with more of "an extremely strange and unfamiliar aspect."[39] Nonetheless, the picture became quite familiar

Knight's frontispiece for Osborn's 1916 book *Men of the Old Stone Age* portrayed Neanderthals as much less active and alert than the Neanderthals of *The Flint Workers*. The Neanderthal standing in this picture, with slumping posture and holding a club, was reproduced in newspapers and advertisements frequently in the 1920s. Image #19545. © American Museum of Natural History

in newspapers in the 1920s. The Neanderthal people in *The Flint Workers* mural were short, squat, and low browed, but they participated in family and social life, made and used tools, and embodied alertness, in distinct contrast to the caveman with the club. But the caveman with the club remained a more familiar type.

Why a club? What evidence was there of Neanderthals bearing clubs? None. There were no papers describing scientific studies of fossil clubs, and no one argued in the scientific literature that Stone Age people used clubs. There was abundant evidence of much more sophisticated tools. An image of a flint implement adorned the cover of *Men of the Old Stone Age,* and the text included many pages of detailed drawings of flint implements that had been found among the remains of early humans, arguing that these were relatively sophisticated weapons, difficult to craft, and that they constituted evidence of the impressive abilities of their makers. This was one of the primary reasons that George Langford contacted Osborn: he hoped above all in *Pic, the Weapon Maker* to demonstrate to readers that early humans were sophisticated toolmakers. The crucial role of flint tools in human evolution was, he told Osborn, the "basic underlying theme" of the book.[40] The flints represented culture. Clubs, by contrast, represented something antithetical to culture.

A study by the anthropologist Stephanie Moser has made a convincing case that the caveman with a club was a relic of a very long historical tradition; the club in the hands or at the side of an ostensibly primitive person is an ancient pictorial motif, beginning with Greek images of Herakles. The ancient Greek tradition, according to this argument, conflated distance in space and distance in time—the past—to arrive at images of non-Greeks as Barbarians, using iconographic signs like fur clothing and clubs paired with physical strength (foreign giants carried clubs) in contrast to the civilized Greeks. The visual tradition continued with medieval images of Adam and Eve leaving the Garden in fur skins instead of fig leaves and through the medieval and Renaissance traditions of the wild man of the woods, feral children, and similar images of primitives with clubs.[41]

The first published image of a Neanderthal, in *Harper's Weekly* in 1873, belonged to a related visual tradition of imagining hirsute primitives which has been traced back at least to the medieval hairy "wild man" images;[42] the wild man of this tradition appeared in *The Faerie Queen* and *The Tempest.* A study by Judith C. Berman postulates that the connotations of hairiness probably included more than mere biogeographical adaptation; culturally, hairiness was associated with wildness.[43]

And why did Knight's 1915 Neanderthal have such stooped shoulders? Was poor posture a transitional stage on the road from apes to humans? Scientific illustrations of monkeys and apes did not portray them with stooped shoulders or curved backs. When geologists began to lengthen the span of deep time, artists responded by employing visual conventions suggesting the primitiveness of the ancient world, but not until after the first discovery of Neanderthal bones in Germany in 1857 and publication of *Origin of Species* in 1859 did they begin to depict early humans as anatomically distinct from modern people.

Knight's club-wielding Neanderthal with stooped shoulders was based on the work of Marcellin Boule. The influential French paleontologist effectively promulgated the idea that Neanderthals represented a divergent, dead-end branch of the human family tree. Boule described Neanderthals as bent kneed, stoop shouldered, brutish, and primitive, and this image was widely adopted. Boule's text describing the anatomy of the Neanderthal specimen contrasted strikingly with his anatomical illustrations. He said in words, for example, that Neanderthals were more similar to chimps than to modern humans, but the pictures he published of their limbs belied that claim. He included an illustration of the outline of a Neanderthal skull superimposed on the outline of the skull of the American paleontologist Edward Drinker Cope, who had bequeathed his skeleton to science. Cope's brain case differed very little in size from that of the Neanderthal, Boule admitted. He insisted, however, that since the face of the Neanderthal protruded farther forward, its cranial endowment relative to the size of its face placed it closer to the anthropoid apes than the sheer size of the skull might suggest: "They are almost equal in cranial capacity, but we see how great is the difference between the size of the face in one of the most intelligent of men and in our savage of quaternary times."[44]

And Boule published some inconsistent series—for example, describing in words a comparison of chimpanzee and Neanderthal thighs but accompanying the text with an illustration of two Neanderthal femora and a modern femur—no chimps. Illustrations of Neanderthals with poor posture expressed this search for a "missing link" anatomy, but in a fashion that made little anatomical sense. Chimps and gorillas walking on all fours still had straight backs. How was slouching posture an intermediate stage? Some anatomical discussions pointed out that the position of the foramen magnum of Neanderthals would have placed the head at an angle on the neck that implied a less erect posture than that of more modern humans; they would have leaned forward in the manner of quadrupedal apes. It has also been suggested that Boule based his reconstructions on an arthritic individual and that, expecting Neanderthals to have crooked backs and bent knees, he saw the morphology he had expected to find. Boule was aware that the specimen was arthritic; his extrapolation from this individual to Neanderthals in general as bent kneed and stoop shouldered may attest to the persistence of the idea of a transitional form that should look like this.[45] The form this restoration took may also have been related to his understanding of human evolution as bushy; he may have exaggerated the primitiveness of Neanderthal out of a conviction that it was very different from modern humans.[46]

Boule's characterizations of Neanderthal anatomy were embedded in a debate about the pattern of primate and human evolution. William King Gregory thought

Kupka and Marcellin Boule's 1909 Neanderthal. As a consultant to the artist for an illustration of the Neanderthal for the *Illustrated London News* in 1909, the influential French paleontologist Marcellin Boule ratified an extremely chimpanzee-like portrayal. This is a reconstruction of the Neanderthal from La Chapelle-aux-Saints, drawn by Frantisek Kupka for the *Illustrated London News*, February 1909.

Boule had exaggerated the differences between anthropoids and Neanderthals: "Nobody could suppose that the anthropoids had entirely stood still in their evolution during the millions of years that have elapsed since their separation from the common man and anthropoid stock," he acknowledged, "but Boule, observing a few characters in which the Neanderthal limbs recall the Cynomorph monkeys, greatly overemphasizes these resemblances." Boule's reconstruction of Neanderthal skeletons, Gregory suggested, reflected his belief that the ancestors of humans must have walked on all fours. Gregory himself, postulating a semierect ancestor, interpreted the configuration of the bones differently.[47]

Looking for intermediate stages, Boule exaggerated the similarities to apes and chose illustrations that confirmed his expectations. As a consultant to the artist for an illustration of the Neanderthal for the *Illustrated London News* in 1909, he ratified an extremely chimpanzee-like portrayal. The caveman needed to be rescued, it seemed, not only from popular images but also from images sponsored and endorsed by scientists. And Osborn worked with Knight to create a stunning contrast to apelike cavemen.

Charles R. Knight's mural of the Cro-Magnon cave artists, painted for the Hall of the Age of Man at the American Museum of Natural History, expressed museum president Henry Fairfield Osborn's vision of the Cro-Magnon people as highly sensitive, artistic, and anatomically modern. This mural was reproduced frequently in newspaper articles and books in the 1920s; it has recently been restored and is again on display at the museum. Image #322602. © American Museum of Natural History

Osborn acknowledged the humanity of Neanderthals, but with a suggestion of ambivalence. He expressed no ambivalence about Cro-Magnon people; in discussions of Cro-Magnon he became eloquent. Knight's Cro-Magnon mural was lyrical. It expressed the idea that, much as Chesterton had said, real humanity resided in art; Neanderthals may have had souls, but it was with the appearance of art that the superior modern soul came into being. Depicting a group of cave artists in the cave of Font-de-Gaume at Les Eyzies-de-Tayac, France, Knight used light to such dramatic effect that on the basis of this painting Osborn was asked to serve as an adviser to a General Electric Company educational film about the history of lighting.[48] Osborn and Knight consulted with the French anthropologist Henri Breuil, the scientist best acquainted with the cave art, about the details of the painting, although they accepted his suggestions only selectively. He agreed to the use of an antler as a palette for the artist but advised against garments—it would have been too hot and moist in the caves, even in winter, for fur clothing.[49] The tension between the dark cave and the light from the lamps created a sense of drama and mystery. The pattern of light illuminated the central figure, a tall, graceful, long-limbed young man whose face resembled Knight's own, perhaps as an homage by the artist to this ancient predecessor. The stance of the artist,

leaning forward and reaching upward, must have pleased Osborn for its evocative power. That the painting in progress was of woolly mammoths could not
have been coincidence. This painting, perhaps more than anything else, expressed Osborn's conviction that "creation of this man of a higher order, known
as the Cro-Magnon, with his moral, spiritual, and intellectual powers, is utterly
incomprehensible as purely a process of survival of the fittest."[50]

Many viewers found Knight's mural inspiring, and the message about the importance of art did not go unnoticed. Henry C. Tracy, a writer of "Stone Age
fiction," credited Osborn with helping to convince people that "later Paleolithic
men were . . . beyond our general public in their essential intelligence. And the
proof is, their art."[51] The museum received numerous requests for copies of the
mural over the years; it was reproduced often, remained on exhibit until the early
1960s, and has recently been restored and put back on exhibit. A silhouette of the
central figure in the painting, and the mammoth he was painting, graced the dust
jacket and newspaper advertisements for *The Earth Speaks to Bryan* and, later, the
cover of a 1936 edition of Stanley Waterloo's 1896 novel *The Story of Ab*.[52]

Osborn and Knight intended the painting to help "redeem the caveman," and
in some quarters, at least, perhaps it did. Osborn and his colleagues at the museum received letters from people who had read, written, or proposed to write
books, especially novels, about the romance of life in the Paleolithic.[53] There were
proposals by artists who had painted or wanted to paint murals depicting the lives
of cavemen, numerous poems about cavemen, and brochures from a man in

*By Henry
Fairfield
Osborn*

**The Earth
Speaks to Bryan**

Dedicated to John Thomas Scopes

This book, with two of its chapters dealing directly with the Tennessee trial, is the most concise and
powerful statement of the case for
the Evolutionists that has yet been
made.

Don't fail to read this timely book.

Artist of 30,000
Years Ago

$1.00 at all bookstore

Charles Scribner's Sons, 5th Ave., New York.

The silhouette of the cave artist
used in this 1925 advertisement for
The Earth Speaks to Bryan comes
from Knight's mural for the Hall
of the Age of Man. Osborn wrote
that he was especially proud of his
achievements in helping to redeem
the popular image of the caveman.
Illustrations like this one were central to that effort. The advertisement appeared in the *New York
Times Book Review*, July 12, 1925.

New Mexico who called himself "The Cave Man" and offered tours of the caves at Carlsbad.[54] For many people, cave people, or at least cavemen, continued to represent the romantic possibilities of an ancient past in contrast to the compromised present. One Ohio student, writing to Osborn to request an autograph, remarked, "I am deeply interested in science and especially in prehistoric man. Although we usually think of science as cold and unfeeling yet there is something romantic in studying about the lives of these ancient men."[55] And Osborn, like other scientists, was asked to speculate about them. One letter, for example, invited him to contribute a paper on "the personality of prehistoric man" to a journal called *Character and Personality: Quarterly for Psychodiagnostic and Allied Studies.*[56]

In 1931 Osborn sent a statement to the *New York Herald:* "Many of the highest sentiments, emotions and impulses are traceable back to our very remote ancestors of the Stone Age, and even of the pre-Stone Age." The "primordial virtues" included "fine primary moral instincts and impulses, especially in relation to home, tribal and family life and to the survival and leadership by the finest members of the tribe." These fine impulses now seemed, however, to be endangered, along with much of the natural world, "laughed out of court and even destroyed by the cynicism of modern times . . . thrown into the discard by civilization."[57]

In the spirit of the "back-to-nature" movements that had expressed Progressive Era ambivalence about modernity, Osborn stretched the old notion of recapitulation, suggesting that if moderns were not careful, the development metaphor might end in senescence. Adopting the perspective of the development, or recapitulation, metaphor—primitive people as "the childhood of the race" bound to recapitulate the development of the species until they reached the final, adult, stage, civilization—late nineteenth- and early twentieth-century images of Neanderthals often borrowed modern ethnographic imagery to depict ancient peoples, assigning European notions of the physiognomy of North American Indians, Australians, or Africans to portrayals of Stone Age people, for example.[58] When William King Gregory placed the Tasmanian woman Trucanini just below the Roman athlete on a family tree, or when Henry Fairfield Osborn imagined Neanderthal people anticipating a life after death in the "happy hunting grounds," they participated in this Victorian tradition. This was the caveman as "noble savage."

For the American Museum anthropologist Nels C. Nelson, answering a query from a writer of prehistoric fiction, the appeal of primitive life lay not in its simplicity, as was commonly supposed, but in its balance, as contrasted with modern life. Although "civilized" life was on the whole "vastly more complicated," individuals in contemporary society functioned, in some sense, in an atrophied environment. In this age of specialization, the individual could no longer function independently. "I am inclined to think that primitive man was on the whole far bet-ter acquainted with his environment than we are with ours."[59]

For Osborn, this loss of individual skills in the face of a complex modern so-ciety represented a lamentable loss of virility. The solution included a disciplined return to the manly virtues of what his old friend Theodore Roosevelt had ex-tolled as "the Strenuous Life" in the outdoors. Modern civilization, Osborn re-peatedly wrote, suffered under an excessively feminizing influence; like Roo-sevelt, he believed that a return to virility would require greater efforts to preserve and live in harmony with primitive nature. In 1925 he contributed an essay to *Col-lier's* on the subject, "The Cave Men Knew," and he reiterated this theme more and more often, in talks, interviews, and essays through the decade of the twen-ties. The *New York Times* featured his statements on this theme prominently. The cave boy had been trained by "that stern master, which we now designate the 'Struggle for Existence,'" but had also been "surrounded on all sides by vibrant nature, full of inspiring and wonderful phenomena, which filled him with rever-ence and awe." Osborn suggested that the disadvantages suffered by the modern boy, in contrast, could be blamed in part on the newspapers: "In our large cities, in the press, and in the minds of teachers who depend upon the press, civiliza-tion has reared a Frankenstein which shuts out the direct vision and inspiration of nature and banishes the struggle for existence."[60] Responding somewhat wryly, the *Times* observed that Osborn's prescription for the modern malady was that children "should spend more time at the American Museum!" and it com-mented, "Dr. Osborn's pessimism is thus coincident with an adroit and pardon-able publicity."[61] In an editorial the same day, citing recent athletic achievements, the editor protested that modern Americans "would seem to be not quite the bi-ological effeminates" Osborn had implied they were.[62]

The cave boy who could teach such important lessons to the modern boy was distinctly a gendered image: the idea was not simply competence but masculine competence. From Osborn's point of view, the flint workers in Knight's mural practiced an exemplary gendered division of labor. Art was apparently also a male activity: in his instructions to Knight for the cave artist mural, he stipulated firmly, "Omit women from the Cro-Magnon picture entirely, and do not make the poses too classical or artistic."[63]

Even when scientists described numerous fossil flints and other tools made by early hominids, depictions of "cavemen" reiterated the tradition of the club as the weapon denoting the primitive, and caveman cartoons would ignore flints in favor of clubs. All efforts at scientific reserve notwithstanding, the caveman re-tained a protean symbolic allure and remained compelling as a vehicle for sensationalist imagination. One artist, hoping for an American Museum commission, wrote to Osborn that, inspired by his reading of *Men of the Old Stone Age,* he had painted six scenes depicting Old Stone Age life which he hoped the museum might choose to put on display. His description of one of the paintings includes "an almost nude Cro-Magnon woman, a look of defiance and fear on her face. . . . Confronting her, his eyes blazing, and accusing left hand with index finger pointing, his right grasping most of the garments torn from the woman's body, a powerfully built cave man."[64] It is safe to speculate that this reading of his book could not have delighted Osborn. The popular interpretation of his work had clearly gotten away from him.

The kidnapping of women by cavemen echoed an older popular motif of abductions of human women by apes. In 1915 Osborn corresponded with Chester K. Field, of *Sunset* magazine, about an outdoor theatrical production called *The Cave Man.* Osborn requested a copy of the play.[65] On reading it, he wrote to Field of "the great pleasure I have derived from reading your original Grove Play 'The Cave Man.'" He did detect a major flaw, however: "The general conception seems to me admirable. I only dissent from the introduction of the gorilla or great ape, because there is no truth in the widely spread stories that the great apes of Africa sometimes attempt to capture women."[66] Many people wrote to Osborn to ask him whether rumors of apes kidnapping women were true. He received so many queries that he saw fit to deny them even in scientific papers, mentioning in particular the notorious 1854 sculpture by Emmanuel Frémiet, *Gorilla Abducting a Negress.* The museum had been given this sculpture as a gift, he revealed, but would never put it on exhibit "because in the Museum exhibits we are trying to present only truth and to eliminate all misrepresentations of ape and human resemblance." Much public misunderstanding of evolution, he asserted, had its source in such "vicious falsifications of natural history," which, unfortunately, had "dominated the stage and literature."[67]

The abduction theme persisted. It had, after all, formidable commercial potential. An especially troubling example, from Osborn's perspective, arrived in the form of the 1932 movie *The Blonde Captive.* When Russell Spaulding, a representative of Columbia Pictures, requested permission to meet with Osborn for advice, a flurry of memos ensued. One of Osborn's secretaries, Miss Tyler, notified him that "the picture, as reported by those who have seen it, is a cheap and commercialized performance."[68] A note to Spaulding declined a meeting with eloquent terseness: "The reports we have received about the picture 'The Blonde Captive,' are distinctly unfavorable and Professor Osborn does not wish to have anything to do with the film in any form. It is unpleasant to learn that his name and one of his books have been exploited in this connection."[69] The movie had cited Osborn's *Men of the Old Stone Age* without his knowledge or permission. Even Spaulding found the film distasteful, it seemed. A memo from a different assistant, Florence Milligan, to the first assistant, Miss Tyler, records: "One Mr. Russell Spaulding desires to go on record as follows: In reply to your letter to him re 'The Blonde Captive,' that he is not responsible for the film and shares none of its 'glory.' He considers it 'drool' (his word) and has no connection with it." Spaulding wanted to make sure everyone understood that he thought "very highly of Professor Osborn and very badly of the film in question" and that he wanted to "clear himself of the malodorousness" attending the movie.[70]

According to a letter from Lillian Crockett Lowder, a woman who wrote to Osborn asking whether he sanctioned the film, the movie located a Neanderthal man in present-day Australia, "untouched by evolution since time began." The distressing thing for Lowder, however, was not so much the implausibility of such a survival; she objected to the racial implications of the film: "The claim is made that this black proto-savage cave man became the father of a blond child by a shipwrecked white woman. It is further claimed that this savage black man was our ancestor along with the ape." She especially hoped to be informed that this idea was not scientific: "May I ask, Dr. Osborn, if it is your conviction, based on scientific deduction, that it is possible for a pure type white woman to become the mother of blond or white children by a pure type black man?"[71]

Osborn answered, again through an assistant, thanking her for calling his attention to the "obvious errors in the film" and noting that the use of his work was unauthorized. Though he was undoubtedly appalled at this use of his name, the most distressing implications of the movie seemed to extend beyond just scientific accountability: "He wishes it known that . . . he is, therefore, in no way responsible for any of the scientific statements in the film. He is strongly opposed to miscegenation or race mixture of any kind."[72]

An advertisement for the movie showed a man meant to look "savage" clinging to a blond woman with bobbed hair, along with the words: "Can a civilized woman find love, happiness, peace in being a Primitive Wife?"[73] *The Blonde Captive* may have been the only movie of this type to appropriate Osborn's reputation, but it was far from being the only movie of the type. In 1930 a movie called *Ingagi*, extremely successful at the box office, drew down the wrath not only of the Will Hays office of the Motion Picture Producers and Distributors of America and the Better Business Bureau but also of the American Society of Mammalogists, holding its annual meeting at the American Museum. Distributors of the movie, which purported to document lusty encounters between gorillas and

"ape-women," ran advertisements reading: "Has the Man Ape Been Found? Gorillas! Wild Women! Apparently Half Ape! Half Human!"[74] A poster for another
movie of the early 1930s, *The Gorilla Woman*, featured an ape carrying a semi-
nude, smiling woman, with the caption, "Giant monsters enthroned as love gods!
Startling in its weird action!"[75]

Although many people undoubtedly shared Osborn's distaste for such movies,
they did find an audience. Whether they seemed scandalous must have been a
matter of perspective. Not every viewer found *The Blonde Captive* appalling. Edna
Briggs, of Brooklyn, New York, wrote to Osborn that she had just seen the film
"and enjoyed it very much." Apparently she was not at all shocked. "It was a rest-
ful change from the jungle pictures we have had so much of lately. In this picture
one of the explorers mentioned your book, 'Men of the Old Stone Age' and I
would like to inquire if I may where I might purchase a copy." Miss Briggs expressed an avid interest in natural history and seemed to be relatively well informed, asking about an error in the film's representation of the geographical
distribution of the dugong.[76]

Reactions to images like those in the film varied widely. Even as ardent a proponent of evolution as Frances Mason found herself put off by unsavory images
of apes. Writing to Conklin in 1929, distraught over an article in the Hearst papers illustrated with a picture of a "big repulsive gorilla," Mrs. Mason pleaded
that scientists should work together to combat such false impressions, which
seemed to her "to caricature God's work."[77] A Staten Island clergyman wrote to
Osborn in 1927 announcing that he was a theistic evolutionist and asking to be
reassured that the common ancestor of apes and humans was "no more an ape
than he was a man . . . one of the nobler animals, manifesting traits which would
at least command such admiration as we give to the intelligent dog or a horse."
When he insisted, "The Wellsian portrait is as false as it is abhorrent. Man's
ancestor was neither an ape nor a hyena!" he may have had in mind Wells's science fiction, especially *The Island of Dr. Moreau*, with its population of human
atavisms.[78]

Osborn shared the minister's point of view, of course. Although he often expressed admiration for the noble qualities of mammals like elephants both ancient and modern and a genuine reverence for nature, his sense of fellowship
with animal life seemed not to extend to the anthropoid apes. He rejected an ape
ancestry even as he declared that no one should be offended at the idea of our
connection with the natural world. He feared that public revulsion from such unpleasant ape-man and caveman images would harm the cause of evolution and
ultimately of science education.[79]

When the Reverend Charles Francis Potter declared an "Evolution Day" at the West Side Unitarian Church in New York in 1924, partly in response to attacks on the American Museum, and announced that the centerpiece of the event would be Carl Akeley's sculpture *Chrysalis*, depicting a man emerging from the skin of a gorilla, Osborn firmly declared that he was "not willing to have Mr. Akeley's 'Chrysalis' connected with the name of the American Museum of Natural History." Through William King Gregory, Osborn informed newspapers, "The museum is a public and municipal institution of an educational character only, which neither by charter nor by municipal sanction has the authority to express opinions on works of art in any way connected with a matter in philosophical dispute." Potter objected, noting that the church suffered from no such qualms about the controversy. Akeley was not only "a great sculptor" but also a "man of science," and his sculpture, Potter maintained, held a "real spiritual message for men of today. . . . The point of the statue is not the gorilla but the man, who has risen above his animal ancestry." John Roach Straton countered with an accusation that *Chrysalis* represented "the glorification of bestiality . . . part of the propaganda of a bestial philosophy."[80] This was not the first time that Osborn demurred when faced with controversy; he often insisted that he was determined to hold the museum aloof from politics. But he avoided controversy only very selectively.

Like many biologists, Osborn attempted to distance humans from apes, ironically in service to the idea that being related to animals should not lead to embarrassment. For conservationists, the human connection to the animal world was cause for humility at least and for celebration at best. And although he seemed to understand viscerally the revulsion of many people for a direct ape ancestry, Osborn, who was also an active conservationist, found the notion of human proximity to nature ennobling. Working with Knight on the plans for murals for the Hall of the Age of Man, Osborn wrote to J. H. McGregor, "I want to make the men in this Hall worthy of the animals." This could have been a statement about nothing more than the quality of the art, of course, but maybe not: Osborn was given to expressions of admiration and even reverence for the large Pleistocene mammals that shared the hall with the hominids.[81]

Evolutionists sometimes protested that popular versions of cavemen exceeded all known animals in beastlike vices. In 1927 the archaeologist Earl Douglass wrote to Osborn objecting to the images of human ancestors in magazines like the *Saturday Evening Post:* "[Early man] is represented as being fiercer, more cruel and more idiotic than any beast that has ever been known." Portraying cavemen in this way made no sense. First of all, "I do not see any reason to believe that man has not been an intelligent animal, and perhaps as peaceable as his struggle for existence would allow, perhaps way back to Oligocene times. How could he have existed if he was not intelligent?" How could natural selection have favored protohumans if they were idiots? Were ancient humans, Douglass wondered facetiously, the only animals entirely lacking in sensitivity? "Did all mammals, as well as birds, tenderly care for their young and were attached to their mates except degraded, brutal man?" Echoing an idea that had become increasingly common since the world war, Douglass added, "Personally I do not believe that man was ever more brutal than he is today but now he is a little more polished and refined in his cruelties."[82]

Missing Links and Irreverent Humor

Cartoon cavemen and cartoon monkeys served to express very similar sentiments about the covert brutality of humankind. Monkeys made a ubiquitous ironic counterpoint to the foolishness of contemporary society. As the historian Jeffrey Moran has persuasively argued, "reverse-monkey" cartoons during the 1920s commented sharply on human failings by depicting monkeys observing cruelties and injustices and disavowing any relationship with humans.[83] During the Scopes trial political commentary by cartoon monkeys proliferated wildly. Even Osborn appeared in a reverse-monkey cartoon, in which a grateful chimpanzee thanked him for denying the familial relationship.[84] Osborn may have liked this cartoon; it is one of the few he preserved in his files. He was not, however, a fan of the comic pages in the newspapers.

In a 1926 address at a Philadelphia high school, reprinted in a book about education, he commented, "It seems a harsh thing to say of the American press, but if I had the power of a Mussolini I would shut it off from our school youth entirely; I would exclude absolutely the irreverent 'funny page.'"[85] He often attributed the confusion wrought by some of his pronouncements about religion and science to misrepresentation by sensation-seeking headlines, and the salacious use of his work in *The Blonde Captive* belonged to a sinister commercialism he associated with the yellow press.

Osborn was right to see the funny pages as irreverent. Cartoons had become regular features of newspapers by the 1920s, and they played important roles in the evolution debates. Cartoons could capture delicious and telling paradoxes— they could be funny and sometimes astute because they made unexpected pairings, using iconic images or metaphors in telling ways. They had the potential to work subversively, and they could be subversive in an unpredictable variety of ways. They could also, however, function to obscure important truths as well as to reveal them. The joke of a cartoon could seem so natural that the reader might simply accept it as true and insightful.[86]

Yet the bemused monkey theme could also work to criticize human priorities and values by inverting the normal hierarchy and allowing monkeys to congratulate themselves on reports that they were not related to humans. Many cartoons in the 1920s made fun of the evolutionary hierarchy. One of the most common cartoon motifs of the decade was to imply the reversal of the evolutionary trajectory from "monkey to man" by variations on the theme of monkeys reading newspapers and hoping that misbehaving humans are not relatives of theirs. The now familiar cartoon joke on the evolutionary sequence "from fish to man" appeared in Scopes trial commentary, as in the *New Yorker* cartoon of the sequence from chimpanzee to Neanderthal to Socrates to William Jennings Bryan, or the *Judge* sequence culminating in the 1920s figure of the "cake-eater."

Defenders of evolution often used caveman references to make fun of antievolutionists, as in the *New York World*'s use of McGregor's Neanderthal juxtaposed with a photograph of Bryan to illustrate Luther Burbank's quip that the shape of Bryan's head resembled "the Neanderthal type." Evolutionists used words like *caveman, simian,* and *Neanderthal* and related terms to sarcastic effect. H. L. Mencken delighted in referring to the people of Dayton, Tennessee, as "simians" and to Bryan and his followers as "bawling primates." A pamphlet by W. H. Rucker remarked, "Even to this day certain types of men, for instance, abhor progress, a reversion perhaps to their stone-age or anthropoid ancestors," and it declared that "a snarling anthropoid or bawling Pithecanthropus in a pulpit venting wrath against a revolving earth, or evolution, resting on the greater proof, is no credit to a church."[87] A cartoon published in *Scientific Monthly* during the summer of the trial played on the cave theme by showing a teacher leading Tennessee schoolchildren into a school housed in a cave.[88]

Scientists—even those as influential as Osborn was—could not control the use of evolutionary images. Osborn encountered a vexing instance of loss of control over the interpretation of a picture that he believed belonged—conceptually as well as legally—to the American Museum when the editors of *Scientific American* used a Charles R. Knight painting of Neolithic hunters, commissioned by the museum and on display in the Hall of the Age of Man, to illustrate an article called provocatively "Which Races Are Best?" Osborn objected strenuously, in particular because he disliked the way the article answered the question. According to *Scientific American,* there was no "best" race. This was not a view that Osborn wanted the museum to appear to endorse, and he demanded and received a public apology in the pages of the magazine.[89] The incident demonstrated clearly, however, that pictures created at the museum became—conceptually, at least—public property, no matter who held the copyrights. Osborn could deny permission to reproduce them, but he could not confine their symbolic relevance.

The most astute anti-evolutionists understood the power of images and of irony and knew how to use them to turn scientists' own pronouncements against them. In a pamphlet titled *Human Evolution and Science,* Father Francis Le Buffe referred to anatomical studies of Neanderthals: "Another argument advanced at times is that of the supra-orbital ridges. 'Prominence of the ridges over the eyes is an indication of nearness to the apes.' Is that so? Then our friends, the evolutionists, including Messrs. Osborn and Conklin, are nearer the apes than the Negroes of Africa."[90] Osborn must have considered this irreverent, indeed.

The paintings by Knight under Osborn's attentive guidance did make a strong impression, but they may have represented a Stone Age for the educated. In the everyday world of the newspapers, cavemen with clubs remained ubiquitous, fixed in a common cultural vocabulary. Even among Knight's illustrations designed in close collaboration with Osborn, the most often reproduced was the stoop-shouldered Neanderthal with a club.

As the editor of the *New Republic* pointed out during the Scopes trial, newspapers' attempts to explain evolution drew heavily on American Museum exhibits for heuristic illustrations. Books about evolution addressed to the nonscientist reader borrowed illustrations from the American Museum much more often than from any other source, and they often included well-informed discussions—many such books were written by scientists or science educators. But the symbolic caveman remained an iconic figure throughout the decade, and even books that used the museum as a source of illustrations often included funny cavemen as well. For example, the familiar hairy caveman with a one-shouldered fur and with club held aloft appeared in an illustration for a 1930 children's book about evolution, *The Earth for Sam,* striding across the top of a globe below the caption "The New Boss!"[91]

Osborn was mistaken in thinking that he could restrict the symbolic meaning of museum images like Knight's Neolithic *Stag Hunters.* Nor could he and his cohorts entirely shake off the pictures of monkeys and apes in their own heads or in the heads of the public. But they could add new images to the store of available ideas, as Osborn did in *Men of the Old Stone Age*—ultimately much more influential than *The Earth Speaks to Bryan*—and as he did in creating his museum. Reactions to the didactic *The Earth Speaks to Bryan* were much less positive than those to Osborn's more detailed, arcane, and technical *Men of the Old Stone Age,* perhaps because the latter conveyed more vividly the really appealing part of science—the science itself.

In 1924 Charles F. Dutton, a Unitarian clergyman from Pennsylvania, wired Osborn: "Wild antievolution propaganda here[.] Rabid evangelist says British American science thrown over evolution[.] . . . Will you wire short statement that science and leaders more convinced of evolution than ever."[92] Osborn responded, "The scientific staff of the American Museum [of] Natural History live in the midst of overwhelming evidence of evolution."[93] This was a straightforward and even eloquently simple statement; but Dutton requested something more. When Dutton wrote at greater length, asking that he elaborate, Osborn responded: "Scientific men the world over as well as all learned Christians and theologians accept the gradual evolution and ascent of life and of man as the natural and divine order of creation." Opposition to evolution, he continued, came only from "willfully or unconsciously ignorant people who are misled by the incomplete or garbled statements of certain scientists."[94] This response, from the author of *The Earth Speaks to Bryan*, would surely have confounded many critics. And it was very much at odds with the elegance of the earlier, simple statement: the people at the museum did live in the midst of overwhelming evidence. But the evidence did not speak in the same way to everyone.

In 1925 the Reverend John Roach Straton published a collection of sermons, *The Old Gospel at the Heart of the Metropolis*, that included a description of "the Cave Man" as the beginning of "the long procession of the ages passing in gorgeous and inspiring pageantry." Inspiring pageantry! Why would this most prominent anti-evolutionist minister, famous antagonist of the American Museum, describe cavemen in this fashion? The passage came from a sermon that began: "Ours is an age of discovery." By "discovery" Straton apparently meant science, and among his list of some of the "wonderful triumphs of science" he included a statement that must have startled some people in his audience: "Men have searched out the hidden secrets of nature as never before. They have . . . read the history of our planet in fossil bone and volcanic rock."[95] The title of the sermon was "Imagination and Life," and in it Straton extolled the imagination as "one of the most constructive and vital of all our powers." What, exactly, did he mean by "Imagination"? It was, he said, a "magical power," and it was visual: "The imagination is the picture-making faculty of man's being."[96] By way of example of the power of this faculty, he noted that "the imagination . . . makes history real to us. . . . Man finds in a cave a few jars and bits of stone and a rude charcoal sketch upon the wall, and this powerful faculty seizes upon these things and bodies forth the conditions, activities and habits of the Cave Man."[97]

The sermon's inclusion in a book published in 1925, the year of the Scopes evolution trial—is striking. Straton's reference to the "Cave Man" at the beginning of "gorgeous and inspiring pageantry" must have piqued acute interest on the part of his auditors, who had good reason to know of his strenuous campaign against evolution.

One of the lessons of the Scopes trial is that in a heated public debate about a scientific issue, symbols, images, metaphors, and iconographic traditions matter a lot. This was so in the case of the Scopes trial because the debate was not simply about the content of science. It was also about the authority of scientists. And it was about what the content of science—cast in terms of a crowded and metaphorically resonant hybrid vocabulary—might imply.

Straton understood this, of course, as did other leading anti-evolutionists. Anti-evolutionists like Straton and William Jennings Bryan well knew the power of visual images, and they invoked and used them. For many anti-evolutionists the vivid contrast between the iconography of Adam and Eve and *any* picture of Neanderthals constituted the most unequivocal possible refutation of evolutionary theory.

But why did Straton include in his "Imagination and Life" sermon this puzzling image of the caveman?

Perhaps because he intended a warning: the imagination could make people empathetic, and therefore better, Christians, and it could give us the "picturable" God that the liberal minister Harry Emerson Fosdick had claimed we yearned for. But the imagination could also—like "science, falsely so-called"—lead us astray, and in subtle and powerful ways. Straton meant it when he called the imagination powerful, even magic—and it was, therefore, potentially dangerous. How to discriminate among images—how to distinguish those that would lead us to God from those that might be snares of the devil? Straton's answer: The heart! The heart, not the intellect, would protect us from the seductions of the imagination, and of science, falsely so-called. It was no accident that anti-evolutionists like Straton targeted visual images of evolutionary ideas.

The imagination was visual and therefore seductive. Straton understood perfectly what Knight's Cro-Magnon cave artist was intended to prove. And he appreciated its seductiveness. Osborn and Knight hoped to vanquish the mythology of the brutish caveman with a visual image of the caveman as noble artist. Straton knew this. While, like many anti-evolutionists, he cited the grotesqueries of the popular caveman, Straton seems also to have perceived *any* image of the caveman as a snare. The cavemen he objected to were not only the cavemen who were obviously brutish and ugly. It is true that many anti-evolutionists found the brutish, nasty, and short caveman offensive. Straton did, too. But he also recognized the seductiveness of the visually compelling caveman. His "Imagination and Life" sermon suggests that *no* image of a human ancestor would mollify him in his opposition to evolution. Even more: like William Jennings Bryan, who insisted that theistic evolution presented a much greater threat to Christianity than did atheistic science because it worked as an anesthetic, Straton comprehended the subtle persuasiveness of appealing images of cavemen. And he understood the power of arguments that were visual. Visual reconstructions of human ancestors were, for scientists, both tools of scientific reasoning—among the "working hypotheses" of paleontology—and heuristic persuasive devices. Science was creative, and scientific method, at least in some of the sciences, included visual tools. For fundamentalists like Straton, recognition of the imagination as a powerful visual faculty demonstrated its perils. The ability to picture the human past was very much at issue. Images of the human past occupied the heart of the debate not only when they caused revulsion but perhaps even more when they were seductive.

Images of cavemen unsettled many people, including some evolutionists; they also supplied persuasive rhetorical devices for anti-evolutionists. But they held a multitude of other cultural meanings. Cavemen in the comics carried clubs because clubs held symbolic meanings that made them funnier than spears or atlatls. Sometimes cavemen in the comics were clichés not because the artists who drew the comics were ignorant or scientifically illiterate but because the accurate portrayal of a scientific idea was not the point.

Just as the Scopes trial was about more than evolution, cavemen were about more than human prehistory. Editors and advertisers undoubtedly sometimes chose Knight's stoop-shouldered, club-carrying caveman not because they believed it was more accurate than his flint workers, or even because it came to hand more readily, but because it better fit the familiar iconic use of the caveman cliché. The conversation they illustrated was about more than simply the content of science; visions of the human past both shaped and reflected the concerns of the present.

Conclusion

In his 1940 autobiography, *Dusk of Dawn*, W. E. B. Du Bois wrote, "I remember once in a museum, coming face to face with a demonstration: a series of skeletons arranged from a little monkey to a tall well-developed white man, with a Negro barely outranking a chimpanzee."[1] Du Bois's reminiscence illustrates the long resonance of images of evolutionary ideas and suggests a reason for his own critical awareness of the distinction between the authority of science in general and the authority of individual scientists.

One of the arguments scientists frequently made in defending evolution was that their years of experience and training conferred on them an authority that gave their judgments special weight. It was ludicrous, from their point of view, for people like William Jennings Bryan to challenge scientific interpretations of the evidence for evolution. When had Bryan ever studied in Germany, worked in a laboratory, or published a scientific discovery? They had a point. Interpretation of the scientific evidence for evolution depended more and more on detailed knowledge of embryology, paleontology, geology, and comparative anatomy. Scientists' years of experience in the lab and the field counted for a great deal. And many of Bryan's challenges were—or at least sounded—disingenuous. But not all of them were. When Osborn began using *Hesperopithecus,* the "Nebraska ape," to ridicule Bryan, his scientific judgment failed him. Bryan understood this, well before the evidence against *Hesperopithecus* as an advanced primate was in. Just look at the flimsiness of the evidence scientists rely on, he told appreciative Dayton audiences, citing Osborn and the Nebraska tooth. In recruiting fossil finds— or anticipations of fossil discoveries—for rhetorical use against fundamentalists, Osborn let his instinct for publicity get the better of him. But this kind of lapse in judgment was not the only difficulty in the way of communication between scientists and a multiform public.

The scientific issues were complex and the scientific community fragmented. Disagreements that scientists saw as internal to science could not be kept under wraps; yet explaining them to a diverse and poorly understood public, at a time when adversaries like Bryan monitored scientific dialogues for inconsistencies, presented a serious challenge. Some of the evidence convincing biologists that evolution was beyond dispute—a law of nature like gravity—did not translate well, as Edwin Grant Conklin learned in trying to revise his contribution to *Creation by Evolution.*

When facing challenges from anti-evolutionists, scientists often responded that their years of training and experience made the evidence they looked at transparent to them in a way that it could never be to the public. This response may in part have been less a conscious rhetorical strategy than a kind of throwing up their hands in despair at the difficulty of communicating with a segment of the public that seemed to them determined not to understand. Yet science popularizers often did an extremely creditable job of explaining the evidence in books for the public, as did some scientists—including Osborn in *Men of the Old Stone Age* and *The Age of Mammals.* This was not enough, however. Scientists and popularizers faced a difficulty beyond that of explaining scientific details to nonscientists. The greatest barrier to communication was much more formidable: it was that the reasons for many people's skepticism about evolution had little to do with the substance of the science after all.

Evolution carried overwhelming symbolic significance. In many cases it represented the irritations and dislocations of modern life and the uncertainties of new roles for groups of people whose amplified assertiveness worried traditional authority. It also stood for science in general, in a way that recent revelations in physical science and mathematics could not, quite. For some religious traditionalists, the implications of science were those so persuasively limned by Walter Lippmann's fictional fundamentalist in *American Inquisitors:* eternal salvation traded for perpetual uncertainty.

Scientists who joined the debate aligned themselves with theological modernists, asserting the compatibility of evolution with Christian faith. This strategy carried no weight with theologically conservative Christians; indeed, it was counterproductive, for the Christianity of the science reconcilers sounded distinctly modernist—even pantheistic—to conservatives. The notion of progress as the organizing principle in evolution, so helpful to nineteenth-century scientists in accommodating evolution to their religious views, did not comfort conservative Christians. The emphasis on progress seemed to them to deny the centrality of the Fall of Adam and Eve in Christian theology, and it offered no solution at all to the problem of evil in a world designed by a loving God. They could not, therefore, dismiss natural selection as a peripheral or technical matter. Natural selection, from their perspective, could never be reconciled with Christian faith.

In claiming that evolution was not a threat to Christian values, scientists failed to change conservative Christian minds. This was true in part because the scientists involved in the debate referred only to their own more or less modernist version of Christianity. It was also a result of professional seclusion. When scientists wrote to their professional organization, suggesting that the American Association for the Advancement of Science take an official stand because the public would find the imprimatur of the organization reassuring, they revealed a degree of parochialism. Similarly, Osborn suggested to Charles Davenport that a publication describing a eugenics conference should include photographs of the participants because "the personality of the men involved" would persuade the public of the legitimacy of their doctrines. These men believed that their status as scientific men and the mastery they had attained through their years of experience should be weighed in as evidence, and they assumed that most of the public would agree.

They had reasons beyond arrogance to assume such a thing. Osborn, for example, received numerous letters from the public asking for his advice about matters of import. In a typical request, John Reid of Spokane requested references on evolution, specifying that he sought "any trustworthy articles by accredited scientists of unquestioned reputation and standing . . . not too technical . . . in a form which can be easily grasped by 'the average man on the street.'" Reid added that he was interested only in "verifiable and verified *facts*, not speculations," and that he hoped one day to visit the "nationally celebrated Museum," especially the "much talked of 'Hall of the Age of Man.'"[2] One man in Coeur D'Alene, Idaho, Charles Hooper, wrote to ask Osborn to use his influence to do something about daylight saving time—an affront to God and to nature. In an enclosed letter to the newspapers Hooper complained, "Ordinary time is not good enough for us modern, advanced, scientific mortals. We must invent a sun of our own to compete with the sun that God Almighty placed in the heavens." Yet it was to Osborn's influence as a scientist that Hooper made his appeal.[3] Will Durant asked Osborn to comment publicly on "the meaning of life," and others requested his opinion as to the possibility of life after death.[4] Scientists were not mistaken in thinking that much of the public turned to them for answers to questions beyond the narrow confines of their disciplinary specialties.

Once they joined the public fray, some scientists tended to extrapolate beyond their own areas of expertise to make pronouncements on matters that many intellectuals defined as extrascientific. They did so partly for obvious reasons of human frailty, a predictable result of hubris on the part of men in their position. But it was also true that, as Kirtley Mather perceived, it was not so easy to define science strictly. Although Osborn's theistic evolution found many appreciative followers among the public, for nonscientist intellectuals science meant, by definition, knowledge untainted by religion. The editor of the *New Republic* assumed that Osborn's pronouncements on the spiritual implications of evolution must have represented prevarication or dishonest rhetorical flourishes; of course the president of the American Museum of Natural History must have known better than to infer religious meaning from the data of science. But Osborn was quite sincere.

In arguing for the spiritual implications of evolution, scientists like Osborn put themselves at odds with some colleagues in the scientific community. By the 1920s, the dictates of scientific objectivity seemed, to many scientists and nonscientists alike, to preclude metaphysics. By the 1930s, Osborn found his footing more and more precarious in such matters. He would answer requests to give sermons with principled notes insisting that he could not, as a scientist, pontificate outside his area of expertise. But pontificate he did. He understood that evolution held enormous symbolic import; but he did not quite understand his audiences, especially when he imagined himself "stooping to conquer." And in trying to assuage the fears of an audience he imagined, he claimed an extended territory for scientific authority and for his own authority as an expert and a man of standing.

Perhaps because they understood so well how much of their knowledge was cumulative, and the degree to which they had learned to trust their scientific judgment based on their years of training and experience, few scientists seemed to recognize that, from the perspective of the public, it was possible to challenge the authority of scientists without denying the authority of science itself. Neither Du Bois nor Bryan, for example, rejected the authority of science; they did challenge the authority of scientists, in very different ways. Bryan expressed confidence that the average, ordinary person could understand the Bible; and he extended this precept to the realm of science, joining the American Association for the Advancement of Science as a gesture intended to announce a form of scientific populism. Du Bois, trained as a social scientist, fought the exploitation of the prestige of science to promote racial doctrines he believed to be unscientific as well as unsavory. In 1910, in the very first issue of the *Crisis,* he featured an article by Franz Boas, dissenting from the doctrines of biological determinism.

Bryan and his followers were not the only members of a variegated public to worry about the implications of the increasing authority of experts—especially experts in arcane disciplines—in a democracy. Many of the most passionate arguments at the Scopes trial centered, for good reasons, on the question of expertise, both the relevance of expert testimony and the participation of "outside" experts. These were very difficult questions, questions that the Scopes trial would not resolve or even illuminate very much.

If the Scopes trial did demonstrate anything, it was that publicity had become an intractable sort of animal, unruly and hard to confine. Osborn seems to have understood that this would be the case at Dayton. Though he had long used publicity adroitly in building his museum, he had good reason, by 1925, to realize that he would not be able to control the media circus in Dayton. No one was. And publicity had become a real force. In this increasingly media-dominated era celebrity conferred, over many and heartfelt objections, a new kind of authority—but only along paths that were very difficult to negotiate. Scientists, in particular, had cause to be wary of it.

Uneasiness about media sensationalism and about the rising professions of public relations and publicity agents was widespread in the 1920s, and probably especially so among people of Osborn's privileged background. Publicity posed a particular dilemma for scientists, caught between a sense of duty to communicate with the public—along with an appreciation of the fact that much of their funding derived ultimately from public sources—and the constraints of scientific caution. Scientists, or "scientific men" as they more often called themselves, while bitterly disagreeing over essentials, had by the 1920s formed something like a separate social class, but their concerns about objectivity were not just psychological or careerist.

They believed in science—it was, as William Diller Matthew so nicely put it, a house that they had built. They lived in it, they worked in it, and they took pride in it. But defending "science" was not really like defending a building. It was more like defending an impalpable like a "home" than a bricks-and-mortar thing like a house. Science carried cultural weight that vastly complicated defending it from assault in an era of proliferating popular media—and in the wake of the Great War and the beginning of the age of Einstein. When scientists fretted about preserving their dignity during the Scopes trial debate, they were not just expressing a general concern, although they were doing that too. The idea of professional dignity for scientists in this difficult decade was wrapped up in notions of scientific objectivity and disinterest. Scientists' fears that objectivity would be hard to maintain in the midst of a contentious public debate were well founded.

The culture of professional science and the culture of the media combined to ensure that the public would not be offered anything like a cross section of scientific opinion. After the Great War, scientists mounted an organized effort to improve communication with the public, complaining that, without their guidance, newspapers only published, as the head of the new Science Service put it, "snippets of sensational science."[5] The formation of the Science Service represented an attempt to promote credible science and a worthy public image of science. It also implied an acknowledgment that not all scientists could expect to gain access to the media. As one persistent journalist protested when Osborn suggested another, less busy scientist for an interview, newspaper chains did not want to interview scientists who were not famous.

Nor did all scientists want to be interviewed. Many of them eschewed the limelight, believing that unseemly public controversies threatened to compromise both their dignity and their objectivity. Museum scientists, engaged in communication with the public and dependent upon the public for the expansion of their institutions, could seldom afford to be so fastidious about publicity.

Osborn's museum took a far more active role in public education than did any other museum. The growth of the American Museum of Natural History during Osborn's tenure was extraordinary—in part, to be sure, because he was able to make use of his position as a member of New York's power elite. But his connections alone could not account for the museum's growth; he also possessed a combination of skills and abilities—including verbal and artistic gifts and the perceptiveness to discern and encourage the talents of students and museum workers—that allowed him to build a museum that piqued the public imagination.

The mandate of the American Museum had always been explicitly educational, and in his strong commitment to that mission, Osborn had expertly promoted the museum's visibility as a public institution. He wrote well, and through years of practice as a successful museum president he had cultivated the ability to craft colorful and pithy statements for use in public addresses and interviews. This skill proved to be a mixed blessing when his statements came under intense scrutiny during the debates about evolution. Adversaries parodied such witticisms as "the earth speaks to Bryan." As the public debate heated up in the 1920s, Osborn's attractiveness as a newspaper representative of science could work against him as well as for him. His denunciations of the "ape-man" theory, stripped of their context and framed as headlines, meant that his ideas about human evolution caused more confusion than enlightenment. Never a fan of media sensationalism, he became increasingly bitter about the exuberant journalism so much in evidence during the evolution debates, warning that newspapers had so much power that, in practice, democracy could degenerate into tyranny by headlines. He told a National Republican Club audience in 1926 that the writer of newspaper headlines was "the only man in the community that I am afraid of."[6]

Osborn's words made frequent headlines during the evolution debates of the

1920s, and he often charged circulation-seeking editors with misrepresentation of his statements. Visual depictions of his ideas in museum exhibits proved to be just as vulnerable. In 1920, Osborn expressed great enthusiasm about the coming together of the new Hall of the Age of Man. As the decade progressed, he found himself having to defend the hall against, as he perceived it, "constant attacks." He and his colleagues suspected the critics of sophistry, a skepticism that sometimes had merit. In asserting that the exhibits displayed simple statements of fact, uncontaminated by theory, however, Osborn adopted an indefensible position. Pictures, as Walter Lippmann could have told him, always implied theories.

As the biologist Theodore Cockerell had noted in an essay in *Science*, the difficulties of explaining increasingly arcane scientific ideas to the uninitiated, especially under the constraints of publishing, meant that scientists sometimes felt as if they were attempting to communicate "by a system of signs."[7] And so they were. When Lippmann wrote that people interpret new ideas in the light of the "pictures in their heads," he was not referring exclusively to visual images, of course. But it was no accident that he put the idea in this form. The decade of the Scopes trial was suffused with awareness of the new prominence of visual images in an age Vachel Lindsay had so memorably described as a "hieroglyphic" civilization.[8] Anti-evolutionists targeted Osborn and the American Museum because they understood the power of visual images and because they understood how protean such images could be.

Osborn also understood the power of visual display, but ultimately he believed that the exhibits told the truth simply and transparently, educating the public not only about the zoology of the past and the fossil record but also about larger truths. He created confusion when, faced with conflicting interpretations of the visual images in the Hall of the Age of Man, he attempted implicitly to invoke his authority as a scientific man in order to reassure the public that the exhibits reaffirmed sturdy traditional values. In the end, the exhibits spoke eloquently, but they did not speak only in Osborn's voice. They added to, but did not efface, the array of other pictures in people's heads.

Osborn did believe in the transparent truth of his exhibits; but he also understood the contingency of scientific hypothesis. In defending his exhibits and his ideas, however, he ultimately took untenable positions. Although he assumed an unperturbed demeanor, he had a great deal at stake in the evolution debates. For one thing, he identified with the museum, calling it a "great temple of nature." He had always been artistically inclined, and he took an active part in exhibit design. After he died, those who were most fond of him recalled his love of nature and his love of art as two aspects of a single instinct, fused in the creation of the museum. Although he produced many influential books and articles, the museum was ultimately his medium and his legacy.

Through the medium of the museum, he attempted to defend the truth as he knew it. Evolution, he insisted, should win the day at the Scopes trial because it was true. The truth of evolution itself was not ultimately the real issue for Bryan, for Darrow, for the ACLU, or for the magazines and newspapers commenting on the debate, however. Even for many scientists, the debate was also about more than evolution itself—it was a debate over reason, scientific method, freedom of thought, and academic freedom. It also addressed the claims of science to authority.

Despite his disclaimers to the contrary, the debate meant all these things to Osborn, and more. When he declared that the only real issue in the Scopes case was the truth of evolution, he was not simply being naive. He was expressing a fervent belief in the transparency of scientific truth, of order, meaning and reason, a faith in his ability to convey the great truths implicit in nature through the scientific display of well-chosen objects in his museum. In putting these truths on display he intended to reaffirm them. He tried to use the museum to uphold larger truths. In teaching and illustrating evolution he meant to preserve a reliable world.

He could not, of course. Scientific truths were not transparent; the very nature of scientific truth was at issue, as were definitions of science. Osborn and his scientific colleagues disagreed about many things, but perhaps their most fundamental disagreement was about what it meant to be scientific. The limits of scientific authority eluded consensus, but in the course of the debate during the Scopes decade, Osborn's interior map of those boundaries began to seem peculiarly dated.

And the debate took place in a context in which the "pictures in people's heads" carried complex and mutable cultural associations. For Osborn, pictures and objects had always served as a medium for education. He had designed visual arguments both in technical publications and in the museum that were dramatic, effective, and often brilliant. He took great pride in the displays at the museum and expressed unswerving confidence that their meanings were clear. By the 1920s, however, as William Jennings Bryan, Alfred Watterson McCann, and John Roach Straton understood, many people had learned to be skeptical of visual arguments. Pictures, they knew, could lie. As symbols, they could also magnify, could act as lenses to reveal multiple, sometimes contradictory meanings.

Du Bois's memory of the visual image of the evolutionary series highlights an irony shading the public face of evolution in the 1920s, and a reason for the central role of images in the debate. Anti-evolutionists like Francis Le Buffe, McCann, and Straton accused the American Museum of exhibiting groups of hominids as sequences based on inadequate evidence. Yet most of the scientists who studied human evolution in this period, while subscribing to a progressive view of evolution in general and of human evolution in particular, nevertheless also perceived primate and human evolution to have occurred in a "bushy," more or less Darwinian pattern. This was especially true of the recent hominids including Neanderthal and Cro-Magnon. Museums, textbooks, and books for the lay public exhibited more linear views of evolution partly as a heuristic device, not so much to explain the pattern of evolution but to make a case for its direction. Showing the net effect at the expense of the complex path could, in theory, make a case more convincing to the uninitiated that evolution had actually occurred. The linear depiction of evolution also served a less obvious purpose, however. Scientists like Osborn absolutely believed that evolution occurred in a meaningful direction and that the direction of evolution implied progress. And they never doubted that the lay public would find this message comforting and therefore convincing. Progress, however, is a very subjective thing. Osborn's version of progress, entangled as it was with racial theories, was far from universally convincing. Racial theories were also implicit, and sometimes explicit, in the proliferation of missing-link images, especially during the evolution debates of the 1920s, a decade sometimes obsessed with race.

Visual images of apes, ape-men, cavemen—missing links of every kind—carried complex associations; they referred to long visual traditions that could not be banished from memory. Scientists could add new images to the store of imaginative possibilities, but they could not control their context or their interpretation. How one pictured the human past was at the heart of the problem for people concerned about evolution. As Harry Emerson Fosdick so astutely observed, if modern science had made it much more difficult to imagine a "picturable" God, the implications were troubling for many people. When Edwin Grant Conklin scoffed at the notion of God as "a big man in the sky," he dismissed too carelessly the conviction of many anti-evolutionists that humans were made, as Bryan so often said, in God's image. To substitute a Neanderthal—or even a Cro-Magnon—for Adam and Eve was no trivial matter. Evolution, even for scientists, was not simply an internal matter of arcane scientific theory. The debate was not simply about whether evolution itself was true, Osborn's insistent use of that word notwithstanding.

Although Osborn sometimes referred to the museum as a "temple of nature," he could not, ultimately, use it effectively as a pulpit. He could not ensure that visitors would come away believing that evolution sanctioned struggle and hard work, or that eugenics measures would save the nation, or that evolution reflected "sublime" Christian values. People who agreed with him on those issues might see them reflected and verified in the exhibits at the museum. People who disagreed would see the same exhibits in a different light. Visitors to the museum might not learn to think in terms of "the gradual evolution and ascent of life and of man as the natural and divine order of creation," but they would undoubtedly remember the graceful caveman painting a mammoth. The museum hosted a multitude of publics, and although it could never control the pictures in their heads, it added new kinds of pictures to the available store, making the past imaginable in new ways. That was no small thing.

Abbreviations

Conklin Correspondence	Edwin Grant Conklin Correspondence, Edwin Grant Conklin Papers, Rare Books and Manuscripts, Princeton University, Princeton, New Jersey
Conklin Papers	Edwin Grant Conklin Papers, Rare Books and Manuscripts, Princeton University, Princeton, New Jersey
Knight Papers	Charles R. Knight Papers, New York Public Library
Osborn Family Papers, NYHS	Osborn Family Papers, New-York Historical Society
Osborn Papers, AMNH	Henry Fairfield Osborn Papers, Library of the American Museum of Natural History, New York

Preface

1. Peter Bowler attributes the phrase "life's splendid drama" to the paleontologist William Diller Matthew. See Peter J. Bowler, "The Moral Significance of 'Life's Splendid Drama' from Natural Theology to Adaptive Scenarios," in Klaas van Berkel and Arjo Vanderjagt, eds., *The Book of Nature in Early Modern and Modern History* (Louvain, Belgium: Pieters, 2006). The phrase "dramatists of evolution" comes from Alfred Watterson McCann, *God—or Gorilla: How the Monkey Theory of Evolution Exposes Its Own Methods, Refutes Its Own Principles, Denies Its Own Inferences, Disproves Its Own Case* (New York: Devin-Adair, 1922), 83.

2. John Roach Straton, "Worldliness and the Vainglory of Life," in Straton, *The Old Gospel at the Heart of the Metropolis* (New York: George H. Doran Co., 1925), 174.

3. John Roach Straton, "The Modern Need of a Great God," in Straton, *Old Gospel at the Heart of the Metropolis*, 25.

4. Straton, "Worldliness and the Vainglory of Life," 174–175.

5. Harry Emerson Fosdick, "How Shall We Think of God?" *Harper's*, July 1926, 229.

6. Ibid., 230.

Chapter One • The Caveman and the Strenuous Life

1. Michael Kazin, *A Godly Hero: The Life of William Jennings Bryan* (New York: Knopf, 2006); Lawrence W. Levine, *Defender of the Faith: William Jennings Bryan, the Last Decade, 1915–1925* (Cambridge: Harvard University Press, 1987); Ronald L. Numbers, *Darwinism Comes to America* (Cambridge: Harvard University Press, 1998); Edward J. Larson, *Summer for the Gods: The Scopes Trial and America's Continuing Debate over Science and Religion* (New York: Basic Books, 1997); Michael Lienesch, *In the Beginning: Fundamentalism, the Scopes Trial, and the Making of the Antievolution Movement* (Chapel Hill: University of North Carolina Press, 2007).

2. *Chicago Defender*, June 20, 1925, sec. 2, p. 12, reprinted in Jeffrey P. Moran, *The Scopes Trial: A Brief History with Documents* (Boston: Bedford/St. Martin's, 2002), 173; for a more extended discussion of this theme, see the excellent essay by Jeffrey Moran, "Reading Race into the Scopes Trial: African American Elites, Science, and Fundamentalism," *Journal of American History* 90 (Dec. 2003): 891–911. There is a variation on this theme in an Art Young cartoon published in *The Masses* during World War I in which the apes' responses are used to express horror over the war: see Mark Pittenger, *American Socialists and Evolutionary Thought, 1870–1920* (Madison: University of Wisconsin Press, 1993), 245.

3. L. Perry Curtis Jr., *Apes and Angels: The Irishman in Victorian Caricature*, rev. ed. (Washington, DC: Smithsonian Institution Press, 1997), xxii–xxiv, 122. See also Henry Fairfield Osborn, "The Influence of Habit in the Evolution of Man and the Great Apes," *Bulletin of the New York Academy of Medicine* 4 (1928): 216–230. See also Osborn to Chester K. Field, Feb. 15, 1916, folder 31, box 7, Osborn Papers, AMNH; Roselyne de Ayala and Jean-Pierre Gueno, eds., *Illustrated Letters: Artists and Writers Correspond* (New York: Harry N. Abrams, 1999), 98. For the longer historical context of similar images, see Stephanie Moser, *Ancestral Images: The Iconography of Human Origins* (Ithaca, NY: Cornell University Press, 1998); Martin Kemp, *The Human Animal in Western Art and Science* (Chicago: University of Chicago Press, 2007); and for the strong racial and ethnic subtexts of such images, see Gustav Jahoda, *Images of Savages: Ancient Roots of Modern Prejudice in Western Culture* (London: Routledge, 1999); Ter Ellingson, *The Myth of the Noble Savage* (Berkeley: University of California Press, 2001); Alan E. Mann, "Imagining Prehistory: Pictorial Reconstructions of the Way We Were," *American Anthropologist* 105 (March 2003): 139–143; Douglas A. Lorimer, "Science and the Secularization of Victorian Images of Race," in Bernard Lightman, ed., *Victorian Science in Context* (Chicago: University of Chicago Press, 1997), 212–235; and Mark Pittenger, "Imagining Genocide in the Progressive Era," *American Studies* 28 (1987): 73–91.

4. *Chicago News*, July 22, 1925, clipping in folder 6, box 92, Osborn Papers, AMNH.

5. Janet Browne, "Charles Darwin as a Celebrity," *Science in Context* 16 (June 2003): 175–194.

6. Judith C. Berman, "Bad Hair Days in the Paleolithic: Modern (Re)Constructions of the Cave Man," *American Anthropologist* 191 (June 1999): 288–304; Roy Pilot and Alvin Rodin, *The Annotated Lost World: The Classic Adventure Novel by Sir Arthur Conan Doyle* (Indianapolis: Wessex Press, 1996).

7. Vachel Lindsay, *The Art of the Motion Picture* (1915; New York: Modern Library, 2000), 116–125.

8. Jack London, *Before Adam* (New York: Macmillan, 1906), 27.

9. Ibid., 39.

10. Ibid., 1, 13–14, 20.

11. Edith Wharton, *The Glimpses of the Moon* (1922; New York: Scribner, 1996), 152. See also Jennie Kassanoff, *Edith Wharton and the Politics of Race* (New York: Cambridge University Press, 2004). Thanks to Jennie Kassanoff and to my colleague Kent Ljungquist for helpful discussions of Wharton's interest in race and in evolution. See Kent Ljungquist, introduction to Edith Wharton, *Ethan Frome and Selected Stories* (New York: Barnes and Noble, 2004). See also Osborn Diary for 1923, June 20, 1923, box 25, Henry Fairfield Osborn Diaries, Osborn Family Papers, NYHS, for mention of a discussion between Osborn and Wharton on the subject of race.

12. Eugene O'Neill, *The Hairy Ape*, in O'Neill, *Three Plays* (New York: Modern Library, 1937), 186.

13. For both the uses of the development metaphor in rationalizing theories of biological determinism and an account of the theory of ontogeny and phylogeny in the history of biology, see Stephen Jay Gould, *Ontogeny and Phylogeny* (Cambridge: Harvard University Press, 1977). For important and subtle treatments of it, see George W. Stocking Jr., *Victorian Anthropology* (New York: Free Press, 1987); George W. Stocking Jr., *Race, Culture and Evolution: Essays in the History of Anthropology* (New York: Free Press, 1968); and Robert J. Richards, *The Meaning of Evolution: The Morphological Construction and Ideological Reconstruction of Darwin's Theory* (Chicago: University of Chicago Press, 1992). For an excellent discussion of the influence of Haeckel's visual images of this idea, see Nick Hopwood, "Pictures of Evolution and Charges of Fraud: Ernst Haeckel's Embryological Illustrations," *Isis* 97 (June 2006): 260–301. See also Gavin R. de Beer, *Embryos and Ancestors* (Oxford: Clarendon Press, 1958), and W. Garstang, "The Theory of Recapitulation: A Critical Restatement of the Biogenetic Law," *Journal of the Linnaean Society of London, Zoology* 35 (1922): 81–102. For the appearance of the notion of stages in social evolution in nineteenth-century American ethnology, see Lewis Henry Morgan, *Ancient Society, or Researches in the Lines of Human Progress from Savagery through Barbarism to Civilization* (Chicago: Charles H. Kerr and Co., 1877). For intriguing examples of the cultural ramifications of the analogy, see Philip J. Deloria, *Playing Indian* (New Haven: Yale University Press, 1998); for an example of the visual display of this idea in American culture, see Robert W. Rydell, *All the World's a Fair: Visions of Empire at American International Expositions, 1876–1916* (Chicago: University of Chicago Press, 1987). For more on the social context of these ideas, see Pittenger, *American Socialists and Evolutionary Thought*; Hamilton Cravens and John Burnham, "Psychology and Evolutionary Naturalism in American Thought, 1890–1940," *American Quarterly* 23 (1971): 635–657; and T. J. Jackson Lears, *No Place of Grace: Antimodernism and the Transformation of American Culture, 1880–1920* (New York: Pantheon, 1981). Gail Bederman points out that though historians refer to this notion as a metaphor, it was used quite literally; her point is well taken, but it was also applied, I think, across such a wide array of thought that it is still useful to see it as having metaphoric application. See Gail Bederman, *Manliness and Civilization: A Cultural History of Gender and Race in the United States, 1880–*

1917 (Chicago: University of Chicago Press, 1995), 92–93. For an especially clear and insightful discussion of metaphor in science, see Gregg Mitman, "Evolution as Gospel: William Patten, the Language of Democracy, and the Great War," *Isis* 81 (Sept. 1990): 446–463.

14. Rydell, *All the World's a Fair;* James Gilbert, *Perfect Cities: Chicago's Utopias of 1893* (Chicago: University of Chicago Press, 1991); Matthew F. Bokovoy, *The San Diego World's Fairs and Southwestern Memory, 1880–1940* (Albuquerque: University of New Mexico Press, 2005), 80–113; G. Blair Nelson, "'Men before Adam!': American Debates over the Unity and Antiquity of Humanity," in David C. Lindberg and Ronald L. Numbers, eds., *When Science and Christianity Meet* (Chicago: University of Chicago Press, 2003); Roslyn Poignant, *Professional Savage: Captive Lives and Western Spectacle* (New Haven: Yale University Press, 2004; Phillips Verner Bradford and Harvey Blume, *Ota Benga: The Pygmy in the Zoo* (New York: St. Martin's, 1992); "The Scandal at the Zoo," *New York Times,* Aug. 6, 2006, 1 (L). Many thanks to Michael Sokal for alerting me to this last article.

15. For more on antimodernism and the notion of "authenticity," see especially Jackson Lears, *No Place of Grace.*

16. For more on the cultural meanings associated with Native Americans, see Deloria, *Playing Indian.*

17. Edgar Rice Burroughs, *Tarzan of the Apes* (1912). For extended discussions of this theme, see Bederman, *Manliness and Civilization,* 219–232; Kasson, *Houdini, Tarzan and the Perfect Man;* and Kemp, *Human Animal in Western Art and Science.*

18. Edgar Rice Burroughs, *The Land That Time Forgot* (1918; New York: Barnes and Noble, 2005).

19. Bederman, *Manliness and Civilization,* 79–101; Dorothy Ross, *G. Stanley Hall: The Psychologist as Prophet* (Chicago: University of Chicago Press, 1972); Jackson Lears, *No Place of Grace,* 144–149; Gould, *Ontogeny and Phylogeny;* Philip J. Pauly, *Biologists and the Promise of American Life: From Meriweather Lewis to Alfred Kinsey* (Princeton: Princeton University Press, 2000).

20. On American Christians' responses to Freud, see the insightful essay by Jon H. Roberts, "Psychoanalysis and American Christianity, 1900–1945," in Lindberg and Numbers, *When Science and Christianity Meet,* 225–244. On the influence of Freud in American popular, literary, and mass culture in the 1920s, see Lynn Dumenil, *The Modern Temper: American Culture and Society in the 1920s* (New York: Hill and Wang, 1995), 145–147.

21. William J. Fielding, *The Caveman within Us: His Peculiarities and Powers; How We Can Enlist His Aid for Health and Efficiency* (New York: E. P. Dutton, 1922).

22. Ibid., xi–xii.

23. Ibid., 109.

24. Ibid., 344.

25. Bruce J. Evenson, "'Cave Man' Meets 'Student Champion': Sports Page Storytelling for a Nervous Generation during America's Jazz Age," *Journalism Quarterly* 70, no. 4 (1993): 767–779.

26. Oscar Ameringer, "From Cave-Man to Edison," *Montana Socialist* (Butte), July 20, 1913, 2. Thanks to John Enyeart for this article.

27. Stanley Waterloo, *The Story of Ab: A Tale of the Time of the Cave Man* (1897; Kessinger Publishers, 2004).

28. Lucy Fitch Perkins, *The Cave Twins* (Cambridge, MA: Riverside Press, 1916).

29. Richard Schickel, *D. W. Griffith: An American Life* (New York: Limelight Editions, 1996), 172.

30. Charlie Chaplin, *His Prehistoric Past* (1914), film available on DVD in the collection "Chaplin: The Legend Lives On" (St. Laurent, Quebec: Madacy Entertainment Group, 2004).

31. The caveman literature of this period was so fundamentally gender-inflected that it is difficult—and perhaps impossible—to avoid the term *caveman*. This literature—including comic strips and caricatures—was specifically about cave*men*, except when, as discussed below, cave women became part of a joke by upsetting expectations.

32. This cartoon appears in *Caricature: Wit and Humor of a Nation in Picture, Song and Story*, 8th ed. (New York: Leslie-Judge Co., 1911), not paginated, and no author's credit was given to either the book or the cartoon.

33. Simeon Strunsky, "About Books, More or Less: Chesterton's Faith," *New York Times Book Review*, Nov. 22, 1925, 4.

34. "He No Cave Man, Wife Willing Bride, He Says," *New York World*, Mar. 3, 1921, 5.

35. "Not Only an Able Lochinvar, 'Bride Stealer' Stars at Bar: Cave-Man Bookkeeper of New Orleans Pleads His Case and Court Finds His Only Weapon a Marriage License," *New York Evening Post*, June 11, 1925, 5. A similar story, this time from Kansas, was reported in the *Post* the next day to have been a hoax: "News from Other Cities," *New York Evening Post*, June 12, 1925, 13. For another example of the association of cavemen and kidnapping, see Sinclair Lewis, *Free Air* (New York: Grosset and Dunlap, 1919), 215.

36. Edgar Rice Burroughs, *Tarzan of the Apes* (New York: Signet Classics, 1990), 198; John Taliaferro, *Tarzan Forever: The Life of Edgar Rice Burroughs, Creator of Tarzan* (New York: Scribner, 1999).

37. John F. Kasson, *Houdini, Tarzan and the Perfect Man: The White Male Body and the Challenge of Modernity in America* (New York: Hill and Wang, 2001).

38. "Goings On: Motion Pictures," *New Yorker*, Dec. 31, 1925, 21. On King Kong, see Cynthia Erb, *Tracking King Kong: A Hollywood Icon in World Culture* (Detroit: Wayne State University Press, 1998), and Mark Cotta Vaz, *Living Dangerously: The Adventures of Merian C. Cooper, Creator of King Kong* (New York: Villard Books, 2005).

39. Frederick H. Martens and Francesco B. De Leone, *Cave-Man Stuff: A Prehistoric Operetta in Two Acts* (New York: G. Schirmer, 1929); advertisement for play *Survival of the Fittest*, *New York World*, Mar. 15, 1921, 9.

40. Helen Bullitt Lowry, "High Art and the Apes," *New York Times Book Review and Magazine*, Feb. 26, 1922, 5.

41. Fillmore Hyde's Grandmother, "Prehistoric New York As I Knew It," *New Yorker*, July 3, 1926, 13–14.

42. Cole Porter, "Find Me a Primitive Man" (1929), in Robert Kimball, ed., *The Complete Lyrics of Cole Porter* (New York: Da Capo Press, 1992), 119.

43. Clarence Day, *This Simian World* (New York: Knopf, 1920), 45.

44. *Judge*, Aug. 8, 1925, 11; another example is *Judge*, Sept. 5, 1925, 6.

45. Moser, *Ancestral Images*, 134.

46. Harry Emerson Fosdick, "How Shall We Think of God?" *Harper's*, July 1926, 229–233.

47. W. E. Dodd, *Christian Index* (Jan. 22, 1925), quoted in Willard T. Gatewood, *Controversy in the Twenties: Fundamentalism, Modernism, and Evolution* (Nashville: Vanderbilt University Press, 1969); for a similar set of associations by Henry Ford, see Roderick Nash, *The Nervous Generation: American Thought, 1917–1930* (1970; Chicago: Ivan R. Dee, 1990), 161–163.

48. Edmund Wilson, *The American Earthquake: A Documentary of the Twenties and Thirties* (New York: Farrar, Straus and Giroux, 1979), 113; Stephen Kern, *The Culture of Time and Space, 1880–1918* (Cambridge: Harvard University Press, 1983), 124.

49. "Jazzing the Scriptures," *Nation*, Feb. 17, 1926, 172.

50. Wilson, *American Earthquake*, 89.

51. See, for example, Billy Sunday, "Back to the Old-Time Religion," *Collier's*, July 10, 1926, 8, 34. The paleontologist Henry Fairfield Osborn's pronouncements on "the jazz mind" are scattered throughout his correspondence of the decade; this correspondence can be found at the American Museum of Natural History Library, Special Collections, and in the Osborn Family Papers, NYHS. On jazz and jazz symbolism in the 1920s, see Kathy J. Ogren, *The Jazz Revolution: Twenties America and the Meaning of Jazz* (New York: Oxford University Press, 1989), and Scott Appelrouth, "Body and Soul: Jazz in the 1920s," *American Behavioral Scientist* 48 (July 2005): 1496–1509.

Chapter Two • The Museum in the Modern Babylon

1. Rev. Henry Sloane Coffin, "Science and Religion," closing address at memorial service at St. Bartholomew's Church, New York, Dec. 18, 1935, reprinted in "Henry Fairfield Osborn: August 8, 1857–November 6, 1935," supplement to *Natural History* 37 (Feb. 1936). For more on Osborn, see also Ronald Rainger, *An Agenda for Antiquity: Henry Fairfield Osborn and Vertebrate Paleontology at the American Museum of Natural History, 1890–1935* (Tuscaloosa: University of Alabama Press, 1991); Brian Regal, *Henry Fairfield Osborn: Race and the Search for the Origins of Man* (Burlington, VT: Ashgate Press, 2002); Charlotte M. Porter, "The Rise to Parnassus: Henry Fairfield Osborn and the Hall of the Age of Man," *Museum Studies Journal* 1 (1983): 26–34; John Michael Kennedy, "Philanthropy and Science in New York City: The American Museum of Natural History, 1868–1968" (Ph.D. diss., Yale University, 1968); Sheila Ann Dean, "What Animal We Came From: William King Gregory's Paleontology and the 1920s Debate on Human Origins" (Ph.D. diss., Johns Hopkins University, 1994); Ronald Rainger, "Vertebrate Paleontology as Biology: Henry Fairfield Osborn and the American Museum of Natural History," in Ronald Rainger, Keith R. Benson, and Jane Maienschein, eds., *The American Development of Biology* (Philadelphia: University of Pennsylvania Press, 1988), 219–256; William King Gregory, "Henry Fairfield Osborn and the American Museum of Natural History," *Nature*, Oct. 31, 1942, 513–515; Gregory, "Henry Fairfield Osborn," *American Philosophical Society Proceedings* 76 (1936): 395–408; Gregory, "Henry Fairfield Osborn, 1857–1935," *Biographical Memoirs of the Na-*

tional Academy of Sciences 19 (1938): 53–119; Douglas Sloan, "Science in New York City, 1867–1907," *Isis* 71 (Mar. 1980): 35–76; Simon Baatz, *Knowledge, Culture, and Science in the Metropolis: The New York Academy of Sciences 1817–1970* (New York: New York Academy of Sciences, 1990), 165–183; and Philip J. Pauly, *Biologists and the Promise of American Life from Meriweather Lewis to Alfred Kinsey* (Princeton: Princeton University Press, 2000).

2. John Roach Straton, *The Old Gospel at the Heart of the Metropolis* (New York: George H. Doran Co., 1925); see also John Roach Straton, *Fighting the Devil in Modern Babylon* (Boston: Stratford, 1929); Ronald L. Numbers, *The Creationists*, expanded ed. (Cambridge: Harvard University Press, 2006); and Ronald L. Numbers, *Darwinism Comes to America* (Cambridge: Harvard University Press, 1998). Clippings from newspapers reporting on attacks on Osborn and the museum are preserved in folder 2, box 86, Osborn Papers, AMNH. The same folder includes news accounts of Straton's sermons about the museum, and Osborn's correspondence and records of his exchanges with Straton are in folder 4, box 21.

3. "Science: American Association," *Time*, Dec. 31, 1928, cover and p. 20; the correspondence with Hughes—likely to have been rather busy in 1921 with the logistics of the Washington Conference—can be found in folder 12, box 28, Osborn Papers, AMNH, especially Osborn to Hughes, June 28, 1921, and an undated telegram from Hughes in response. The letter of reference from Theodore Roosevelt is reprinted in Henry Fairfield Osborn, "The Elephants and Mastodonts Arrive in America," *Natural History* 25 (Feb. 5, 1925): 3.

4. Henry Fairfield Osborn's brother William Church Osborn is depicted in Edmund Wilson, "An Estate on the Hudson," in Wilson, *The Higher Jazz*, ed. Neale Reinitz (Iowa City: University of Iowa Press, 1998), 39–58. See also the introduction by Neale Reinitz, xxiii.

5. George T. Brett to Henry Fairfield Osborn, Nov. 15, 1904, folder 11, box 86, Osborn Papers, AMNH.

6. Henry Fairfield Osborn, *The Age of Mammals in Europe, Asia, and North America* (New York: Macmillan, 1910), 1.

7. *The American Museum School Service: Fifty-eighth Annual Report of the Trustees* (New York: American Museum of Natural History, 1927), 3. According to the annual report for 1927, the museum sold 20,036 popular publications in that year, the best-selling one being the *Guide to the Hall of the Age of Man*, closely followed by the *Guide to the Hall of Dinosaurs*. See *Building the American Museum, 1869–1927: Fifty-ninth Annual Report of the Trustees* (New York: American Museum of Natural History, 1928), 35. In that report curator George H. Sherwood claimed that if all outreach and public programs were added together with attendance, the figures would show that the museum had, in a single year, "reached more than 9,900,000 school children." George H. Sherwood, "The Museum and School Service," in *Building the American Museum*, 91. On the increase in the number of visitors, see Frederic A. Lucas, *General Guide to the Exhibition Halls*, 16th ed. (New York: American Museum of Natural History, 1931), 14.

8. Porter, "Rise to Parnassus"; see also Rainger, *Agenda for Antiquity*; Regal, *Henry Fairfield Osborn;* and Sylvia Massey Czerkas and Donald F. Glut, *Dinosaurs, Mammoths, and Cavemen: The Art of Charles R. Knight* (New York: E. P. Dutton, 1982). A long draft of Knight's autobiography is available in folder 4, box 5, Knight Papers. An edited edition with

appreciations by several scientists and the special-effects artist Ray Harryhausen was published in 2005. See Charles R. Knight, *Autobiography of an Artist* (Ann Arbor, MI: G. T. Labs, 2005). Letters to the American Museum requesting copies of Knight's paintings and permission to reproduce them are scattered throughout the Osborn Papers, AMNH, and the Knight Papers.

9. Osborn frequently complained—and sometimes boasted—that Wells had cribbed images and ideas from his *Men of the Old Stone Age*. The correspondence in the Osborn Papers, AMNH, includes numerous such complaints.

10. Knight unpublished autobiography, folder 4, box 5, Knight Papers; Knight, *Autobiography of an Artist*. For more on Knight, see Czerkas and Glut, *Dinosaurs, Mammoths, and Cavemen;* on exhibits at the American Museum of Natural History, see the biography of Osborn by Rainger, *Agenda for Antiquity*, 152–181; John Michael Kennedy, "Philanthropy and Science in New York City: The American Museum of Natural History, 1868–1968" (Ph.D. diss., Yale University, 1968); Sheila Ann Dean, "What Animal We Came From: William King Gregory's Paleontology and the 1920s Debate on Human Origins" (Ph.D. diss., Johns Hopkins University, 1994); Douglas J. Preston, *Dinosaurs in the Attic: An Excursion into the American Museum of Natural History* (New York: St. Martin's, 1986); Joseph Wallace, *A Gathering of Wonders: Behind the Scenes at the American Museum of Natural History* (New York: St. Martin's, 2000); Geoffrey Hellman, *Bankers, Bones and Beetles: The First Century of the American Museum of Natural History* (Garden City, NY: Natural History Press, 1968); and Chris Beard, *The Hunt for the Dawn Monkey: Unearthing the Origins of Monkeys, Apes, and Humans* (Berkeley: University of California Press, 2004).

11. Martin J. S. Rudwick, *Scenes from Deep Time: Early Pictorial Representation of the Prehistoric World* (Chicago: University of Chicago Press, 1992), 135–172. See also Steve McCarthy and Mick Gilbert, *The Crystal Palace Dinosaurs: The Story of the World's First Prehistoric Sculptures* (London: Croydon, n.d.).

12. On the history of museums, generally, see Sally Gregory Kohlstedt, "Essay Review: Museums: Revisiting Sites in the History of the Natural Sciences," *Journal of the History of Biology* 28 (1995): 151–166; Steven Conn, *Museums and American Intellectual Life, 1876–1926* (Chicago: University of Chicago Press, 1998); William Leach, *Land of Desire: Merchants, Power, and the Rise of a New American Culture* (New York: Pantheon, 1993); and Lynn K. Nyhart, "Science Art, and Authenticity in Natural History Displays," in Soraya de Chadarevian and Nick Hopwood, eds., *Models: The Third Dimension of Science* (Stanford: Stanford University Press, 2004), 307–335.

13. Ronald Rainger, "Just before Simpson: William Diller Matthew's Understanding of Evolution," *Proceedings of the American Philosophical Society* 130, no. 4 (1986): 453–474; Rainger, "Vertebrate Paleontology as Biology"; Leo F. Laporte, "George G. Simpson: Vertebrate Paleontologist as Biologist," in Keith R. Benson, Jane Maienschein, and Ronald Rainger, eds., *The Expansion of American Biology* (New Brunswick, NJ: Rutgers University Press, 1991), 80–106; Leo F. Laporte, *George Gaylord Simpson: Paleontologist and Evolutionist* (New York: Columbia University Press, 2000); Dean, "What Animal We Came From"; Regal, *Henry Fairfield Osborn*.

14. Clark Wissler, Survey of the American Museum of Natural History Made at the Re-

quest of the Management Board in 1942–3. Copy in Special Collections, Library, American Museum of Natural History.

15. See, for example, Paul H. Bade to Osborn, Nov. 29, 1933, folder 4, box 27, Osborn Papers, AMNH; Frederick Morse Cutler to Osborn, June 28, 1929, ibid.; and many other examples in boxes 27, 19, 99, and 92, ibid.

16. Walter C. Kraatz to Osborn, Apr. 13, 1924, folder 6, box 86, ibid. Kraatz was an assistant professor of zoology at Miami University in Oxford, Ohio.

17. Edith Wharton, postcard to Professor Fairfield Osborn, June 29, 1923, folder for 1923, box 9, Osborn Correspondence, Osborn Family Papers, NYHS. The fact that Wharton addressed the postcard to Fairfield Osborn rather than to Henry Fairfield Osborn suggests that she may have been a family friend or at least that they traveled in the same social circles, which is not unlikely; they certainly had friends—Theodore Roosevelt, for one—in common. Reviews of the book, sales figures, and correspondence about it can be found in box 99, Osborn Papers, AMNH. The book was also adopted as a required course textbook at the University of California for courses with enrollments of some thirteen hundred students: Osborn to Nicholas Murray Butler, Nov. 23, 1915, folder 7, box 99, Osborn Papers, AMNH.

18. See sales figures in folder 9, box 56, Osborn Papers, AMNH.

19. Osborn to Scribner, Nov. 6, 1924, ibid.

20. Osborn to Scribner, July 18, 1923, ibid.

21. On racial and eugenics themes at the American Museum, see Rainger, *Agenda for Antiquity*; Regal, *Henry Fairfield Osborn*; Donna J. Haraway, "Teddy Bear Patriarchy: Taxidermy in the Garden of Eden, New York City, 1908–36," in Haraway, *Primate Visions: Gender, Race and Nature in the World of Modern Science* (New York: Routledge, 1989), 26–58; Porter, "Rise of Parnassus"; and Timothy Luke, "Museum Pieces: Politics and Knowledge at the American Museum of Natural History," *Australasian Journal of American Studies* 16, no. 2 (1997): 1–28.

22. Alfred Watterson McCann, *God—or Gorilla: How the Monkey Theory of Evolution Exposes Its Own Methods, Refutes Its Own Principles, Denies Its Own Inferences, Disproves Its Own Case* (New York: Devin-Adair, 1922).

23. "How 'Missing Links' Are Made," *Sunday School Times*, Nov. 18, 1922, article in folder 5, box 86, Osborn Papers, AMNH.

24. William King Gregory, "Physical Evidence of the Origin of Man," *New York Evening Post*, Apr. 1, 1922, 9.

25. "Dr. Straton Assails Museum of History," *New York Times*, Mar. 9, 1924, E1.

26. "Pulpits Continue Doctrinal Dispute," *New York Times*, Mar. 9, 1924, 20.

27. See letters and clippings about the Osborn-Straton exchange, folder 4, box 21, Osborn Papers, AMNH. See especially Henry Fairfield Osborn to John Roach Straton, Mar. 8, 1924, ibid; John Roach Straton, "Making Poison Plausible," sermon, 1924, ibid; John Dickenson Sherman, " 'Treason to God Almighty': Rev. Dr. J. R. Straton Denounces American Museum of Natural History," newspaper clipping, *Fort Bragg (CA) News*, May 3, 1924, ibid; and Henry Fairfield Osborn, "Evolution and Daily Living," *Forum* 73 (Feb. 1925): 169. See also John Roach Straton and Charles Francis Potter, *Evolution versus Creation: Second in the*

Series of Fundamentalist-Modernist Debates (1924), in Ronald L. Numbers, ed., *Creationism in Twentieth-Century America: A Ten-Volume Anthology of Documents, 1903–1961*, vol. 2: *Creation-Evolution Debates* (New York: Garland, 1995), 21–131.

28. Osborn to Straton, Mar. 8, 1924, folder 2, box 86, Osborn Papers, AMNH; also cited in Henry Fairfield Osborn, *Evolution and Religion in Education: Polemics of the Fundamentalist Controversy of 1922 to 1926* (New York: Charles Scribner's Sons, 1926), 46.

29. Henry Fairfield Osborn, "The Annual Meeting of the Board of Trustees of the American Museum of Natural History," *Science*, Feb. 20, 1920, 182–184.

30. Henry Fairfield Osborn to J. H. McGregor, Mar. 20, 1924, folder Correspondence 1924, box 9, Osborn Correspondence, Osborn Family Papers, NYHS; Henry Fairfield Osborn, memo to George Sherwood, Apr. 2, 1924, folder 209C, Central Archives, American Museum of Natural History.

31. "How 'Missing Links' Are Made," *Sunday School Times*.

32. Osborn, *Age of Mammals*, 7.

33. For recent discussions of the visual historiography of science, see the bibliographic essay in Jennifer Tucker, *Nature Exposed: Photography as Eyewitness in Victorian Science* (Baltimore: Johns Hopkins University Press, 2005), and the special section "Focus: Science and Visual Culture," in *Isis* 97 (Mar. 2006): 75–132. See also Luc Pauwels, ed., *Visual Cultures of Science: Rethinking Representational Practices in Knowledge Building and Scientific Communication* (Hanover, NH: Dartmouth College Press, 2006), and Nick Hopwood, "Pictures of Evolution and Charges of Fraud: Ernst Haeckel's Embryological Illustrations," *Isis* 97 (June 2006): 260–301. The seminal article for historians of science is Martin J. S. Rudwick, "The Emergence of a Visual Language for Geological Science, 1760–1840," *History of Science* 14 (Sept. 1975): 149–195. See also Martin J. S. Rudwick, "Encounters with Adam, or at Least the Hyaenas: Nineteenth-Century Visual Representations of the Deep Past," in James R. Moore, ed., *History, Humanity and Evolution: Essays for John C. Greene* (Cambridge: Cambridge University Press, 1989), 231–251, and Rudwick, *Scenes from Deep Time*. See also "Special Issue: Seeing Science," *Representations* 40 (Fall 1992); "Special Issue on Pictorial Representation in Biology," *Biology and Philosophy* 6 (Apr. 1991); the special issue "Science and the Visual," *British Journal for the History of Science* 31 (June 1998); Brian S. Baigre, ed., *Picturing Knowledge: Historical and Philosophical Problems Concerning the Use of Art in Science* (Toronto: University of Toronto Press, 1996); Caroline A. Jones and Peter Galison, eds., *Picturing Science, Producing Art* (New York: Routledge, 1998); Michael Lynch and Steve Woolgar, eds., *Representation in Scientific Practice* (Cambridge, MA: MIT Press, 1990); Thomas L. Hankins and Robert J. Silverman, *Instruments and the Imagination* (Princeton: Princeton University Press, 1995); Bruno Latour, "Visualization and Cognition: Thinking with Eyes and Hands," *Knowledge and Society* 6 (1986): 1–40; Massimiano Bucchi, "Images of Science in the Classroom: Wallcharts and Science Education, 1850–1920," *British Journal of the History of Science* 31 (1998): 161–184; Martin Kemp, "Taking It on Trust: Form and Meaning in Naturalistic Representation," *Archives of Natural History* 17, no. 2 (1990): 127–188; Gregg Mitman, *Reel Nature: America's Romance with Wildlife on Film* (Cambridge: Harvard University Press, 1999); and Charlotte M. Porter, "Essay Review: The History of Scientific Illustration," *Journal of the History of Biology* 28 (1995): 545–550. For a review of

studies of visual culture by historians generally, see George Roeder Jr., "Filling in the Picture: Visual Culture," *Reviews in American History* 26 (Mar. 1998): 275–293.

34. Adolph H. Schultz to Charles R. Knight, Apr. 18, 1948, folder 2, box 2, Knight Papers. The Knight Papers include other letters describing "experiments" of this sort. See, for example, a letter Knight wrote to a newspaper (which paper is not disclosed) dated July 18, 1948, describing the method of reconstruction he used in testing an interpretation by the paleontologist Franz Weidenreich of the discovery of large fossil teeth of hominid type, folder 2, box 2. There are similar letters in the archives of the American Museum of Natural History about the use of illustrations and reconstructions by William King Gregory, William Diller Matthew, and J. H. McGregor. For other examples of reasoning about the reconstruction of fossil animals by artistic means and about exhibits at the museum, see Rainger, *Agenda for Antiquity*, 152–181.

35. H. E. Le Grand, "Is a Picture Worth a Thousand Experiments?" in Le Grand, ed. *Experimental Inquiries: Historical, Philosophical and Social Studies of Experimental Science*, vol. 8 in *Australasian Studies in History and Philosophy of Science* (Dordrecht, Netherlands: Kluwer Academic Publishers, 1990), 241–270. Le Grand argues that earth scientists, who historically had not engaged in experiments in the traditional sense, used illustrations as a kind of experimentation, a way to test hypotheses and predictions related to the theories of continental drift and plate tectonics.

36. Tucker, *Nature Exposed*.

37. Stephen Jay Gould has called this "the most familiar of all illustrations" of horse evolution. Stephen Jay Gould, "Life's Little Joke," in Gould, *Bully for Brontosaurus* (New York: W. W. Norton, 1991), 174.

38. According to the museum's annual reports, in 1924 some 546,160 slides went out on loan to the public schools as part of the museum's active education program; the number increased to 672,479 in 1925 and to 786,115 by 1926. The museum's figures for 1926 show a total of 808,789 slides lent altogether. Between 1923 and 1927 the number increased from 440,315 to 921,811, possibly an index of the museum's prominent role in the evolution debates of the period. The museum received numerous requests for permission to reproduce images, as did the artist Charles R. Knight, whose murals painted for the museum achieved lasting fame. George H. Sherwood, "Public Education in the Museum and in the Schools," in *The American Museum and Education, Fifty-sixth Annual Report of the Trustees* (New York: American Museum of Natural History, 1925), 128; *American Museum School Service: Fifty-eighth Annual Report*, 19, 99; *Building the American Museum*, 98. For an insightful discussion of another example of the longevity of the influence of scientific diagrams, see Hopwood, "Pictures of Evolution and Charges of Fraud."

39. Robert T. Bakker, pers. comm.; Stephen Jay Gould, foreword to Charles R. Knight, *Life through the Ages*, Commemorative Edition (Indianapolis: Indiana University Press, 2001), vii; Philip J. Currie, introduction to Knight, *Life through the Ages*, xi–xii; Rainger, *Agenda for Antiquity*, 180.

40. George Langford to Osborn, Mar. 15, 1920, folder 25, box 12, Osborn Papers, AMNH.

41. Nicolaas Rupke, " 'The End of History' in the Early Picturing of Geological Time," *History of Science* 36 (Mar. 1998): 61–90.

42. Henry Fairfield Osborn, *The Origin and Evolution of Life: On the Theory of Action, Reaction and Interaction of Energy* (New York: Charles Scribner's Sons, 1925), vii.

43. Rainger, *Agenda for Antiquity,* 139.

44. George A. Dorsey, "Selling Evolution to the Masses: Mrs. Osborn's 'Chain of Life' Suffers by Dependence on Husband's Writings," *New York World,* Dec. 13, 1925, copy on file in folder 26, box 16, Osborn Papers, AMNH. This is a review of Lucretia Perry Osborn, *The Chain of Life* (New York: Charles Scribner's Sons, 1925).

45. Dorsey, "Selling Evolution to the Masses."

46. Osborn to Frances Mason, Sept. 13, 1929, folder 13, box 14, Osborn Papers, AMNH.

47. See especially the letters in box 27, ibid. "Cosmic Christianity" was the topic of an aspiring writer named Leon H. Barnett, of New York, who sent an undated letter preserved in folder 7, box 27, ibid.. Mrs. Julia W. Molina wrote: "I believe that I have the true and great secret of life . . . and would appreciate an audience with you to consult as to what you believe would be the way to bring it out upon the world." Julia W. Molina to Osborn, Dec. 29, 1929, folder 6, box 27, ibid. The Conklin Correspondence includes numerous similar letters; Conklin kept them in a file labeled "crank letters."

48. Dorothy Giles to Osborn, Dec. 31, 1925, folder 1, box 27, Osborn Papers, AMNH. Will Durant wrote that he was requesting statements on "the meaning of life" from a select group of eminent men: Will Durant to Osborn, folder 15, box 7, ibid.

49. Secretary to Osborn to Dorothy Giles, Jan. 6, 1925, folder 1, box 27, ibid. Osborn believed that "unfeminine fashions" threatened modern evolutionary health.

50. Rhoda E. McColloch to Osborn, July 28, 1925, folder 1, box 27, ibid.

51. Secretary to Osborn to Rhoda E. McColloch, Oct. 2, 1925, folder 1, box 27, ibid.

52. It was much more common for the scientists in Osborn's circle, and for those who were published in the journal *Science*—two groups with a lot of overlap—to refer to themselves as "scientific men" than as "scientists." The newspapers and outsiders generally called them scientists.

53. This paragraph and the next: editorial, *New York Evening Post,* reprinted in *Literary Digest,* Oct. 15, 1927, 80–81.

54. Dorsey, "Selling Evolution to the Masses."

55. George A. Dorsey, *Why We Behave Like Human Beings* (New York: Harper and Brothers, 1925).

56. "Churches Urged to Advertise More," *New York Herald,* Jan. 30, 1921, sec. 2, p. 4. See also Douglas Carl Abrams, *Selling the Old-Time Religion: American Fundamentalists and Mass Culture, 1920–1940* (Athens: University of Georgia Press, 2001).

57. J. C. Grinnan to Osborn, Sept. 21, 1925, folder 1, box 92, Osborn Papers, AMNH. Similar criticisms of Straton regularly appeared in newspapers.

58. Charles W. Wood, "Religion Becomes News," *Nation,* Aug. 19, 1925, 204.

59. Science Service pamphlet, folder 4, box 56, Osborn Papers, AMNH. In addition to photographs of contemporary scientists, the pamphlet offered pictures of Aristotle and Archimedes. Osborn and his colleagues also commonly exchanged photographs among themselves; Osborn posted photographs of important naturalist acquaintances—including John Scopes—on a wall in his office.

60. David Starr Jordan, "Discussion and Correspondence: Concerning 'Species-Grinding,'" *Science*, July 1, 1927, 14–15.

61. Conklin to Frances Mason, Apr. 19, 1929, folder 10, box 15, Conklin Correspondence.

62. Roy Chapman Andrews, *Across Mongolian Plains: A Naturalist's Account of China's "Great Northwest"* (New York: D. Appleton, 1921); Roy Chapman Andrews, *On the Trail of Ancient Man: A Narrative of the Field Work of the Central Asiatic Expeditions* (New York: G. P. Putnam's Sons, 1926); Roy Chapman Andrews, *Under a Lucky Star: A Lifetime of Adventure* (New York: Viking Press, 1943); Charles Gallenkamp, *Dragon Hunter: Roy Chapman Andrews and the Central Asiatic Expeditions* (New York: Viking Press, 2001).

63. Walter Tittle, "Personalities: A Champion of Evolution. President Osborn of the American Museum of Natural History," *World's Work* 56 (May 1928): 85.

64. J. Walker McSpadden, *To the Ends of the World and Back: Scouting for a Great Museum* (New York: Thomas Y. Crowell, 1931), 14. According to the book jacket, McSpadden was also the author of *Pioneer Heroes, Indian Heroes* ("Boys will be thrilled!" according to the jacket blurb), and *Boys' Book of Famous Soldiers*.

65. "Camera" cartoon in "Christmas luncheon place card" collection, Paleontology Department Library, American Museum of Natural History. Margaret Mead's statement was for an interview with the *New Yorker*, quoted in Gallenkamp, *Dragon Hunter*, 40.

66. Wissler, Survey of the American Museum of Natural History.

67. Osborn, memo to Miss Ethel Newman, July 26, 1928, Osborn Papers, AMNH; Osborn, memo of May 3, 1921, folder for 1921, box 19, Osborn Correspondence, Osborn Family Papers, NYHS.

68. Osborn to his daughter Josephine, Jan. 22, 1922, folder for 1922, box 9, H. F. Osborn Correspondence, Osborn Family Papers, NYHS.

69. "American Association," *Time*, Dec. 31, 1928, 20, and cover drawing of Osborn.

70. Rae Goodell, *The Visible Scientists* (Boston: Little, Brown, 1977). See also Marcel C. LaFollette, *Making Science Our Own: Public Images of Science, 1910–1955* (Chicago: University of Chicago Press, 1990); Ronald C. Tobey, *The American Ideology of National Science, 1919–1930* (Pittsburgh: University of Pittsburgh Press, 1971); and John C. Burnham, *How Superstition Won and Science Lost: Popularizing Science and Health in the United States* (New Brunswick, NJ: Rutgers University Press, 1987). Peter Bowler has demonstrated a similar pattern among scientists writing for the public in Britain. See Peter J. Bowler, *Reconciling Science and Religion: The Debate in Early-Twentieth-Century Britain* (Chicago: University of Chicago Press, 2001). For many examples of teachers seeking advice from Osborn, see especially box 27, containing requests for advice, opinions, and information; box 19, containing material pertaining to the Scopes trial; and box 86, a collection of correspondence and news clippings about evolution, Osborn Papers, AMNH.

Chapter Three • Nineteen Twenty-two or Thereabouts

1. See Osborn Diary for 1922, box 25, Henry Fairfield Osborn Diaries, Osborn Family Papers, NYHS. Bryan's article was William Jennings Bryan, "God and Evolution," *New York*

Times, Feb. 26, 1922, sec. 7, p. 1. For Osborn's response, see Henry Fairfield Osborn, "Evolution and Religion," *New York Times*, Mar. 5, 1922, sec. 7, pp. 2, 14. In telling the story of his involvement in the controversy in an essay in *Evolution and Religion*, Osborn pinpointed Bryan's entry as the stimulus to his own involvement. Henry Fairfield Osborn, *Evolution and Religion in Education: Polemics of the Fundamentalist Controversy of 1922 to 1926* (New York: Charles Scribner's Sons, 1926), 3.

2. Osborn, *Evolution and Religion in Education*, 3.

3. James McKeen Cattell to Osborn, Feb. 4, 1922, folder 4, box 4, Osborn Papers, AMNH.

4. Winterton C. Curtis, *Fundamentalism vs. Evolution at Dayton, Tennessee: Abstracts from the Autobiographical Notes of Winterton C. Curtis* (Columbia, MO: Winterton C. Curtis, 1956), 8.

5. William Jennings Bryan, "The Prince of Peace," in *Speeches of William Jennings Bryan* (New York: Funk and Wagnalls, 1909), 2:266–267.

6. See especially Ronald L. Numbers, *The Creationists: The Evolution of Scientific Creationism*, 2d ed. (Cambridge: Harvard University Press, 2006); Ronald L. Numbers, *Darwinism Comes to America* (Cambridge: Harvard University Press, 1998); Michael Lienesch, *In the Beginning: Fundamentalism, the Scopes Trial, and the Making of the Antievolution Movement* (Chapel Hill: University of North Carolina Press, 2007); Jon H. Roberts, *Darwinism and the Divine in America: Protestant Intellectuals and Organic Evolution, 1859–1900* (Madison: University of Wisconsin Press, 1988); Jon H. Roberts, "Conservative Evangelicals and Science Education in American Colleges and Universities, 1890–1940," *Journal of the Historical Society* 3 (Fall 2005): 297–329; Edward J. Larson, *Summer for the Gods: The Scopes Trial and America's Continuing Debate over Science and Religion* (New York: Basic Books, 1997); James R. Moore, *The Post-Darwinian Controversies: A Study of the Protestant Struggle to Come to Terms with Darwin in Great Britain and America, 1870–1900* (Cambridge: Cambridge University Press, 1979); and Peter J. Bowler, *Reconciling Science and Religion: The Debate in Early-Twentieth-Century Britain* (Chicago: University of Chicago Press, 2001).

7. *The Fundamentals: A Testimony to the Truth*, 12 vols. (Chicago: Testimony Publishing Co., 1910–12). Numbers, *The Creationists;* Numbers, *Darwinism Comes to America;* Larson, *Summer for the Gods;* Lienesch, *In the Beginning;* Paul K. Conkin, *When All the Gods Trembled: Darwinism, Scopes, and American Intellectuals* (Lanham, MD: Rowman and Littlefield, 1998); Jeffrey P. Moran, *The Scopes Trial: A Brief History with Documents* (Boston: Bedford/St. Martin's, 2002); Garry Wills, *Under God: Religion and American Politics* (New York: Simon and Schuster, 1990); George M. Marsden, *Fundamentalism and American Culture: The Shaping of Twentieth-Century Evangelicalism, 1870–1925* (New York: Oxford University Press, 1980). For Bryan's involvement, see also Michael Kazin, *A Godly Hero: The Life of William Jennings Bryan* (New York: Knopf, 2006), and Lawrence W. Levine, *Defender of the Faith: William Jennings Bryan, the Last Decade, 1915–1925* (Cambridge: Harvard University Press, 1987).

8. Edwin Grant Conklin, "Evolution and the Bible," *New York Times*, Mar. 5, 1922, sec. 7, p. 14, reprinted as a pamphlet in the series Popular Religion Leaflets by American Institute of Sacred Literature (Chicago: First Impression, 1922). A copy of this pamphlet can be found in folder 20, box 5, Osborn Papers, AMNH.

9. Vernon Kellogg, *Headquarters Nights: A Record of Conversations and Experiences at the Headquarters of the German Army in France and Belgium* (Boston: Atlantic Monthly Press, 1917). See also Larson, *Summer for the Gods*, 40–42.

10. On the vogue for Nietzsche, see Henry F. May, "The Rebellion of the Intellectuals, 1912–1917," *American Quarterly* 8 (1956): 114–116.

11. James Leuba, *The Belief in God and Immortality* (Boston: Sherman, French, 1916).

12. Robert S. Lynd and Helen Merrell Lynd, *Middletown: A Study in Modern American Culture* (New York: Harcourt, Brace and World, 1929); Warren I. Susman, *Culture as History: The Transformation of American Society in the Twentieth Century* (New York: Pantheon, 1984); Lynn Dumenil, *The Modern Temper: American Culture and Society in the 1920s* (New York: Hill and Wang, 1995); Roderick Nash, *The Nervous Generation: American Thought, 1917–1930* (1970; Chicago: Ivan R. Dee, 1990); David J. Goldberg, *Discontented America: The United States in the 1920s* (Baltimore: Johns Hopkins University Press, 1999); Stanley Coben, *Rebellion against Victorianism* (New York: Oxford University Press, 1991); Frederic J. Hoffman, *The Twenties: American Writing in the Postwar Decade* (1949; New York: Free Press, 1965).

13. The phrase "acids of modernity" comes from Walter Lippmann, *A Preface to Morals* (New York: Macmillan, 1929). See also Dumenil, *Modern Temper*.

14. Roscoe Pound, "The Cult of the Irrational," quoted in Michael Schudson, *Discovering the News: A Social History of American Newspapers* (New York: Basic Books, 1978), 126–127.

15. Joseph Wood Krutch, *The Modern Temper* (1929; New York: Harcourt Brace Jovanovich, 1957), 7.

16. Ibid., 9.

17. *Chicago Evening Post* quoted in "Is Scientific Advance Impeding Human Welfare?" *Literary Digest*, Oct. 1, 1927, 32–33.

18. Andrew Dickson White, *A History of the Warfare of Science with Theology in Christendom* (New York: D. Appleton, 1896); John William Draper, *History of the Conflict between Religion and Science* (New York: D. Appleton, 1874). For more on the influence of these two books in the 1920s, see Larson, *Summer for the Gods*, 21–23.

19. Jon H. Roberts, "Psychoanalysis and American Christianity, 1900–1945," in David C. Lindberg and Ronald L. Numbers, eds., *When Science and Christianity Meet* (Chicago: University of Chicago Press, 2003), 225–244. On the vogue for Freud and behaviorism, see Dumenil, *Modern Temper*, 145–146.

20. Paul Carter, *Politics, Religion and Rockets: Essays in Twentieth-Century American History* (Tucson: University of Arizona Press, 1991), 7.

21. Jozsef Illy, ed., *Albert Meets America: How Journalists Treated Genius during Eisntein's 1921 Travels* (Baltimore: Johns Hopkins University Press, 2006); Edward A. Purcell Jr., *The Crisis of Democratic Theory: Scientific Naturalism and the Problem of Value* (Lexington: University of Kentucky Press, 1973); Dumenil, *Modern Temper*, 147–149. See also Arthur I. Miller, *Einstein, Picasso: Space, Time, and the Beauty That Causes Havoc* (New York: Basic Books, 2001), and Stephen Kern, *The Culture of Time and Space, 1880–1918* (Cambridge: Harvard University Press, 1983). Robert Millikan's response to Lucretia Perry Osborn's query is Millikan to Osborn, June 3, 1926, folder 10, box 15, Osborn Papers, AMNH.

22. Purcell, *Crisis of Democratic Theory.*

23. Conklin, "Evolution and the Bible," 14.

24. Frank E. E. Germann, "Cooperation in Research," *Science,* Mar. 26, 1926, 324–327; Michael M. Sokal, "Promoting Science in a New Century: The Middle Years of the AAAS," in Sally Gregory Kohlstedt, Michael M. Sokal, and Bruce V. Lewenstein, *The Establishment of Science in America: 150 Years of the American Association for the Advancement of Science* (New Brunswick, NJ: Rutgers University Press, 1999), 50–102.

25. Marcel C. LaFollette, *Making Science Our Own: Public Images of Science, 1910–1955* (Chicago: University of Chicago Press, 1990); John C. Burnham, *How Superstition Won and Science Lost: Popularizing Science and Health in the United States* (New Brunswick, NJ: Rutgers University Press, 1987); Sokal, "Promoting Science in a New Century"; Ronald C. Tobey, *The American Ideology of National Science, 1919–1930* (Pittsburgh: University of Pittsburgh Press, 1971); Charles William Heywood, "Scientists and Society in the United States, 1900–1940: Changing Concepts of Social Responsibility" (Ph.D. diss., University of Pennsylvania, 1954), 7; David J. Rhees, "A New Voice for Science: Science Service under Edwin E. Slosson, 1921–29" (M.A. thesis, University of North Carolina, 1979). See also Bernard Lightman, "The Visual Theology of Victorian Popularizers of Science from Reverent Eye to Chemical Retina," *Isis* 91 (Dec. 2000): 651–680; Katherine Pandora, "Knowledge Held in Common: Tales of Luther Burbank and Science in the American Vernacular," *Isis* 92 (Sept. 2001): 484–516; and Bowler, *Reconciling Science and Religion.*

26. The Osborn Papers, AMNH, contain many reviews of Osborn's and others' books of this type. For materials relating to Osborn's *Men of the Old Stone Age,* see box 99; to *Earth Speaks to Bryan,* box 92; and to Frances Mason, ed., *Creation by Evolution,* box 14. Sales figures for Osborn's books during the 1920s are in folders 7–11, box 56, ibid; see also F. F. Van de Water, "Books for Babbitt," *World's Work* 58 (June 1929): 68–71; Susman, *Culture as History,* 105–121; James Steel Smith, "The Day of the Popularizers: The 1920's," *South Atlantic Quarterly* 62 (Spring 1963): 297–309; Janice A. Radway, *A Feeling for Books: The Book-of-the-Month Club, Literary Taste, and Middle-Class Desire* (Chapel Hill: University of North Carolina Press, 1997); and Joan Shelley Rubin, *The Making of Middlebrow Culture* (Chapel Hill: University of North Carolina Press, 1992).

27. "Evolution Books Popular. Princeton Students Taking Great Interest in the Subject," *New York Times,* Jan. 31, 1926, sec. 2, p. 3. These books continued to be popular through the decade.

28. See F. F. Van de Water, "Books for Babbitt," *World's Work* 58 (June 1929); Susman, *Culture as History,* 105–21; Smith, "Day of the Popularizers"; Matthew D. Whalen, "The Public and American Culture: A Preface to the Study of Popular Science," *Journal of American Culture* 4 (Winter 1981): 14–26; Matthew D. Whalen and Mary F. Tobin, "Periodicals and the Popularization of Science in America, 1860–1910," *Journal of American Culture* 3 (Spring 1980): 195–203; Radway, *Feeling for Books;* Rubin, *Making of Middlebrow Culture;* Hoffman, *The Twenties;* and Joan Shelley Rubin, "Between Culture and Consumption: The Mediations of the Middlebrow," in Richard Wightman Fox and T. J. Jackson Lears, eds., *The Power of Culture: Critical Essays in American History* (Chicago: University of Chicago Press, 1993), 163–191. For a longer list of popularizations of evolution in the 1920s, see Con-

stance Areson Clark, "Evolution for John Doe: Pictures, the Public, and the Scopes Trial Debate," *Journal of American History* 87 (Mar. 2001): 1283n13.

29. For scientists' responses to the Kentucky initiative, see Arthur M. Miller, "Kentucky and the Theory of Evolution," *Science*, Feb. 17, 1922, 178–179, and Arthur M. Miller, "The Vote on the Evolution Bill in the Kentucky State Legislature," *Science*, Mar. 24, 1922, 317. See also correspondence between Osborn and Arthur Miller, a former student of Osborn's, who taught at the University of Kentucky; this file also includes a series of articles about evolution penned by Miller for the *Lexington (KY) Herald*, in folder 5, box 15, Osborn Papers, AMNH.

30. Bateson's address at the Toronto meeting of American Association for the Advancement of Science was reprinted in *Science:* William Bateson, "Evolutionary Faith and Modern Doubts," *Science*, Jan. 29, 1922, 55–61.

31. James Gilbert, *Redeeming Culture: American Religion in an Age of Science* (Chicago: University of Chicago Press, 1997), 37–61.

32. Osborn to Sir Arthur Shipley, Feb. 10, 1922, folder 4, box 15, Osborn Papers, AMNH; Osborn to Conklin, Nov. 22, 1922, folder 31, box 17, Conklin Correspondence; Conklin to Osborn, Nov. 25, 1922, folder 1, box 86, Osborn Papers, AMNH. See also E. B. Poulton to Osborn, Feb. 23, 1922, folder for 1922, box 9, Osborn Correspondence, Osborn Family Papers, NYHS.

33. Osborn to E. B. Wilson, Mar. 8, 1922, folder 9, box 86, Osborn Papers, AMNH.

34. Henry Fairfield Osborn, "William Bateson on Darwinism," *Science*, Feb. 24, 1922, 194–197.

35. Osborn, *Evolution and Religion in Education*, 3.

36. Osborn, "William Bateson on Darwinism."

37. Henry Fairfield Osborn, "William Bateson on Darwinism," *Science*, Apr. 7, 1922, 373; William Bateson, "Genetical Analysis and the Theory of Natural Selection," *Science*, Apr. 7, 1922, 373. See also correspondence between Osborn and Conklin in folder 31, box 17, Conklin Correspondence. Bateson's letter to Curtis, dated Dec. 11, 1922, is reprinted in Curtis, *Fundamentalism* vs. *Evolution at Dayton, Tennessee*, 12.

38. Conklin to David Starr Jordan, July 14, 1922, folder 23, box 12, Conklin Correspondence.

39. Edwin Linton, "The Scientific Method and Authority," *Science*, Feb. 12, 1926, 174.

40. This and following paragraphs: Ira D. Cardiff to Burton E. Livingston, Mar. 27, 1922, folder 1, box 86, Osborn Papers, AMNH.

41. Osborn to Livingston, May 22, 1922, folder 1, box 86, ibid.

42. The American Institute of Sacred Literature published pamphlets by Conklin, "Evolution and the Bible" (Chicago: First Impression, 1922); by Harry Emerson Fosdick, "Evolution and Mr. Bryan" (Chicago: First Impression, 1922); and by the University of Chicago modernist theologian Shailer Mathews, "How Science Keeps Our Faith" (Chicago: First Impression, 1922). These three pamphlets are preserved in folder 20, box 5, Osborn Papers, AMNH. For more on them, see Edward B. Davis, "Science and Religious Fundamentalism in the 1920s," *American Scientist* 93 (May–June 2005): 253–260, and Edward B. Davis, "Fundamentalist Cartoons, Modernist Pamphlets, and the Religious Image

of Science in the Scopes Era," in Charles L. Cohen and Paul S. Boyer, eds., *Religion and the Culture of Print in Modern America* (Madison: University of Wisconsin Press, forthcoming). I wish to thank Ted Davis for giving me an advance copy of this excellent essay.

43. Michael I. Pupin, *From Immigrant to Inventor* (1922; New York: Charles Scribner's Sons, 1930), 378–387. See also Michael I. Pupin, *The New Reformation: From Physical to Spiritual Realities* (New York: Charles Scribner's Sons, 1927).

44. The statement drafted by Millikan, a Nobel laureate in physics, and known as the Millikan Manifesto was Robert Andrews Millikan, "Joint Statement upon the Relations of Science and Religion, by Religious Leaders and Scientists," *Science*, June 1, 1923, 630–631. It was reprinted in Eldred C. Vanderlaan, ed., *Fundamentalism versus Modernism* (New York: H. W. Wilson, 1925), 294–296.

45. Davis, "Science and Religious Fundamentalism in the 1920s," 256.

46. Lienesch, *In the Beginning*, 77–79.

47. Walter C. Kraatz to Osborn, Apr. 13, 1924, folder 6, box 86, Osborn Papers, AMNH.

48. Miller's letters to Osborn and his newspaper articles are preserved in folder 5, box 15, ibid.; on H. H. Lane's concerns about losing his job, see Conklin to Osborn, June 27, 1925, folder 10, box 19, and H. H. Lane to Osborn, July 1, 1925, folder 11, box 19, both in Osborn Correspondence, Osborn Papers, AMNH.

49. Albert Edward Wiggam, *The New Decalogue of Science* (Indianapolis: Bobbs-Merrill, 1923), 115.

50. On eugenics, see Daniel J. Kevles, *In the Name of Eugenics: Genetics and the Uses of Human Heredity* (New York: Knopf, 1985); Edward J. Larson, *Evolution: The Remarkable History of a Scientific Theory* (New York: Modern Library, 2004), 177–197; Martin S. Pernick, *The Black Stork: Eugenics and the Death of "Defective" Babies in American Medicine and Motion Pictures since 1915* (New York: Oxford University Press, 1996); and Philip J. Pauly, *Biologists and the Promise of American Life: From Meriwether Lewis to Alfred Kinsey* (Princeton: Princeton University Press, 2000), 214–227.

51. Carroll Lane Fenton, "The Pathology of Rationalism," *Haldeman-Julius Monthly* 3 (May 1926): 704–711 (thanks to Mark Pittenger for telling me about this paper); Kirtley F. Mather, "Evolution and Religion," *Scientific Monthly* 21 (Sept. 1925): 322–328; Winterton C. Curtis, "The Fact, the Course, and the Causes of Organic Evolution," *Scientific Monthly* 21 (Sept. 1925): 295–302; Curtis, *Fundamentalism* vs. *Evolution at Dayton, Tennessee*; Kennard Baker Bork, *Cracking Rocks and Defending Democracy: Kirtley Fletcher Mather, Scientist, Teacher, Social Activist, 1888–1978* (San Francisco: Pacific Division, AAAS, 1994).

52. Correspondence between Raymond Pearl and H. L. Mencken in Raymond Pearl Papers, American Philosophical Society, Philadelphia.

53. Pauly, *Biologists and the Promise of American Life*, 194–213; Garland Allen, "Naturalists and Experimentalists: The Genotype and the Phenotype," *Studies in the History of Biology* 3 (1979): 179–209; Jane Maienschein, *Transforming Traditions in American Biology, 1880–1915* (Baltimore: Johns Hopkins University Press, 1991).

54. Vernon Kellogg, "Where Evolution Stands Today," *New Republic*, Apr. 11, 1923, 179–181.

55. "Calls John T. King to Finance Straton," *New York Times*, Jan. 28, 1924, 17.

56. Lienesch, *In the Beginning*, 86–87, 102, 105–106; John Roach Straton and Charles

Francis Potter, *Evolution versus Creation: Second in the Series of Fundamentalist-Modernist Debates* (1924), in Ronald L. Numbers, ed., *Creationism in Twentieth-Century America: A Ten-Volume Anthology of Documents, 1903–1961*, vol. 2: *Creation-Evolution Debates* (New York: Garland, 1995); "Akeley Bronze Ape to Stand in Church," *New York Times*, Apr. 9, 1924, 14.

57. Wendy McElroy, *Queen Silver: The Godless Girl*. (Amherst, NY: Prometheus Books, 2000), 177, 191. I thank Rickie Solinger for telling me about this book.

58. William Green to Osborn, folder 5, box 86, Osborn Papers, AMNH.

59. Dale M. Herder, "Haldeman-Julius, the Little Blue Books, and the Theory of Popular Culture," *Journal of Popular Culture* 4 (Spring 1971): 881–891; Dale M. Herder, "American Values and Popular Culture in the Twenties: The Little Blue Books," *Canadian Historical Association, Historical Papers* (1971): 289–299; Stuart McConnell, "E. Haldeman-Julius and the Little Blue Bookworms: The Bridging of Cultural Styles, 1919–1951," *Prospects* 11 (1987): 59–79.

60. Ernest Rice McKinney, "This Week," *New York Amsterdam News*, Apr. 15, 1925, sec. 2, p. 1.

61. William Pickens, "Bryan and Evolution," *New York Amsterdam News*, June 3, 1925. For more on African American responses to the evolution debate, see especially the excellent article by Jeffrey Moran, "Reading Race into the Scopes Trial: African American Elites, Science, and Fundamentalism," *Journal of American History* 90 (Dec. 2003): 891–911.

62. Maynard Shipley, *The War on Modern Science: A Short History of the Fundamentalist Attacks on Evolution and Modernism* (New York: Knopf, 1927).

63. Miriam Allen De Ford, *Up-Hill All the Way: The Life of Maynard Shipley* (Yellow Springs, OH: Antioch Press, 1956). See also Nick Salvatore, *Eugene V. Debs: Citizen and Socialist* (Urbana: University of Illinois Press, 1982), 331. On scientists' ambivalence about Shipley and the Science League, see, for example, Holmes to E. G. Conklin, Mar. 26, 1929; Conklin to H. S. Jennings, Oct. 26, 1925; and Jennings's reply to Conklin, Oct. 27, 1925, all in folder 14, box 11, Conklin Correspondence in the Conklin Papers. See also Conklin to Burton Livingston, Mar. 4, 1929, in Conklin Papers, organizations, AAAS folder; Shipley folder in Raymond Pearl Papers at the American Philosophical Society.

64. Jennings to Conklin, Oct. 27, 1925, folder 15, box 12, Conklin Papers; Conklin to Jennings, Oct. 26, 1925, ibid; Conklin to Jennings, Nov. 9, 1925, ibid. See also letters in the Raymond Pearl Papers and the Herbert Spencer Jennings Papers at the American Philosophical Society.

65. Clarence Darrow, introduction to Clarence Darrow and Wallace Rice, eds., *Infidels and Heretics: An Agnostic Anthology* (Boston: Stratford Co., 1929), i–x. For the popularity and influence of the books by Draper and White, see Larson, *Summer for the Gods*, 21–22.

66. Clarence Darrow, "Why I Am an Agnostic," reprinted in Arthur and Lila Weinberg, eds., *Clarence Darrow, Verdicts out of Court* (Chicago: Ivan R. Dee, 1989), 434.

67. Clarence Darrow, reprinted in Arthur Weinberg, ed., *Attorney for the Damned: Clarence Darrow in the Courtroom* (Chicago: University of Chicago Press, 1989), 56. This published version of the address was thoroughly edited by Darrow after the fact, however. Simon Baatz, pers. comm., Nov. 2, 2007. See the forthcoming book by Simon Baatz on the Leopold and Loeb case (HarperCollins).

68. Ibid., 44.

69. Ibid., 86–87.

70. Morrill Goddard to Osborn, May 24, 1923, folder 5, box 86, Osborn Papers, AMNH.

71. *Detroit Free Press* quoted in "Banishing Evolution in the South, *Literary Digest,* Apr. 3, 1926, 30.

72. For more on the use of the term *Puritan* in the 1920s, see Frederick J. Hoffman, "Philistine and Puritan in the 1920s," *American Quarterly* 1 (1949): 247–263.

73. "Tennessee vs. Truth," *Nation,* July 8, 1925, 58.

74. Rollin Lynde Hartt, "What Lies beyond Dayton," *Nation,* July 22, 1925, 111.

75. "Tennessee vs. Truth," 58.

76. Mather, "Evolution and Religion." Scientists' statements for the Scopes trial were published in *Scientific Monthly* as well as in the trial transcript; see "Evidences for Evolution: Statements Prepared for the Defense Counsel, State of Tennessee v. John Thomas Scopes," *Scientific Monthly* 21 (Sept. 1925): 291–328. See also Bork, *Cracking Rocks and Defending Democracy.*

Chapter Four • Saving the Phenomena

1. Charles Robert Darwin, *On the Origin of Species by Means of Natural Selection: Or the Preservation of Favoured Races in the Struggle for Life* (London: John Murray, 1859; facsimile of the first edition with an introduction by Ernst Mayr, Cambridge: Harvard University Press, 1964); John Ray, *The Wisdom of God Manifested in the Works of the Creation* (London: W. Innys and R. Manby, 1691); C. E. Raven, *John Ray: Naturalist* (1942; New York: Cambridge University Press, 1960); William Paley, *Natural Theology: Or, Evidences of the Existence and Attributes of the Deity Collected from the Appearances of Nature* (London: R. Faulder, 1802); Klaas van Berkel and Arjo Vanderjagt, eds, *The Book of Nature in Early Modern and Modern History* (Louvain, Belgium: Pieters, 2006), especially the essays by Peter Harrison, "'The Book of Nature' and Early Modern Science," 1–26; Edward B. Davis, "The Word and the Works: Concordism in American Evangelical Thought," 195–207; and Ronald L. Numbers, "Reading the Book of Nature through American Lenses," 261–274.

2. Stephen Jay Gould, *The Structure of Evolutionary Theory* (Cambridge: Harvard University Press, 2002); Dov Ospovat, *The Development of Darwin's Theory: Natural History, Natural Theology, and Natural Selection, 1838–1859* (New York: Cambridge University Press, 1981); Janet Browne, *Voyaging: Charles Darwin, a Biography,* vol. 1 (New York: Knopf, 1995), and Janet Browne, *The Power of Place: Charles Darwin, a Biography,* vol. 2 (New York: Knopf, 2002). Adrian Desmond and James Moore, *Darwin* (New York: Viking Penguin, 1991); Robert M. Young, *Darwin's Metaphor: Nature's Place in Victorian Culture* (New York: Cambridge University Press, 1985); Sandra Herbert, "Darwin, Malthus, and Selection," *Journal of the History of Biology* 4 (1971): 209–218; Peter J. Bowler, *Fossils and Progress: Paleontology and the Idea of Progressive Evolution in the Nineteenth Century* (New York: Science History Publications, 1976); Peter J. Bowler, "Darwinism and the Argument from Design: Suggestions for a Reevaluation," *Journal of the History of Biology* 10 (1977): 29–43; Adrian Desmond, *Archetypes and Ancestors: Palaeontology in Victorian London, 1850–1875* (London: Blond and Briggs, 1982); Martin J. S. Rudwick, *The Meaning of Fossils: Episodes in the History of Palaeontology* (London: Macdonald, 1972); Michael Ruse, *The Darwinian Revolution:*

Science Red in Tooth and Claw (Chicago: University of Chicago Press, 1979); David Kohn, ed., *The Darwinian Heritage* (Princeton: Princeton University Press, 1985).

3. Ernst Mayr, *The Growth of Biological Thought: Diversity, Evolution and Inheritance* (Cambridge: Harvard University Press, 1982); Ernst Mayr, "Prologue: Some Thoughts on the History of the Evolutionary Synthesis," in Mayr and William Provine, eds., *The Evolutionary Synthesis: Perspectives on the Unification of Biology* (Cambridge: Harvard University Press, 1980), 1–48; Peter J. Bowler, *Evolution: The History of an Idea* (Berkeley: University of California Press, 1989); Gould, *Structure of Evolutionary Theory.*

4. Ernst Mayr, *One Long Argument: Charles Darwin and the Genesis of Modern Evolutionary Thought* (Cambridge: Harvard University Press, 1991).

5. John Herschel quoted in David Hull, *The Metaphysics of Evolution* (Albany: State University of New York Press, 1989), 29.

6. Jon H. Roberts, *Darwin and the Divine in America: Protestant Intellectuals and Organic Evolution, 1859–1900* (Madison: University of Wisconsin Press, 1988); James R. Moore, *The Post-Darwinian Controversies: A Study of the Protestant Struggle to Come to Terms with Darwin in Great Britain and America, 1870–1900* (Cambridge: Cambridge University Press, 1979); Ronald L. Numbers and John Stenhouse, eds., *Disseminating Darwinism: The Role of Place, Race, Religion and Gender* (New York: Cambridge University Press, 1999); Gould, *Structure of Evolutionary Theory;* Peter J. Bowler, *The Eclipse of Darwinism: Anti-Darwinian Evolution Theories in the Decades around 1900* (Baltimore: Johns Hopkins University Press, 1983); Peter J. Bowler, *The Non-Darwinian Revolution: Reinterpreting a Historical Myth* (Baltimore: Johns Hopkins University Press, 1988); Peter J. Bowler, "Darwinism and Modernism: Genetics, Paleontology, and the Challenge to Progressionism, 1880–1930," in Dorothy Ross, ed., *Modernist Impulses in the Human Sciences, 1870–1930* (Baltimore: Johns Hopkins University Press, 1994), 236–54. For an examination of expressions of the wider influence on nonscientists of evolutionary thought, see Mark Pittenger, *American Socialists and Evolutionary Thought, 1870–1920* (Madison: University of Wisconsin Press, 1993). For clear discussions of the elements of chance implied in Darwin's theory by one of the preeminent biologists of the twentieth century, see Mayr, *One Long Argument;* see also Stephen Jay Gould, *Full House: The Spread of Excellence from Plato to Darwin* (New York: Harmony Books, 1996).

7. See, for example, Joseph LeConte, *Evolution and Its Relation to Religious Thought* (New York: D. Appleton, 1891).

8. Joseph LeConte, *Evolution: Its Nature, Its Evidences, and Its Relation to Religious Thought* (New York: D. Appleton, 1896); David Starr Jordan, *The Factors in Organic Evolution: A Syllabus of a Course of Elementary Lectures Delivered in Leland Stanford Junior University* (Boston: Ginn, 1895).

9. Bowler, *Eclipse of Darwinism;* Bowler, *Non-Darwinian Revolution.* Numbers, *Darwinism Comes to America,* 33–36; Roberts, *Darwin and the Divine in America,* 85–87.

10. LeConte, *Evolution;* Ronald Rainger, "The Continuation of the Morphological Tradition: American Paleontology, 1880–1919," *Journal of the History of Biology* 14 (1981): 129–158.

11. For instance, Samuel Butler, *Evolution, Old and New* (London: Harwicke and Bogue, 1879), and St. George Jackson Mivart, *On the Genesis of Species* (London: Macmillan, 1871).

For a perspective criticizing the notion that the fossil record demonstrates directionality and progress, see the elegant arguments in Stephen Jay Gould, *Wonderful Life: The Burgess Shale and the Nature of History* (New York: W. W. Norton, 1989), and Gould, *Full House*.

12. On the *scala naturae*, or great chain of being, see the classic text by Arthur O. Lovejoy, *The Great Chain of Being: A Study of the History of an Idea* (Cambridge: Harvard University Press, 1936).

13. Glenn L. Jepsen, "Selection, 'Orthogenesis,' and the Fossil Record," *Proceedings of the American Philosophical Society* 93, no. 6 (1949): 479–500.

14. Henry Fairfield Osborn, "The Hereditary Mechanism and the Search for the Unknown Factors of Evolution," *American Naturalist* 29 (1895): 435.

15. James McKeen Cattell, "The Material and Efficient Causes of Evolution," *Science*, n.s., 3 (1896): 668.

16. Henry Fairfield Osborn, "Organic Selection," *Science*, Oct. 15, 1897, 583–587; Henry Fairfield Osborn, "The Limits of Organic Selection," *American Naturalist* 31 (1897): 944–951; James Mark Baldwin, "A New Factor in Evolution," *American Naturalist* 30 (1896): 441–451, 536–553; James Mark Baldwin, "Consciousness and Evolution," *Psychological Review* 3 (1896): 300–308; James Mark Baldwin, "Organic Selection," *Nature*, Apr. 15, 1897, 558; James Mark Baldwin, "Organic Selection," *Science*, Apr. 23, 1897, 634–636; George Gaylord Simpson, "The Baldwin Effect," *Evolution* 7 (1953): 110–117; Robert J. Richards, *Darwin and the Emergence of Evolutionary Theories of Mind and Behavior* (Chicago: University of Chicago Press, 1987), 457–495.

17. For recent discussions of the Baldwin effect and its historiography, see Bruce H. Weber and David Depew, eds., *Evolution and Learning: The Baldwin Effect Reconsidered* (Cambridge, MA: MIT Press, 2003).

18. Robert Olby, *The Origins of Mendelism*, 2d ed. (Chicago: University of Chicago Press, 1985); William B. Provine, *The Origins of Theoretical Population Genetics* (Chicago: University of Chicago Press, 1971); Diane B. Paul and Barbara A. Kimmelman, "Mendel in America: Theory and Practice, 1900–1909," in Ronald Rainger, Keith R. Benson, and Jane Maienschein, eds., *The American Development of Biology* (Philadelphia: University of Pennsylvania Press, 1988), 281–310.

19. Ernst Mayr has said this in many places. See, for example, Mayr, *Growth of Biological Thought*, and Mayr, "Prologue," 1–48.

20. See Garland Allen, *Life Science in the Twentieth Century* (Cambridge: Cambridge University Press, 1978); Garland Allen, "Naturalists and Experimentalists: The Genotype and the Phenotype," *Studies in History of Biology* 3 (1979): 179–209; Garland Allen, "The Transformation of a Science: Thomas Hunt Morgan and the Emergence of a New American Biology," in Alexandra Oleson and John Voss, eds., *Organization of Knowledge in Modern America, 1860–1920* (Baltimore: Johns Hopkins University Press, 1979), 123–210; and the issue of the *Journal of the History of Biology* devoted to a debate about Allen's model, "Special Section on American Morphology at the Turn of the Century," *Journal of the History of Biology* 14 (1981), especially the introduction: Jane Maienschein, Ronald Rainger, and Keith R. Benson, "Introduction: Were American Morphologists In Revolt?" 83–87.

21. *Science*, July 29, 1927, 104.

22. Peter Bowler has established that in Britain theistic evolution appeared more often

in the literature aimed at the public, so that it seemed to be more common among scientists than it actually was, in part because it was espoused by older and more established scientists who wrote for the public more than did the younger ones. My survey of the U.S. literature suggests that this was true in the United States, too. Peter J. Bowler, *Reconciling Science and Religion: The Debate in Early-Twentieth-Century Britain* (Chicago: University of Chicago Press, 2001).

23. John O'Connell to Osborn, Nov. 28, 1930, folder 3, box 86, Osborn Papers, AMNH.

24. Conklin to Thomas Hunt Morgan, Jan. 6, 1924, folder 15, box 16, Conklin Correspondence.

25. Osborn to Morgan, Jan. 24, 1924, June 2, 1924, folder 1, box 16, Osborn Papers, AMNH.

26. Michael Lienesch, *In the Beginning: Fundamentalism, the Scopes Trial, and the Making of the Antievolution Movement* (Chapel Hill: University of North Carolina Press, 2007), 214–216; Christopher Toumey, *God's Own Scientists: Creationists in a Secular World* (New Brunswick, NJ: Rutgers University Press, 1994). I am very grateful to Michael Lienesch for a suggestion that helped me to think more clearly about this issue.

27. Steven Conn has argued that museums had been foci for the production of knowledge in the late nineteenth century but that, with the advent of graduate and professional schools founded on the German model of the research university, beginning with Johns Hopkins in 1876, universities increasingly dominated science as the places where knowledge was created. Steven Conn, *Museums and American Intellectual Life, 1876–1926* (Chicago: University of Chicago Press, 1998). On the intriguing and vexed relationship between Osborn and Morgan, see Allen, "Transformation of a Science."

28. Osborn to Butler, May 28, 1932, folder 5, box 4, Osborn Papers, AMNH. See also Osborn's correspondence with Morgan in folder 1, box 16, ibid.

29. Morgan to Osborn, folder 1, box 16, ibid.

30. See also Kathleen Pandora, "Knowledge Held in Common: Tales of Luther Burbank and Science in the American Vernacular," *Isis* 92 (Sept. 2001): 484–516.

Chapter Five • Unlikely Infidels

1. Editor, "The Scientist Bends the Knee," *New Republic*, Aug. 5, 1925, 280–281.

2. "House Reviews," *Variety*, July 15, 1925, 31.

3. The movie caption mistakenly claimed, for example, that the tapir was "the link between the rhinoceros and the elephant." It also misidentified several animals. "Evolution Film for Rivoli," *New York Evening Post*, July 3, 1925, 9.

4. Editor, "Scientist Bends the Knee." The film quoted the last two lines of a poem by William Herbert Carruth first published in 1902. The complete poem reads: "A fire-mist and a planet, / A crystal and a cell, / A jelly-fish and a saurian, / And caves where the cavemen dwell. / Then a glimpse of law and beauty / And a face turned *from* the sod:— / Some call it Evolution / And others call it God." It was among the most frequently reprinted among the plentiful poems about evolution in the 1920s. See, for example, John Roach Straton and Charles Francis Potter, *Evolution versus Creation: Second in the Series of Fundamentalist-Modernist Debates* (1924), in Ronald L. Numbers, ed., *Creationism in Twentieth-*

Century America: A Ten-Volume Anthology of Documents, 1903–1961, vol. 2: *Creation-Evolution Debates* (New York: Garland, 1995), 49, and the pamphlet in the Haldeman-Julius series, Langdon Smith, *Poems of Evolution* (Girard, KS: Haldeman-Julius, 1924), 3.

5. Henry Fairfield Osborn, *The Earth Speaks to Bryan* (New York: Charles Scribner's Sons, 1925), 3, reprinted in Henry Fairfield Osborn, *Evolution and Religion in Education: Polemics of the Fundamentalist Controversy of 1922 to 1926* (New York: Charles Scribner's Sons, 1926), 93–112. The book began as an article, "Evolution and Religion," in the *New York Times*, Mar. 5, 1922, sec.7, pp. 2, 14, in response to an article by Bryan the previous week: William Jennings Bryan, "God and Evolution," *New York Times*, Feb. 26, 1922, sec. 7, p. 1. The *Times* review of *The Earth Speaks to Bryan* was featured prominently on the first page of the Book Review section: Van Buren Thorne, "Nature Rebukes Mr. Bryan," *New York Times Book Review*, June 28, 1925, 1, 22. Other reviews of *The Earth Speaks to Bryan* are preserved in box 92 of the Osborn Papers, AMNH.

6. Bryan, "God and Evolution."

7. Osborn to Charles Scribner, Mar. 17, 1922, folder 1, box 92, Osborn Papers, AMNH. Osborn was extremely wealthy, and his acquaintance with people of even middle-class status, other than museum employees, appears to have been minimal. His income in 1922 was $88,258, in 1922 dollars, three-quarters of that from his father's estate. He lived on an estate on the Hudson and kept an apartment on Fifth Avenue in New York. See Osborn correspondence with Charles Scribner, folder 1, box 92, and folders 9 and 10, box 56, ibid. For his income, see Osborn Diary for 1923, Mar. 9, 1923, box 25, Henry Fairfield Osborn Diaries, Osborn Family Papers, NYHS.

8. Osborn to Scribner, Mar. 25, 1924, folder 10, box 56, Osborn Papers, AMNH.

9. Osborn, *Earth Speaks to Bryan*, 3, reprinted in Osborn, *Evolution and Religion in Education*, 93–112.

10. See especially Edward J. Larson, *Summer for the Gods: The Scopes Trial and America's Continuing Debate over Science and Religion* (New York: Basic Books, 1997), and Ronald L. Numbers, *Darwinism Comes to America* (Cambridge: Harvard University Press, 1998).

11. Conklin to Osborn, June 27, 1925, folder 10, box 19, Osborn Papers, AMNH. For more on Conklin, see J. W. Atkinson, "E. G. Conklin on Evolution: The Popular Writings of an Embryologist," *Journal of the History of Biology* 18 (1985): 31–50, and Jane Maienschein, *Transforming Traditions in American Biology, 1880–1915* (Baltimore: Johns Hopkins University Press, 1991).

12. Conklin to J. H. McGregor, Dec. 15, 1920, and other letters, folder 24, box 15, Conklin Correspondence.

13. Henry Fairfield Osborn to Scribner, May 22, 1922, folder 1, box 86, Osborn Papers, AMNH.

14. Osborn was quoted in the *New York Times*, Nov. 29, 1925. His statement in the *Times* subsequently stirred up a controversy in *Science*. A letter from a professor at the University of Nebraska denied Osborn's claim, rather hotly. Franklin D. Barker, "Evolution and the University of Nebraska," *Science*, Jan. 15, 1926, 71; Henry Fairfield Osborn, "Evolution and the University of Nebraska," *Science*, Feb. 19, 1926, 209.

15. Henry Fairfield Osborn, *The Age of Mammals in Europe, Asia, and North America* (New York: Macmillan, 1910), 7.

16. For more on Piltdown, see Frank Spencer, ed., *A History of American Physical Anthropology, 1930–1980* (New York: Academic Press, 1982), and Peter J. Bowler, *Theories of Human Evolution: A Century of Debate, 1844–1944* (Baltimore: Johns Hopkins University Press, 1986), esp. 35–38.

17. For more details on Osborn's disagreements with colleagues about human evolution and their relationship to the debates of the 1920s, see especially Brian Regal, *Henry Fairfield Osborn: Race and the Search for the Origins of Man* (Burlington, VT: Ashgate Press, 2002); Ronald Rainger, *An Agenda for Antiquity: Henry Fairfield Osborn and Vertebrate Paleontology at the American Museum of Natural History, 1890–1935* (Tuscaloosa: University of Alabama Press, 1991); Chris Beard, *The Hunt for the Dawn Monkey: Unearthing the Origins of Monkeys, Apes, and Humans* (Berkeley: University of California Press, 2004); Sheila Ann Dean, "What Animal We Came From: William King Gregory's Paleontology and the 1920s Debate on Human Origins" (Ph.D. diss., Johns Hopkins University, 1994); Misia Landau, *Narratives of Human Evolution* (New Haven: Yale University Press, 1991); and Bowler, *Theories of Human Evolution*.

18. Osborn, *Earth Speaks to Bryan*, 24.

19. Ibid., 87. The Millikan Manifesto was Robert Andrews Millikan, "Joint Statement upon the Relations of Science and Religion, by Religious Leaders and Scientists," *Science*, June 1, 1923, 630–31, reprinted in Eldred C. Vanderlaan, ed., *Fundamentalism versus Modernism* (New York, 1925): 294–296.

20. Kirtley F. Mather, "The Religion of a Geologist," popular leaflet, Religion and Science Series (American Institute of Sacred Literature, Chicago, 1931), copy preserved in box 87, Osborn Papers, AMNH; for more on Mather and on the pamphlets in this series, see Edward B. Davis, "Fundamentalist Cartoons, Modernist Pamphlets, and the Religious Image of Science in the Scopes Era," in Charles L. Cohen and Paul S. Boyer, eds., *Religion and the Culture of Print in Modern America* (Madison: University of Wisconsin Press, forthcoming).

21. Arthur M. Miller, "Evolution Identifies Creation with Growth Process," *Lexington (KY) Herald*, Feb. 1922, in folder 5, box 15, Osborn Papers, AMNH.

22. W. Maxwell Reed, *The Earth for Sam: The Story of Mountains, Rivers, Dinosaurs and Men* (New York: Harcourt, Brace and Co., 1930), 93.

23. David Meredith Seares Watson, "Evolution of the Bird," in Frances Mason, ed., *Creation by Evolution: A Consensus of Present-Day Knowledge As Set Forth by Leading Authorities in Non-Technical Language That All May Understand* (New York: Macmillan, 1928), 242–254.

24. J. Arthur Thomson, "Why We Must Be Evolutionists," in Mason, *Creation by Evolution*, 20. This is just one example of the frequently cited argument that all the lines of evidence converged in evolution and that this must be taken as evidence of the truth of evolution.

25. Arthur M. Miller, "The Evidence of Organic Evolution—Article III," *Lexington (KY) Herald*, Mar. 5, 1922, clipping in folder 5, box 15, Osborn Papers, AMNH.

26. Bryan, "God and Evolution."

27. Benjamin C. Gruenberg, "Unifying the Aims of High School Science Teaching," *School and Society*, Jan. 31, 1925, 121–192.

28. Edwin Grant Conklin, "Bryan and Evolution," *New York Times,* Mar. 5, 1922, sec. 7, p. 14.

29. "Skull of Bryan Is Neanderthal Type, Luther Burbank Avers," *New York World,* Dec. 23, 1924, 1. Copy preserved in folder 16, box 86, Osborn Papers, AMNH. It is mislabeled "New York Post" in that archive.

30. Conklin, "Bryan and Evolution."

31. Conklin to Osborn, Jan. 16, 1923, folder 31, box 17, Conklin Correspondence; Clifford W. Francis to Osborn, May 22, 1924, folder 31, box 7, Osborn Papers, AMNH.

32. "Skull of Bryan Is Neanderthal Type, Luther Burbank Avers," *New York World,* Dec. 23, 1924, 1.

33. Osborn, *Evolution and Religion in Education,* 45.

34. Osborn to Straton, Mar. 7, 1924, folder 4, box 21, Osborn Papers, AMNH; Osborn to Butler, July 17, 1925, folder 11, box 19, ibid.

35. "Evolution Museum Dedicated at Yale," *New York Times,* Dec. 30, 1925, 9.

36. "Banishing Evolution in the South," *Literary Digest,* Apr. 3, 1926, 30.

37. Mason, *Creation by Evolution,* See Frances Mason's correspondence with Osborn in folders 13 and 14, box 14, Osborn Papers, AMNH, and in folder 10, box 15, Conklin Correspondence.

38. Mason to Conklin, Jan. 13, 1927, folder 10, box 15, Conklin Correspondence.

39. Conklin to Mason, Jan. 14, 1927, ibid.

40. Mason to Conklin, no date, folder 10, box 15, ibid.

41. Conklin to Mason, Mar. 15, 1927, ibid.

42. *Creation by Evolution* was listed among the best sellers of the decade in "Books for Babbitt," *World's Work* 58 (June 1929), copy in folder 14, box 14, Osborn Papers, AMNH.

43. Review by John Gould Curtis, "The Bookmakers," *Erie Daily Times,* n.d., and advertising fliers in folder 14, box 14, Osborn Papers, AMNH.

44. Henry Skinner, American Entomological Society, to Osborn, July 21, 1925, folder 2, box 92, Osborn Papers, AMNH.

45. McAlister Coleman, "Dayton Re-Echoes: A Flood of Books on Evolution, Religion and Biology," *New York City,* Sept. 12, 1925, clipping in folder 6, box 92, Osborn Papers, AMNH. See also other clippings and correspondence on *The Earth Speaks to Bryan* in box 92 of the Osborn Papers, AMNH.

46. Edward W. Berry, "Antaeus, or the Future of Geology," *Science,* May 7, 1926, 475–476.

47. Vernon Kellogg, "Some Things Science Doesn't Know," *World's Work* 51 (Mar. 1926): 523–529.

48. Henshaw Ward, *Evolution for John Doe* (Indianapolis: Bobbs-Merrill, 1925).

49. Michael I. Pupin, quoted in Albert Edward Wiggam, "Science Is Leading Us Closer to God: An Interview with Michael Pupin, the Distinguished Scientist," *American Magazine* (Sept. 1927), reprinted in "The Faith of a Great Scientist," *Literary Digest,* Oct. 1, 1927, 33.

50. Ira D. Cardiff, "Evolution and the Bible," *Science,* July 31, 1925, 111.

51. James J. Porter, letter to the editor, *New Republic,* Aug. 12, 1925, 323.

52. Editorial, *Truth Seeker,* July 11, 1925, clipping in folder 6, box 92, Osborn Papers, AMNH.

53. Jacob Benjamin, "The Pious Intellectuals," *Truth Seeker*, Aug. 15, 1925, 519–520.

54. Maynard Shipley, *The War on Modern Science: A Short History of the Fundamentalist Attacks on Evolution and Modernism* (New York: Knopf, 1927); Miriam Allen De Ford, *Up-Hill All the Way: The Life of Maynard Shipley* (Yellow Springs, OH: Antioch Press, 1956).

55. Joseph McCabe, *The Story of Religious Controversy* (Boston: Stratford Co., 1929), 18. For more on McCabe, see Bill Cooke, *A Rebel to His Last Breath: Joseph McCabe and Rationalism* (Amherst, NY: Prometheus Books, 2001).

56. McCabe, *Story of Religious Controversy*, 21.

57. Ibid., 57 and 557.

58. Ibid., 84.

59. Walter Lippmann, *American Inquisitors: A Commentary on Dayton and Chicago* (New York: Macmillan, 1928); John Crowe Ransom, *God without Thunder: An Unorthodox Defense of Orthodoxy* (New York: Harcourt, Brace and Co., 1930). See also Paul Jerome Croce, "Beyond the Warfare of Science and Religion in American Culture—And Back Again," *Religious Studies Review* 26 (Jan. 2000): 29–35.

60. For sympathetic and insightful discussions of religious objections to theistic evolutionism in this scientific corollary of theological modernism, see Michael Lienesch, *In the Beginning: Fundamentalism, the Scopes Trial, and the Making of the Antievolution Movement* (Chapel Hill: University of North Carolina Press, 2007); Numbers, *Darwinism Comes to America;* Larson, *Summer for the Gods;* and Paul K. Conkin, *When All the Gods Trembled: Darwinism, Scopes, and American Intellectuals* (Lanham, MD: Rowman and Littlefield, 1998).

61. Francis P. Le Buffe, S.J., "Evolution and Religion," *America*, Oct. 27, 1923, 29–30.

62. Conklin to Le Buffe, May 19, 1932, box 13, Conklin Correspondence. For more on Conklin and religion, see also Conklin, "Edwin Grant Conklin," in Louis Finklestein, ed., *Thirteen Americans: Their Spiritual Biographies* (Port Washington, NY: Kennikat Press, 1953), 47–76; Maienschein, *Transforming Traditions in American Biology;* and Atkinson, "E. G. Conklin on Evolution."

63. Edmund Wilson, *The American Earthquake: A Documentary of the Twenties and Thirties* (New York: Farrar, Straus and Giroux, 1979), 75–76.

64. Henry Fairfield Osborn, interview with Alva Johnson, *New York Herald*, June 10, 1923, quoted in Le Buffe, "Evolution and Religion," 29–30.

65. Edwin Grant Conklin, *The Direction of Human Evolution* (New York: Charles Scribner's Sons, 1921), 213, quoted in Francis P. Le Buffe, S.J., "Human Evolution and Science," pamphlet, copy in box 87, Osborn Papers, AMNH.

66. Francis P. Le Buffe, S. J., "Human Evolution and Science," pamphlet, copy in box 87, Osborn Papers, AMNH.

67. Alfred Watterson McCann, *God—or Gorilla: How the Monkey Theory of Evolution Exposes Its Own Methods, Refutes Its Own Principles, Denies Its Own Inferences, Disproves Its Own Case* (New York: Devin-Adair, 1922); "How Missing Links Are Made," *Sunday School Times*, Nov. 18, 1922, in folder 5, box 86, Osborn Papers, AMNH. There is also a collection of pamphlets by Father Francis Le Buffe in box 87, Osborn Papers, AMNH. See also William King Gregory, "Physical Evidence of the Origin of Man," *New York Evening Post*, Apr. 1, 1922, 9; "Says Soul of Man Is God's Creation. Pastor Asserts That Many Catholics Believe in Doctrine of Evolution," *New York Times*, Jan. 14, 1926, 6; and newspaper ac-

counts of Osborn's exchanges with Straton and O'Connell in folder 2, box 86, Osborn Papers, AMNH.

Chapter Six • *Stooping to Conquer, and a Hall Full of Elephants*

1. O. Farneur to Osborn, Feb. 1, 1926, folder 3, box 86, Osborn Papers, AMNH. Cardinal O'Connell's criticisms of the Hall of the Age of Man were widely reported in the press; see the collection of correspondence and clippings in folder 2, box 86, ibid.

2. Henry Fairfield Osborn, *The Age of Mammals in Europe, Asia, and North America* (New York: Macmillan, 1910); Barnum Brown, "Tyrannosaurus, the Largest Flesh-Eating Animal That Ever Lived," *American Museum Journal* 15 (1915): 271–290; Henry Fairfield Osborn, "The 'Ostrich' Dinosaur and the 'Tyrant' Dinosaur," *American Museum Journal* 17 (1917): 5–13; William H. Ballou, "Strange Creatures of the Past: Gigantic Saurians of the Reptile Age," *Century Magazine* 55 (1897): 15–23; Walter L. Beasley, "A Carnivorous Dinosaur: A Reconstructed Skeleton of a Huge Saurian," *Scientific American*, Dec. 14, 1907, 446. The *New York Times* carried an illustrated article about *Tyrannosaurus rex*, Dec. 3, 1905, sec. 3, p. 1. For more on Osborn's naming of *T. rex* and his work on popularizing dinosaurs, see also Ronald Rainger, *An Agenda for Antiquity: Henry Fairfield Osborn and Vertebrate Paleontology at the American Museum of Natural History, 1890–1935* (Tuscaloosa: University of Alabama Press, 1991), 94–99.

3. Henshaw Ward, *Evolution for John Doe* (Indianapolis: Bobbs-Merrill, 1925), 15; see also Benjamin C. Gruenberg, *The Story of Evolution: Facts and Theories on the Development of Life* (1919; Garden City, NY: Garden City Publishing Co., 1929), a book that effectively adopted the approach recommended by Farneur.

4. Henry Fairfield Osborn, "The Museum a New Force in Education," in Osborn, *Creative Education in School, College, University and Museum* (New York: Scribner's, 1927), 252–253.

5. Clippings from newspapers reporting on Cardinal O'Connell's attacks on Osborn and the museum are preserved in folder 2, box 86, Osborn Papers, AMNH. For attacks on Osborn and the museum and Osborn's exchanges with Straton, see sources in Chap. 2, nn. 2 and 27. Osborn's correspondence with Father Francis P. Le Buffe includes similar material, as does the collection of papers related to *The Earth Speaks to Bryan*, in box 92, ibid.

6. Henry Fairfield Osborn, *Mastodons and Mammoths of North America*, Guide Leaflet No. 62 (New York: American Museum of Natural History, 1926), 4, reprinted from *Natural History* 23 (Feb. 5, 1923): 3–24, and *Natural History* 25 (Feb. 27, 1925): 3–23.

7. The story of Uncle Pierpont and the hall full of elephants is told in John Michael Kennedy, "Philanthropy and Science in New York City: The American Museum of Natural History, 1868–1968" (Ph.D. diss., Yale University, 1968), 183–186.

8. Henry Fairfield Osborn, "The Hall of the Age of Man in the American Museum," *Natural History* 20 (May–June 1920): 238. For an account of Osborn's work on elephants that emphasizes its orthogenetic interpretation, see Claudine Cohen, *The Fate of the Mammoth: Fossils, Myth, and History*, trans. William Rodarmor (Chicago: University of Chicago Press, 2002), 176–179.

9. Peter J. Bowler, *Theories of Human Evolution: A Century of Debate, 1844–1944* (Baltimore: Johns Hopkins University Press, 1986); Misia Landau, *Narratives of Human Evolu-*

tion (New Haven: Yale University Press, 1991); Erik Trinkhaus and Pat Shipman, *The Neandertals: Of Skeletons, Scientists, and Scandal* (New York: Vintage Books, 1992).

10. See Henry Fairfield Osborn, *Creative Education in School, College, University and Museum* (New York: Charles Scribner's Sons, 1927), 252–253.

11. These letters occur throughout the Osborn Papers, AMNH, but see especially box 27, which contains miscellaneous requests for information and opinions; box 19, the Scopes trial box; and the boxes of correspondence, pamphlets, and clippings on evolution, boxes 85, 86, and 87.

12. Osborn's paper mentioning Frémiet and the rumors of gorilla abductions is Henry Fairfield Osborn, "The Influence of Habit in the Evolution of Man and the Great Apes," *Bulletin of the New York Academy of Medicine* 4 (1928): 216–230. See also the sources listed in Chap. 1, n. 3. For the story of Frémiet's gorilla sculpture and a photograph of it, see Roselyne de Ayala and Jean-Pierre Gueno, eds., *Illustrated Letters: Artists and Writers Correspond* (New York: Harry N. Abrams, 1999), 98.

13. See also Sheila Ann Dean, "What Animal We Came From: William King Gregory's Paleontology and the 1920s Debate on Human Origins" (Ph.D. diss., Johns Hopkins University, 1994).

14. Henry Fairfield Osborn, *New York Times*, Mar. 5, 1922, sec. 7, p. 2.

15. Henry Fairfield Osborn, "Why Central Asia?" *Natural History* 26 (June 25, 1926): 266–267.

16. Madison Grant, *The Passing of the Great Race, or the Racial Basis of European History* (New York: Charles Scribner's Sons, 1916). See also the extensive correspondence between Osborn and Grant in folders 38–40, box 8, and folders 1–5, box 9, Osborn Papers, AMNH; Osborn's correspondence with the anthropologist Robert Lowie, especially Lowie to Osborn, June 19, 1922, calling Grant a charlatan, folder 6, box 13, ibid; and a 1924 exchange of letters between Osborn and Walter Lippmann, folder 1, box 13, ibid.

17. Raymond Dart, "*Australopithecus africanus:* The Man-Ape of South Africa," *Nature* 115 (1925): 195–199; Davidson Black, "Asia and the Dispersal of Primates," *Bulletin of the Geological Society of China* 4 (1925): 133–183.

18. G. H. Findlay, *Dr. Robert Broom, F.R.S. Palaeontologist and Physician 1866–1951* (Cape Town: A. A. Balkema, 1972), 31–32, 52–55.

19. Arthur Keith, "Whence Came the White Race?" *New York Times Sunday Magazine,* Oct. 12, 1930, SM1. See also Peter J. Bowler, "Darwinism and Modernism: Genetics, Paleontology, and the Challenge to Progressionism, 1880–1930," in Dorothy Ross, ed., *Modernist Impulses in the Human Sciences, 1870–1930* (Baltimore: Johns Hopkins University Press, 1994), 236–254.

20. Nelson to Henry C. Tracy, no date, 1928, folder 3, box 99, Osborn Papers, AMNH.

21. Gregory, Jan. 29, 1920, memo to Osborn, in old correspondence folder, Hall of the Age of Man, Library, American Museum of Natural History, New York. Osborn had published this idea in *Men of the Old Stone Age: Their Environment, Life, and Art* (1915; New York: Charles Scribner's Sons, 1916), 61; he retained it in the 1930 edition (also p. 61).

22. William Diller Matthew, "The Value of Palaeontology," *Natural History* 25 (Mar.– Apr. 1925): 167–168.

23. Osborn to Charles Scribner, June 3, 1925, folder 1, box 92, Osborn Papers, AMNH.

24. Osborn to Alexander Ghigi, Sept. 23, 1925, folder 3, box 92, ibid.

25. Osborn to Scribner, June 3, 1925, and June 10, 1925, folder 1, box 92, ibid.

26. Henry Fairfield Osborn, "*Hesperopithecus*, the First Anthropoid Primate Found in America," *Proceedings of the National Academy of Sciences* 8 (1922): 246; the *Hesperopithecus* announcement was also published under the same title in *American Museum Novitates* 37 (1922): 1–5, and in *Nature* as "*Hesperopithecus*, the Anthropoid Primate of Western Nebraska," *Nature* 110 (1922): 281–283.

27. Edwin Slosson to Osborn, May 4, 1922, folder 4, box 56, Osborn Papers, AMNH.

28. Grafton Elliott Smith, "*Hesperopithecus*: the Ape-Man of the Western World," *Illustrated London News*, June 24, 1922, 944.

29. Osborn, "*Hesperopithecus*, the Anthropoid Primate of Western Nebraska."

30. William King Gregory, "*Hesperopithecus* Apparently Not An Ape nor a Man," *Science*, n.s., 66 (1927): 579–581.

31. Straton to Osborn, Apr. 12, 1928, folder 4, box 21, Osborn Papers, AMNH.

32. Stephen Jay Gould, "An Essay on a Pig Roast." *Natural History* 89 (Jan. 1989): 14–24. Reprinted in Gould, *Bully for Brontosaurus: Reflections in Natural History* (New York: W. W. Norton, 1991), 432–447.

33. Osborn's species diagnoses of fossil elephants are documented in Henry Fairfield Osborn, *Proboscidea* (Washington, DC: U.S. Government Printing Office, 1936). The species descriptions in *Proboscidea* include magnificent illustrations and detailed descriptions of the fossils used in naming new species.

34. Henry Fairfield Osborn, *The Earth Speaks to Bryan* (New York: Charles Scribner's Sons, 1925), 40.

35. "Nebraska's 'Ape Man of the Western World,'" *New York Times*, Sept. 17, 1922, sec. 7, p. 2.

36. Turpin to Osborn, Sept. 4, 1922, folder 1, box 50, Osborn Papers, AMNH.

37. Conklin to Osborn, Jan. 16, 1923, folder 1, box 86, ibid.; George Rappleyea to Osborn, box 19, ibid.

38. Forrest Davis, "Bryan Tells Dayton Trial Is Death Duel," *New York Herald Tribune*, July 8, 1925, clipping in folder 15, box 19, Osborn Papers, AMNH.

39. Gregory to Lucretia Perry Osborn, June 6, 1928, folder 11, box 9, Library, American Museum of Natural History, New York.

40. Osborn to Lankester, Oct. 10, 1921, folder 13, box 28, Osborn Papers, AMNH.

41. Osborn to Prof. William M. Goldsmith, Dec. 28, 1922, folder 5, box 86, ibid.

42. McGregor to Osborn, June 9, 1925, folder 23, box 13, ibid.; Alfred Watterson McCann, *God—or Gorilla: How the Monkey Theory of Evolution Exposes Its Own Methods, Refutes Its Own Principles, Denies Its Own Inferences, Disproves Its Own Case* (New York: Devin-Adair, 1922).

43. "How 'Missing Links' Are Made," *Sunday School Times*, Nov. 18, 1922, clipping in folder 5, box 86, Osborn Papers, AMNH.

44. "Prof. Osborn Defends 'Age of Man' Exhibit. Natural History Museum Head Makes Brief Reply to Attack by Cardinal O'Connell," *New York Times*, Feb. 2, 1926, 10, 2. Stories of O'Connell's attacks and Osborn's responses were also reported by the wire ser-

vices and therefore carried nationwide. For correspondence and news clippings, see folder 2, box 86, Osborn Papers, AMNH.

45. Dean, "What Animal We Came From," 259–409.

46. There are examples in the Osborn correspondence in the Osborn Papers, AMNH; see especially box 27, requests for information, and box 86, evolution.

47. Conklin to Morgan, Jan. 2, 1929, folder 15, box 16, Conklin Correspondence.

48. Osborn, "Influence of Habit," 249. McGregor's comment was made during the discussion following Osborn's paper. The discussion was published with the paper.

49. William K. Gregory and J. Howard McGregor, "A Dissenting Opinion as to Dawn Men and Ape Men," *Natural History* 26 (1926): 270–271.

50. A. S. Eve, "Discussion and Correspondence: What Did Darwin Write?" *Science,* Dec. 31, 1926, 649.

51. Osborn to William H. Witte, Jan. 26, 1926, folder 9, box 86, Osborn Papers, AMNH.

52. Ibid.

53. *New York Evening Post,* cartoon, July 14, 1925. The cartoon showed Bryan carrying a weapon labeled "Duel to the Death" glaring at a scarecrow labeled "Science—Imaginary Enemy of Religion."

54. "Evolutionist Logic False, Divine Says: Exhibits of Missing Links Are Attacked," *Tampa Daily Times,* June 29, 1925, clipping in folder 14, box 19, Osborn Papers, AMNH.

55. Wilfred Parson, S.J., "Hypothesis, Theory, or Fact?" *America,* July 25, 1925, 344–345.

56. Francis P. Le Buffe, S.J., "An Evolutionist's Broadside," *America,* July 25, 1925, 346–347.

57. Nelson to Osborn, Nov. 24, 1924, folder 6, box 16, Osborn Papers, AMNH.

58. Ibid.

59. Byron Cummings, "Problems of a Scientific Investigator," *Science,* Mar. 26, 1926, 321.

60. Osborn to Davenport, June 13, 1922, folder 1, box 86, Osborn Papers, AMNH.

61. "The Book Column: Man and Ape—Bryan, Osborn, and a Battle of Books," *New York City Sun,* July 9, 1925, clipping in folder 6, box 92, Library, American Museum of Natural History, New York.

62. T. D. A. Cockerell, "The Duty of Biology," *Science,* Apr. 9, 1926, 367–371.

63. Harry Hansen, "Ee—volution," *Chicago News,* July 1, 1925, clipping in folder 15, box 19, Osborn Papers, AMNH.

64. Editorial, *Truth Seeker,* July 25, 1925, 467.

65. *Life,* July 30, 1925, clipping in folder 15, box 19, Osborn Papers, AMNH.

66. Walter Lippmann, *Public Opinion* (1922; New York: Free Press, 1997).

Chapter Seven • The Pictures in Our Heads

1. H. L. Mencken, "Malone the Victor, Even Though Court Sides with Opponents, Says Mencken," *Baltimore Evening Sun,* July 17, 1925, reprinted in Marion Elizabeth Rodgers, ed., *The Impossible H. L. Mencken. A Selection of His Best Newspaper Stories* (New York: An-

chor Books, Doubleday, 1991), 595. Joseph Wood Krutch, *More Lives Than One* (New York: William Sloane Associates, 1962), 153; Joseph Wood Krutch, "The Monkey Trial," *Commentary* 43 (May 1967): 84.

2. Arthur Garfield Hays, quoted in Lawrence Mark Bernabo, "The Scopes Myth: The Scopes Trial in Rhetorical Perspective" (Ph.D. diss., University of Iowa, 1990), 262.

3. The transcript was published as *Tennessee Evolution Case: A Complete Stenographic Report of the Famous Court Case of the Tennessee Anti-Evolution Act, at Dayton, July 10 to 21, 1925, Including Speeches and Arguments of Attorneys* (Cincinnati: National Book Co., 1925), 174–177; the diagram appeared in George William Hunter, *A Civic Biology Presented in Problems* (New York: American Book Co., 1914), 194.

4. William Jennings Bryan and Mary Baird Bryan, *The Memoirs of William Jennings Bryan* (Philadelphia: John C. Winston, 1925), 535.

5. "'Monkey War' Echoed in Fairfax," *Washington Daily News*, May 30, 1925, folder 3, box 44, Science Service Archives, Smithsonian Institution, Washington, D.C.

6. Howard E. Gruber, "Darwin's 'Tree of Nature' and Other Images of Wide Scope," in Judith Wechsler, ed., *On Aesthetics in Science* (Cambridge, MA: MIT Press, 1978), 121–140.

7. Ernst Haeckel, *The Evolution of Man* (1866; New York: H. L. Fowle, 1896), 189. On Haeckel's tree diagrams, see also Jane M. Oppenheimer, "Haeckel's Variations on Darwin," in Henry M. Hoenigswald and Linda F. Wiener, eds., *Biological Metaphor and Cladistic Classification: An Interdisciplinary Perspective* (Philadelphia: University of Pennsylvania Press, 1987), 123–135; Stephen Alter, *Darwinism and the Linguistic Image: Language, Race, and Natural Theology in the Nineteenth Century* (Baltimore: Johns Hopkins University Press, 1999), 108–145; Stephen Jay Gould, "Ladders and Cones: Constraining Evolution by Canonical Icons," in Robert B. Silvers, ed., *Hidden Histories of Science* (New York: New York Review of Books, 1995), 37–67; Martin Kemp, "Haeckel's Hierarchies," in Kemp, *Visualizations: The Nature Book of Art and Science* (Berkeley: University of California Press, 2000), 90–91; and Peter J. Bowler, "Darwinism and Modernism: Genetics, Paleontology and the Challenge to Progressionism, 1880–1930," in Dorothy Ross, ed., *Modernist Impulses in the Human Sciences, 1870–1930* (Baltimore: Johns Hopkins University Press, 1994), 236–254. On racial hierarchies in Haeckel's diagrams, and on their influence, see Nick Hopwood, "Pictures of Evolution and Charges of Fraud: Ernst Haeckel's Embryological Illustrations," *Isis* 97 (June 2006): 260–301. For more on the concept of progress in evolutionary thought, see Michael Ruse, *Monad to Man: The Concept of Progress in Evolutionary Biology* (Cambridge, MA: Harvard University Press, 1996).

8. The art historian Martin Kemp has pointed out that two trees from Haeckel's *History of Creation*, translated into English by E. Ray Lankester in 1875–76, were "transformed into fern-like structures, with a degree of impressionistic suggestiveness commensurate with the uncertainties of the fossil record." Kemp, "Haeckel's Hierarchies," 91. Haeckel's fernlike trees were published in Ernst Haeckel, *The History of Creation*, trans. E. Ray Lankester (London: John Murray, 1875–76), vol. 2, opposite p. 223. For more on the earlier history of tree and treelike diagrams, see William Coleman, "Morphology between Type Concept and Descent Theory," *Journal of the History of Medicine and Allied Sciences* 31, no. 2 (1976): 149–175; Robert J. O'Hara, "Representations of the Natural System in the Nineteenth Century," *Biology and Philosophy* 6 (Apr. 1991): 255–274; H. J. Lam, "Phylogenetic

Symbols, Past and Present," *Acta Biotheoretica* 2 (Oct. 1936): 153–194; Edward G. Voss, "The History of Keys and Phylogenetic Trees in Systematic Biology," *Journal of the Scientific Laboratories of Denison University* 43 (Dec. 1953): 1–25; Gruber, "Darwin's 'Tree of Nature' "; Gould, "Ladders and Cones"; and Theodore D. McCown and Kenneth A. R. Kennedy, eds., *Climbing Man's Family Tree: A Collection of Major Writings on Human Phylogeny, 1699 to 1971* (Englewood Cliffs, NJ: Prentice-Hall, 1972).

9. Sir Arthur Keith, *The Construction of Man's Family Tree* (London: Watts and Co., 1934), 2–3.

10. William King Gregory, "The Origin, Rise and Decline of *Homo sapiens,*" *Scientific Monthly* 39 (Dec. 1934): 488.

11. Keith, *Construction of Man's Family Tree,* 10.

12. Benjamin C. Gruenberg, *The Story of Evolution: Facts and Theories on the Development of Life* (1919; Garden City, NY: Garden City Publishing Co., 1929), 71; Adam Gowans Whyte, *The Wonder World We Live In* (New York: Knopf, 1921), frontispiece.

13. Thomas Henry Huxley, *Man's Place in Nature* (1863; Ann Arbor: University of Michigan Press, 1959), frontispiece, 129–130. See also Nicolaas Rupke, *Richard Owen: Victorian Naturalist* (New Haven: Yale University Press, 1994), and Adrian Desmond, *Huxley: From Devil's Disciple to Evolution's High Priest* (Reading, MA: Addison-Wesley, 1997).

14. *New Yorker,* June 6, 1925, 3; *Judge,* July 18, 1925, 2. A similar sequence appeared on the cover of *Amazing Stories* (Apr. 1928). Michelle Hope Herwald, "Amazing Artifact: Cultural Analysis of *Amazing Stories* 1926–1938" (Ph.D. diss., University of Michigan, 1977).

15. Ronald Rainger argues that horse diagrams at the museum reflected Osborn's linear, progressive model of evolution. See Ronald Rainger, *Agenda for Antiquity: Henry Fairfield Osborn and Vertebrate Paleontology at the American Museum of Natural History, 1890–1935* (Tuscaloosa: University of Alabama Press, 1991), 164–165, 208–210. W. D. Matthew, "The Evolution of the Horse: A Record and Its Interpretation," *Quarterly Review of Biology* 1, no. 2 (1926): 139–185; American Museum of Natural History, *Evolution of the Horse* (New York: American Museum of Natural History, 1924); Thomas Henry Huxley, *American Addresses* (London: Macmillan, 1877); Stephen Jay Gould, *Bully for Brontosaurus* (New York: W. W. Norton, 1991), 168–181; George Gaylord Simpson, *Horses: The Story of the Horse Family in the Modern World and Through Sixty Million Years of History* (New York: Oxford University Press, 1951).

16. W. Maxwell Reed, *The Earth for Sam: The Story of Mountains, Rivers, Dinosaurs and Men* (New York: Harcourt, Brace and Co., 1930), 258.

17. "House Reviews," *Variety,* July 15, 1925, 31. For the account of Straton's sermon, see John L. Low to Osborn, Mar. 24, 1924, folder 6, box 86, Osborn Papers, AMNH.

18. S. R. Mayer-Oakes to Osborn, Oct. 21, 1930, folder 1, box 14, Osborn Papers, AMNH.

19. William King Gregory and Marcelle Roigneau, *Introduction to Human Anatomy: Guide to Section I of the Hall of Natural History of Man* (New York: American Museum of Natural History, 1934), 27.

20. For Gregory's discussions of human evolution, tree diagrams, and palimpsest evolution, see, for example, the following works by him: "Studies on the Evolution of the Primates," *Bulletin of the American Museum of Natural History* 35 (1916): 239–355; *Our Face*

from Fish to Man: A Portrait Gallery of Our Ancient Ancestors and Kinsfolk Together with a Concise History of Our Best Features (New York: G. P. Putnam's Sons, 1929); "Origin, Rise and Decline of *Homo sapiens*," 481–496; and "Supra-specific Variation in Nature and Classification: A Few Examples from Mammalian Paleontology," *American Naturalist* 71 (1937): 268–276. See also the study of Gregory by Sheila Ann Dean, "What Animal We Came From: William King Gregory's Paleontology and the 1920s Debate on Human Origins" (Ph.D. diss., Johns Hopkins University, 1994).

21. Gregory, "Studies on the Evolution of the Primates."

22. Vivienne Rae-Ellis, "Representing Trucanini," in Elizabeth Edwards, ed., *Anthropology and Photography, 1860–1920* (New Haven: Yale University Press, 1992), 230–233; James Ryan, "Images and Impressions: Printing, Reproduction and Photography," in John M. MacKenzie, ed., *The Victorian Vision: Inventing New Britain* (London: V & A Publications, 2001), 214–239. There is evidence of the lasting influence of Trucanini's image in the fact that a contemporary Australian rock group, "Midnight Oil," has recorded a song about her. I thank Allison Wickens for pointing this out to me.

23. See Philip J. Deloria, *Playing Indian* (New Haven: Yale University Press, 1998); Philip J. Deloria, *Indians in Unexpected Places* (Lawrence: University of Kansas Press, 2004); and Judith C. Berman, "A Note on the Paintings of Prehistoric Ancestors by Charles R. Knight," *American Anthropologist* 105 (Mar. 2003): 143–146.

24. On racial and eugenics themes at the American Museum, see the sources in Chap. 2, n. 21. For the history of eugenics, see especially Daniel J. Kevles, *In the Name of Eugenics: Genetics and the Uses of Human Heredity* (New York: Knopf, 1985). For a review of the voluminous literature on eugenics, see Philip J. Pauly, "Essay Review: The Eugenics Industry—Growth or Restructuring?" *Journal of the History of Biology* 26 (1993): 131–145, and Barry Alan Mehler, "A History of the American Eugenics Society, 1921–1940" (Ph.D. diss., University of Illinois at Urbana-Champaign, 1988).

25. Hunter, *Civic Biology*; Henry Fairfield Osborn, "Our Ancestors Arrive in Scandinavia," *Natural History* 22 (Mar.–Apr. 1922): 116–134.

26. "Ascent to Utopia Evolution's Aim, Says Savant, Decrying Quibbling," *Poughkeepsie Enterprise*, June 22, 1925, clipping in folder 14, box 19, Osborn Papers, AMNH; *New York Times*, July 12, 1925, sec. 8, p. 1. This diagram also appeared in George Grant MacCurdy, "Old Problems and New Methods in Prehistory," *Scientific American* 134 (May 1926): 308–309, and William King Gregory, "Did Man Originate in Central Asia?" *Scientific Monthly* 24 (May 1927): 385–401. See also correspondence between Osborn and Gregory about the preparation of the exhibit in folder 209C, Central Archives, American Museum of Natural History; Dean, "What Animal We Came From," 220, 291; and Alter, *Darwinism and the Linguistic Image*, 128.

27. Henry Fairfield Osborn, "The Influence of Habit in the Evolution of Man and the Great Apes," *Bulletin of the New York Academy of Medicine* 4 (1928): 224.

28. Henry Fairfield Osborn, *Evolution and Religion in Education* (New York: Charles Scribner's Sons, 1925), 206; Dean, "What Animal We Came From," 289.

29. Albert G. Ingalls, "Evolution from the Nebula to 1925," *Scientific American* (Oct. 1925): 234–235.

30. Lucretia Perry Osborn, *The Chain of Life* (New York: Charles Scribner's Sons, 1925),

145; Henry Fairfield Osborn, *Man Rises to Parnassus: Critical Epochs in the Prehistory of Man* (Princeton: Princeton University Press, 1927).

31. Unsigned memo from one of Osborn's secretaries to Mr. John Hall Wheelock, Charles Scribner's Sons, June 9, 1929, Hall of the Age of Man Old Correspondence Folder, Central Archives, American Museum of Natural History, New York.

32. McGregor to Conklin, Dec. 22, 1920, folder 24, box 15, Conklin Correspondence.

33. McDougall to Osborn, July 15, 1925, folder 2, box 92, Osborn Papers, AMNH.

34. Unsigned editorial, *Lexington (KY) Herald*, Jan. 29, 1922, in folder 16, box 86, ibid.

35. Albert G. Ingalls, "Did Man Descend From Monkeys? The Unqualified Denial of Man's Ape-like Origin Is Misleading and Lacks Candor," *Scientific American* (Sept. 1925): 157.

36. Henry Fairfield Osborn, "'Dawn Man' Appears as Our First Ancestor," *New York Times*, Jan. 9, 1927, sec. 8, p. 3.

37. Francis Arthur Bather, "The Record of the Rocks," 109–110; Richard Swann Lull, "Connecting and Missing Links in the Ascent to Man," 261–262; and Herbert Spencer Jennings, "Can We See Evolution Occurring?" 25, all in Frances Mason, ed., *Creation by Evolution: A Consensus of Present-Day Knowledge As Set Forth by Leading Authorities in Non-Technical Language That All May Understand* (New York: Macmillan, 1928).

38. The Osborn Papers, AMNH and the Conklin Papers include extensive correspondence with Mason about the book. Osborn's correspondence with Mason is in folders 13 and 14, box 14, Osborn Papers, AMNH; Conklin's correspondence with Mason is preserved in folder 10, box 15, Conklin Correspondence.

39. Hopwood, "Pictures of Evolution and Charges of Fraud."

40. Frederic A. Lucas, *Guide to the Hall of Mammals*, Guide Leaflet No. 57 (New York: American Museum of Natural History, 1923), 1.

41. Osborn, "'Dawn-Man' Appears as Our First Ancestor."

Chapter Eight • Scientists and the Monkey Trial

1. "Offers for Scopes," *Variety*, July 22, 1925, 1; "Modest Evolutionist Scopes Scorns Money; Wants Post-Graduate Course," *Variety*, July 29, 1925, 4.

2. Michael Schudson, *Discovering the News: A Social History of American Newspapers* (New York: Basic Books, 1978), 128.

3. *New York Evening Post*, June 11, 1925, 10.

4. "In the Day's News," *New York Evening Post*, June 1, 1925, 10.

5. The Butler Act, quoted in Sheldon Norman Grebstein, ed., *Monkey Trial: The State of Tennessee vs. John Thomas Scopes* (Boston: Houghton Mifflin, 1960), 3. The wording of the act was also reprinted in the transcript of the Scopes trial, published as *Tennessee Evolution Case: A Complete Stenographic Report of the Famous Court Case of the Tennessee Anti-Evolution Act, at Dayton, July 10 to 21, 1925, Including Speeches and Arguments of Attorneys* (Cincinnati: National Book Co., 1925).

6. Austin Peay, quoted in *Nashville Tennessean*, Mar. 24, 1925, 1, quoted in Jeffrey P. Moran, *The Scopes Trial: A Brief History with Documents* (Boston: Bedford/St. Martin's, 2002), 23.

7. Ray Ginger, *Six Days or Forever? Tennessee v. John Thomas Scopes* (New York: Oxford University Press, 1958), 66. For more on the American Civil Liberties Union, see Samuel Walker, *In Defense of American Liberties: A History of the ACLU* (New York: Oxford University Press, 1990).

8. Several recent histories have done much to improve historians' understanding of the trial and its context: see especially Ronald L. Numbers, *Darwinism Comes to America* (Cambridge: Harvard University Press, 1998); Michael Lienesch, *In the Beginning: Fundamentalism, the Scopes Trial, and the Making of the Antievolution Movement* (Chapel Hill: University of North Carolina Press, 2007); Edward J. Larson, *Summer for the Gods: The Scopes Trial and America's Continuing Debate over Science and Religion* (New York: Basic Books, 1997); Paul K. Conkin, *When All the Gods Trembled: Darwinism, Scopes, and American Intellectuals* (Lanham, MD: Rowman and Littlefield, 1998); Susan Harding, "Representing Fundamentalism: The Problem of the Repugnant Cultural Other," *Social Research* 58 (Summer 1991): 373–393; Moran, *Scopes Trial*; Garry Wills, *Under God: Religion and American Politics* (New York: Simon and Schuster, 1990); George E. Webb, *The Evolution Controversy in America* (Lexington: University Press of Kentucky, 1994); James Gilbert, *Redeeming Culture: American Religion in an Age of Science* (Chicago: University of Chicago Press, 1997); George M. Marsden, *Fundamentalism and American Culture: The Shaping of Twentieth-Century Evangelicalism, 1870–1925* (New York: Oxford University Press, 1980); Lawrence W. Levine, *Defender of the Faith: William Jennings Bryan, the Last Decade, 1915–1925* (Cambridge: Harvard University Press, 1987); Lawrence Levine, *The Unpredictable Past: Explorations in American Cultural History* (New York: Oxford University Press, 1993), 20–28; and Michael Kazin, *A Godly Hero: The Life of William Jennings Bryan* (New York: Knopf, 2006).

9. Russell D. Owen, "The Significance of the Scopes Trial. I. Issues and Personalities," *Current History* 22 (1925): 878.

10. John Thomas Scopes, "Reflections—Forty Years After," in Jerry R. Tompkins, ed., *D-Days at Dayton* (Baton Rouge: Louisiana State University Press, 1965), 19–20.

11. H. L. Mencken, *Heathen Days: 1890–1936* (New York: Knopf, 1943), 222.

12. Mencken quoted in Wills, *Under God*, 109.

13. John Thomas Scopes and James Presley, *Center of the Storm: Memoirs of John T. Scopes* (New York: Holt, Rinehart and Winston, 1967), 59.

14. *The World's Most Famous Court Trial: Tennessee Evolution Case* (Cincinnati: National Book Co., 1925), 287–303. This is the complete transcript of the trial. On the acceptance by most fundamentalists of the 1920s of the possibility of an ancient earth, see Numbers, *Darwinism Comes to America*, 79–84.

15. Mencken, *Heathen Days*, 216. For the extent of media coverage of the trial, see especially Numbers, *Darwinism Comes to America*, 78–79, and Larson, *Summer for the Gods*, 112–115.

16. "Ballyhooed Hullabaloo," *Variety*, July 15, 1925, 7.

17. On press coverage of the trial, see Edward Caudill, *Darwinism in the Press: The Evolution of an Idea* (Hillsdale, NJ: Lawrence Erlbaum Associates, 1989), 94–113; Edward Caudill, "A Content Analysis of Press Views of Darwin's Evolution Theory, 1860–1925," *Journalism Quarterly* 64, no. 4 (1987): 782–786, 946; Edward Caudill, "The Roots of Bias: An Empiricist Press and Coverage of the Scopes Trial," *Journalism Monographs* 114 (1989), 1–37; Donald F. Brod, "The Scopes Trial: A Look at Press Coverage after Thirty Years," *Jour-*

nalism Quarterly 42, no. 2 (1965): 219–226; Carl E. Hatch, "New Jersey's Reaction to the Tennessee Evolution Trial," *New Jersey History* 90, no. 4 (1972): 226–241; Moran, *Scopes Trial;* James Walter Wesolowski, "Before Canon 35: WGN Broadcasts the Monkey Trial," *Journalism History* 2 (1975): 76–79, 86; and J. Woodfin Wilson, "Northwest Louisiana Newspaper Coverage of the Scopes Trial," *Journal of the North Louisiana Historical Association* 21, no. 1 (1990): 43–48.

18. William Manchester, *Disturber of the Peace: The Life of H. L. Mencken* (Amherst: University of Massachusetts Press, 1986), 180; H. L. Mencken, "Inquisition," in Mencken, *Heathen Days*, 214–238.

19. Joseph Wood Krutch, *More Lives Than One* (New York: William Sloan Associates, 1962), 147.

20. Mencken, *Heathen Days*, 234–235.

21. H. L. Mencken in the *Baltimore Evening Sun*, July 15, 1925, quoted in Tompkins, *D-Days at Dayton*, 43.

22. Ibid., 48.

23. H. L. Mencken, "W.J.B.: In Memoriam," in Mencken, *The American Scene: A Reader* (New York: Vintage Books, 1925), 227–231.

24. H. L. Mencken, " 'The Monkey Trial': A Reporter's Account," in Tompkins, *D-Days at Dayton*, 41.

25. "Decision Today Is Likely," *New York Times*, July 14, 1925, 1.

26. Russell D. Owen, "Dayton's Remote Mountaineers Fear Science," *New York Times*, July 19, 1925, XX3.

27. S. L. Harrison, "The Scopes 'Monkey Trial' Revisited: Mencken and the Editorial Art of Edmund Duffy," *Journal of American Culture* 17 (Winter 1994): 55–63. Any review of newspapers and magazines of the period will yield many cartoons making fun of Bryan.

28. Russell D. Owen, "Significance of the Scopes Trial," 876, 880.

29. Ibid., 881.

30. "Scopes Found Guilty by Jury after Five Minutes, Fined $100 in Evolution Trial," *New York Evening Post*, July 21, 1925, 1.

31. Ibid., 7.

32. "Scopes Jury Barred as Attorneys Argue Evolution Exclusion," *New York Evening Post*, July 16, 1925, 1, 6.

33. Clarence Darrow, *The Story of My Life* (1932; New York: Da Capo Press, 1996), 275–276.

34. Marcet Haldeman-Julius, "Impressions of the Scopes Trial" (excerpts from *Clarence Darrow's Two Great Trials*, pamphlet published in 1927 by Haldeman-Julius, Girard, Kansas), www.law.umkc.edu/faculty/projects/ftrials/scopes/haldeman-julius.html.

35. On the mythology of a simple dichotomy between northern and southern attitudes, see Ronald L. Numbers and Lester Stephens, "Darwinism in the American South," in Numbers and John Stenhouse, *Disseminating Darwinism: The Role of Place, Race, Religion and Gender* (Cambridge: Cambridge University Press, 1999), 123–144.

36. Numbers, *Darwinism Comes to America;* Lienesch, *In the Beginning;* Larson, *Summer for the Gods;* Conkin, *When All the Gods Trembled;* Moran, *Scopes Trial;* Kazin, *Godly Hero;* Levine, *Defender of the Faith;* Levine, *Unpredictable Past*, 20–28; Wills, *Under God;*

Stephen Jay Gould, "William Jennings Bryan's Last Campaign," in Gould, *Bully for Brontosaurus* (New York: W. W. Norton, 1991), 416–431.

37. Felix Frankfurter, "Democracy and the Expert," *Atlantic Monthly* 146 (Nov. 1930): 649. For an introduction to the voluminous literature on eugenics, see Daniel J. Kevles, *In the Name of Eugenics: Genetics and the Uses of Human Heredity* (New York: Knopf, 1985); Hamilton Cravens, *The Triumph of Evolution: American Scientists and the Heredity-Environment Controversy, 1900–1941* (Philadelphia: University of Pennsylvania Press, 1978); John S. Haller Jr. *Outcasts from Evolution: Scientific Attitudes of Racial Inferiority, 1859–1900* (Urbana: University of Illinois Press, 1971); Stephen J. Gould, *The Mismeasure of Man* (New York: W. W. Norton, 1981); Kathy J. Cooke, "The Limits of Heredity: Nature and Nurture in American Eugenics before 1915," *Journal of the History of Biology* 31 (June 1998): 263–287; and Philip J. Pauly, "Essay Review: The Eugenics Industry—Growth or Restructuring?" *Journal of the History of Biology* 26 (1993): 131–145. For a useful list of the scientists involved in a prominent eugenics organization, see Barry Alan Mehler, "A History of the American Eugenics Society, 1921–1940" (Ph.D. diss., University of Illinois at Urbana-Champaign, 1988).

38. Walter Lippmann, *American Inquisitors: A Commentary on Dayton and Chicago* (New York: Macmillan, 1928); J. Gresham Machen, *Christianity and Liberalism* (Grand Rapids, MI: Eerdmans, 1923); Reinhold Niebuhr, *Does Civilization Need Religion?* (New York: Macmillan, 1927); Reinhold Niebuhr, "Can Christianity Survive?" *Atlantic Monthly* 135 (Jan., 1925): 84–88; Reinhold Niebuhr, "Our Secularized Civilization," *Christian Century* (Apr. 22, 1926): 508–510. See also D. G. Hart, *Defending the Faith: J. Gresham Machen and the Crisis of Conservative Protestantism in Modern America* (Baltimore: Johns Hopkins University Press, 1994).

39. Ronald L. Numbers, *The Creationists: From Scientific Creationism to Intelligent Design*, expanded ed. (Cambridge: Harvard University Press, 2006); Numbers, *Darwinism Comes to America;* Lienesch, *In the Beginning;* George Marsden, *Understanding Fundamentalism and Evangelicalism* (Grand Rapids, MI: Eerdmans, 1991); Larson, *Summer for the Gods.* For corrections to the "warfare" myth, see especially Jon H. Roberts, *Darwinism and the Divine in America: Protestant Intellectuals and Organic Evolution, 1859–1900* (Madison: University of Wisconsin Press, 1988); Jon H. Roberts, "Conservative Evangelicals and Science Education in American Colleges and Universities, 1890–1940," *Journal of the Historical Society* 3 (Fall 2005): 297–329; James R. Moore, *The Post-Darwinian Controversies: A Study of the Protestant Struggle to Come to Terms with Darwin in Great Britain and America, 1870–1900* (Cambridge: Cambridge University Press, 1979); James Gilbert, *Redeeming Culture: American Religion in an Age of Science* (Chicago: University of Chicago Press, 1997); David N. Livingstone, *Darwin's Forgotten Defenders: The Encounter between Evangelical Theology and Evolutionary Thought* (Grand Rapids, MI: Eerdmans, 1987); and Ronald L. Numbers, "Science and Religion," *Osiris*, 2d ser., 1 (1985): 59–80.

40. These affidavits are on file in box 19, Osborn Papers, AMNH. They were not, however, allowed to be introduced into evidence at the trial; the judge ruled that only statements by scientists who were present and could be sworn in would be introduced into evidence in the official record of the trial.

41. One of these copies, in its original dust jacket, remains on display in the Scopes Trial Museum in the Rhea County Courthouse in Dayton.

42. Osborn's correspondence related to the Scopes trial is collected in box 19, Osborn Papers, AMNH; additional correspondence about the evolution debates is found in box 86 and in box 92, which contains correspondence pertaining to *The Earth Speaks to Bryan*. Osborn to Butler, July 7, 1925, folder 11, box 19, ibid. See also "Dr. Osborn Advises Scopes on Defense," *New York Times*, June 9, 1925, 6, and "Seek Osborn's Help for Scopes in Trial," *New York Times*, June 12, 1925, 3. Osborn to Neal, June 13, 1925, in folder 9, box 19, Osborn Papers, AMNH.

43. Variations of the same story ran in "Scientific Witnesses Picked for Scopes Trial," *St. Louis Post Dispatch*, June 27, 1925; "Many Noted Scientists Enlisted for Defense in Evolution Trial," *St. Paul (MN) Press*, June 25, 1925; "List of Scopes Witnesses Reads Like Page of Science Who's Who," *Philadelphia Public Ledger*, June 26, 1925, clippings in folder 14, box 19, Osborn Papers, AMNH.

44. "A Country Schoolmaster Stirs the World," *Popular Science Monthly* (Aug. 1925): 27.

45. John Wolf and James S. Mellett, "The Role of 'Nebraska Man' in the Creation/Evolution Debate," *Creation/Evolution* 16 (1985): 31–43.

46. Osborn to Dr. J. M. Coleman, Oct. 14, 1925, folder 1 (second copy in folder 4), box 92, Osborn Papers, AMNH; Henry Fairfield Osborn, "Recent Discoveries Relating to the Origin and Antiquity of Man," *Proceedings of the American Philosophical Society* 66 (1927): 373–389, diagram on 375; Henry Fairfield Osborn, "Recent Discoveries Relating to the Origin and Antiquity of Man," *Science*, May 20, 1927, 481–488, diagram on 482.

47. Osborn's instruction to the museum's public relations staff to play down Darrow's "criminal connections" is in Osborn, memorandum to G. N. Pindar, July 3, 1925, folder 11, box 19, Osborn Papers, AMNH; see also Conklin to Frances Mason, Apr. 9, 1929, folder 10, box 15, Conklin Correspondence; Conklin to Herbert Spencer Jennings, Oct. 26, 1925, Conklin Correspondence; Conklin to Jennings, Nov. 9, 1925, folder 15, box 12, Conklin Correspondence; S. J. Holmes to Conklin, Mar. 26, 1929, folder 14, box 11, Conklin Correspondence; Burton Livingston to Conklin, Apr. 30, 1929, AAAS folder, in Organizations, Conklin Papers.

48. Winterton C. Curtis, *Fundamentalism* vs. *Evolution at Dayton, Tennessee: Abstracts from the Autobiographical Notes of Winterton C. Curtis* (Columbia, MO: Winterton C. Curtis, 1956), 57.

49. Winterton C. Curtis, "The Evolution Controversy," in Tompkins, *D-Days At Dayton*, 76.

50. "Scientists Desert Scopes Trial, Seeing No Evolution Test," *New York Evening Post*, July 9, 1925, 1, 7.

51. Curtis, *Fundamentalism* vs. *Evolution at Dayton, Tennessee*, 57.

52. Conklin to Osborn, June 27, 1925, folder 10, box 19, Osborn Papers, AMNH.

53. For a substantial collection of such communications, see folder 1, box 44, Science Service Archives, Smithsonian Institution, Washington, D.C. This archive also includes many press clippings about the trial and correspondence between the trial lawyers and Watson Davis, who covered the trial for Science Service.

54. Lamont Cole, "Current Thoughts on Biological Evolution," in Tompkins, *D-Days at Dayton*, 105.

55. Curtis, "Evolution Controversy," 76.

56. Kirtley F. Mather, "Geology and Genesis," in Tompkins, *D-Days at Dayton*, 92. See also Kennard Baker Bork, *Cracking Rocks and Defending Democracy: Kirtley Fletcher Mather, Scientist, Teacher, Social Activist, 1888–1978* (San Francisco: Pacific Division, AAAS, 1994); Edward B. Davis, "Fundamentalist Cartoons, Modernist Pamphlets, and the Religious Image of Science in the Scopes Era," in Charles L. Cohen and Paul S. Boyer, eds., *Religion and the Culture of Print in Modern America* (Madison: University of Wisconsin Press, forthcoming); and Correspondence 1925 folder in the the Kirtley Mather Papers at Harvard University.

57. Mencken, "'Monkey Trial,'" 45.

58. "Scopes Jury Barred as Attorneys Argue Evolution Exclusion," 1, 6.

59. *World's Most Famous Court Trial*, 172–177.

60. Ibid., 183–187.

61. Ferenc M. Szasz, "The Scopes Trial in Perspective," *Tennessee Historical Quarterly* 30, 3 (1971): 288–298, 297; Scopes, "Reflections—Forty Years After," 27.

62. "Bryan Defends Tennessee and Its Law; Calls Evolution Attack on Church; Spirited Debate on Expert Evidence" and "Fervid Appeals in Court," both in *New York Times*, July 17, 1925, 1.

63. Winterton C. Curtis, "The Fact, the Course, and the Causes of Organic Evolution," *Scientific Monthly* 21 (Sept. 1925): 296. This article and the other affidavits were also reprinted in the transcript of the trial and were excerpted in newspapers. In *Scientific Monthly* they were all collected under the heading "Evidences for Evolution: Statements Prepared for the Defense Counsel, State of Tennessee, vs. John T. Scopes."

64. Charles Hubbard Judd, "Evolution and Mental Life," *Scientific Monthly* 21 (Aug. 1925): 316.

65. H. H. Newman, "Evidences from Comparative Anatomy," *Scientific Monthly* 21 (Sept. 1925): 302.

66. Ibid., 304.

67. Fay-Cooper Cole, "The Evolution of Man," *Scientific Monthly* 21 (Sept. 1925): 318.

68. Ibid., 320.

69. Curtis, "Fact, the Course, and the Causes," 301.

70. Wilbur A. Nelson, "Geology and Evolution," *Scientific Monthly* 21 (Sept. 1925): 312.

71. Kirtley F. Mather, "Evolution and Religion," *Scientific Monthly* 21 (Sept. 1925): 322. Numbers, *Darwinism Comes to America*, 80–84.

72. Jacob G. Lipman, "Organic Evolution from the Point of View of the Soil Investigator," *Scientific Monthly* 21 (Sept. 1925): 312.

73. Maynard M. Metcalf, "The Truth of Evolution," *Scientific Monthly* 21 (Sept. 1925): 291.

74. Ibid., 292.

75. Newman, "Evidences from Comparative Anatomy," 306.

76. Curtis, "Fact, the Course, and the Causes," 301.

77. Metcalf, "Truth of Evolution," 291.

78. Curtis, "Fact, the Course, and the Causes," 299; Newman, "Evidences from Comparative Anatomy," 304.

79. Metcalf, "Truth of Evolution," 292.

80. "One Compensation," *New York Times*, July 12, 1925, E6.

81. Cole, "Current Thoughts on Biological Evolution," 112.

82. "Run at Library Caused by Trial," *Newark Sunday Call,* quoted in Hatch, "New Jersey's Reaction to the Tennessee Evolution Trial," 237.

83. W. O. McGeehan, "Why Pick on Dayton?" *Harper's,* Oct. 1925, 623–627.

84. "Retrial of Scopes Put to High Court," *New York Evening Post,* July 22, 1925, 2.

85. "Tennessee's Ghastly Travesty," *New York Evening Post,* July 22, 1925, 8.

86. "Tennessee and the Constitution," *New Republic,* July 8, 1925, 166–168.

87. "The Week," *New Republic,* July 15, 1925, 191.

88. W. E. B. Du Bois, "Scopes," *Crisis* (Sept. 1925): 218.

89. Ibid.

90. Lippmann, *American Inquisitors,* 61–62, 63, 87.

91. Arthur Garfield Hays, "The Strategy of the Scopes Defense," *Nation,* Aug. 5, 1925, 157–158.

92. W. T. Brown, letter to the editor, *Nation,* Aug. 5, 1925, 168.

93. "It Is Just as Well to Be Frank," *New York Times,* Dec. 16, 1925, 24.

94. Marcel La Follette found, in a content analysis of science stories in popular magazines from 1910 to 1955, a significant peak in 1926 and a falling off in the early 1930s. La Follette suggests, intriguingly, that after 1926 the proportion of science popularizations written by scientists declined, with science journalists partly but not completely taking up the slack. Marcel La Follette, *Making Science Our Own: Public Images of Science, 1910–1955* (Chicago: University of Chicago Press, 1990). A complete account of La Follette's data and methods can also be found in Marcel Evelyn Chotkowski La Follette, "Authority, Promise and Expectation: The Images of Science and Scientists in American Popular Magazines, 1910–1955" (Ph.D. diss., Indiana University, 1979). See also John C. Burnham, *How Superstition Won and Science Lost: Popularizing Science and Health in the United States* (New Brunswick, NJ: Rutgers University Press, 1987), and Charles William Heywood, "Scientists and Society in the United States, 1900–1940: Changing Concepts of Social Responsibility" (Ph.D. diss., University of Pennsylvania, 1954).

95. Thomas F. Gieryn, George M. Bevins, and Stephen C. Zehr, "Professionalization of American Scientists: Public Science in the Creation/Evolution Trials," *American Sociological Review* 50 (June 1985): 392–409.

96. "The Week," *New Republic,* July 22, 1925, 219.

97. Osborn to Bond, June 13, 1925, folder 10, box 19, Osborn Papers, AMNH.

98. Osborn to Colby, June 13, 1925, folder 10, box 19, ibid.

99. Allene Summer, " 'Monkey Town' Has Trade Boom," *Victoria Daily Times,* June 27, 1925, in folder 14, box 19, ibid.

100. *New York Evening Post,* June 10, 1925, 24.

101. "Billy Sunday Wallops the Devil and Gorillas," *Denver Post,* July 1, 1925, 1, 19.

102. On the use of monkeys to criticize humans in cartoons, see Moran, *Scopes Trial.*

Chapter Nine • Redeeming the Caveman, and the Irreverent Funny Pages

1. Alfred Watterson McCann, *God—or Gorilla: How the Monkey Theory of Evolution Exposes Its Own Methods, Refutes Its Own Principles, Denies Its Own Inferences, Disproves Its*

Own Case (New York: Devin-Adair, 1922). The advertisement appeared in the *New York Evening Post,* June 3, 1925, 11.

2. Osborn F. Hevener, "The Kingdom of the Mind," *Forum* 67 (Apr. 1922): 367.

3. R. Maxwell Bradmer to Osborn, Mar. 6, 1922, folder 3, box 86, Osborn Papers, AMNH.

4. John Roach Straton, "Making Poison Plausible," sermon, 1924, folder 4, box 21, ibid.

5. Henry Fairfield Osborn, "Evolution and Daily Living," *Forum* 73 (Feb. 1925): 171; he repeated the same statement in *Evolution and Religion in Education: Polemics of the Fundamentalist Controversy of 1922 to 1926* (New York: Charles Scribner's Sons, 1926), 51.

6. Harold O. Whitnall, "In Defense of the Cave-Man," *Scientific Monthly* 20 (June 1926): 522–532.

7. McCann, *God—or Gorilla,* 19–20.

8. Ibid., 19; Henry Fairfield Osborn, *Men of the Old Stone Age: Their Environment, Life, and Art* (1915; New York: Charles Scribner's Sons, 1916), 73.

9. McCann, *God—or Gorilla,* 29; James Shreve, *The Neandertal Enigma: Solving the Mystery of Modern Human Origins* (New York: William Morrow, 1995); Pat Shipman, *The Man Who Found the Missing Link: Eugene Dubois and His Lifelong Quest to Prove Darwin Right* (New York: Simon and Schuster, 2001); Peter J. Bowler, *Theories of Human Evolution: A Century of Debate, 1844–1944* (Baltimore: Johns Hopkins University Press, 1986); Frank Spencer, ed., *A History of American Physical Anthropology, 1930–1980* (New York: Academic Press, 1982); John G. Fleagle, *Primate Adaptation and Evolution,* 2d ed. (San Diego: Academic Press, 1999), 511–549; Michael Hammond, "The Expulsion of the Neanderthals from Human Ancestry: Marcellin Boule and the Social Context of Scientific Research," *Social Studies of Science* 12 (1982): 1–36. See also Osborn's letters to McGregor expressing frustration with Dubois, folder 22, box 13, Osborn Papers, AMNH.

10. McGregor to Conklin, Dec. 13, 1920, folder 24, box 15, Conklin Correspondence.

11. Osborn to McGregor, Oct. 27, 1915, folder 22, box 13, Osborn Papers, AMNH.

12. Osborn to McGregor, Jan. 24, 1920, ibid.

13. McCann, *God—or Gorilla,* 10. For a complete account of the Piltdown episode, see Frank Spencer, *Piltdown: A Scientific Forgery* (London: Oxford University Press, 1990).

14. McCann, *God—or Gorilla,* 11.

15. "Evolutionist Logic False, Divine Says: Exhibits of Missing Links Are Attacked," *Tampa Daily Times,* June 29, 1925, clipping in folder 14, box 19, Osborn Papers, AMNH.

16. Osborn to J. H. McGregor, Apr. 25, 1927, folder 23, box 13, Osborn Papers, AMNH.

17. For example, Anne Papp to Osborn, Feb. 9, 1931, folder 5, box 27, Osborn Papers, AMNH; Robert E. Fatherly to Osborn, Mar. 3, 1927, folder 31, box 7, ibid.

18. Osborn, through a secretary, to Robert E. Fatherly, Earlham College, Earlham Indiana, Apr. 14, 1927, in response to a letter from Fatherly, folder 31, box 7, ibid.

19. Osborn to McGregor, Nov. 17, 1921, folder 22, box 13, ibid.

20. J. H. McGregor, "Restoring Neanderthal Man," *Natural History* 26 (May–June 1926): 288–293. See also Ronald Rainger, *Agenda for Antiquity: Henry Fairfield Osborn and Vertebrate Paleontology at the American Museum of Natural History, 1890–1935* (Tuscaloosa: University of Alabama Press, 1991), 170–173.

21. Osborn, *Evolution and Religion in Education,* 51.

22. Ibid., 41.

23. Osborn to Boule, Jan. 14, 1916, folder 7, box 99, Osborn Papers, AMNH.

24. Osborn, *Evolution and Religion in Education*, 223.

25. G. K. Chesterton, *The Everlasting Man* (1925; San Francisco: Ignatius Press, 1993), 34–35.

26. Osborn to Knight, Feb. 9, 1915, folder 8, box 12, Osborn Papers, AMNH.

27. Osborn to Edgar Palmer, Esq., Feb. 2, 1925, folder 9, box 12, ibid.

28. Osborn to Knight, July 28, 1919, folder 5, box 12, ibid.

29. Charles R. Knight, *Before the Dawn of History* (New York: McGraw Hill, 1935), 112.

30. Osborn, *Evolution and Religion in Education*, 127. For recent discussions of the presentation of Neanderthals in this painting, see Stephanie Moser, *Ancestral Images: The Iconography of Human Origins* (Ithaca, NY: Cornell University Press, 1998); Alan E. Mann, "Imagining Prehistory: Pictorial Reconstructions of the Way We Were," *American Anthropologist* 105 (Mar. 2003): 139–148; and Judith C. Berman, "A Note on the Paintings of Prehistoric Ancestors by Charles R. Knight," *American Anthropologist* 105 (Mar. 2003): 143–146.

31. Osborn to Scribner, Apr. 29, 1924, folder 9, box 56, Osborn Papers, AMNH.

32. Osborn, *Evolution and Religion in Education*, 127.

33. Henry C. Tracy to Osborn, Aug. 20, 1928, folder 3, box 99, Osborn Papers, AMNH.

34. Langford to Osborn, Jan. 28, 1920, folder 25, box 12, ibid.

35. Langford to Osborn, July 18, 1920, ibid.

36. Joseph Gilpin Pyle, "Pre-Adamite Man," *St. Paul Pioneer Press*, Sept. 26, 1920, clipping in folder 15, box 12, ibid.

37. Harry Hansen, "Ee—volution," *Chicago News*, July 1, 1925, clipping in folder 15, box 19, ibid.

38. *Fort Bragg News*, May 3, 1924, in folder 4, box 21, ibid.

39. Gregory to Osborn, Jan. 6, 1919, folder 8, box 12, ibid.

40. Langford to Osborn, Jan. 28, 1920, folder 25, box 12, ibid.

41. Moser, *Ancestral Images*.

42. *Harper's Weekly*, July 19, 1873, 617.

43. Judith C. Berman, "Bad Hair Days in the Paleolithic: Modern (Re)Constructions of the Cave Man," *American Anthropologist* 191 (June 1999): 294; Martin Kemp, *The Human Animal in Western Art and Science* (Chicago: University of Chicago Press, 2007).

44. Marcellin Boule, *Fossil Men: Elements of Human Palaeontology*, trans. Jessie Elliott Ritchie and James Ritchie (Edinburgh: Oliver and Boyd, Tweeddale Court, 1923), 196.

45. This is the interpretation of Erik Trinkhaus and Pat Shipman, *The Neandertals: Of Skeletons, Scientists, and Scandal* (New York: Vintage Books, 1992); see also Shreve, *Neandertal Enigma*, and Brian Regal, *Henry Fairfield Osborn: Race and the Search for the Origins of Man* (Burlington, VT: Ashgate Press, 2002).

46. In a comprehensive and elegant essay, Michael Hammond has argued that Boule's commitment to a nonlinear view of human evolution, a commitment growing out of complex concerns having to do with professionalization, scientific and epistemological methodologies, and national loyalties, shaped his interpretations of the important La Chapelle-aux-Saints specimen, on which he based his reading of Neanderthal anatomy. See Michael

Hammond, "The Expulsion of the Neanderthals from Human Ancestry: Marcellin Boule and the Social Context of Scientific Research," *Social Studies of Science* 12 (1982): 1–36.

47. Gregory to Osborn, Jan. 6, 1919, folder 8, box 12, Osborn Papers, AMNH.

48. Anita Maris Boggs to Osborn, Mar. 3, 1916, folder 7, box 99, ibid.

49. Breuil to Osborn, Jan. 27, 1919, box 1, Knight Papers. See also Alan Houghton Brodrick, *Father of Prehistory: The Abbe Henri Breuil. His Life and Times* (New York: William Morrow, 1963).

50. Osborn, "Evolution and Daily Living," 171. I am grateful to Richard Milner for the observation that the artist resembles Knight.

51. Tracy to Osborn, Aug. 20, 1928, folder 3, box 99, Osborn Papers, AMNH.

52. For more on the mural, its composition, and its popularity, see Moser, *Ancestral Images*, 159–60; Rainger, *Agenda for Antiquity*; Sylvia Massey Czerkas and Donald F. Glut, *Dinosaurs, Mammoths, and Cavemen: The Art of Charles R. Knight* (New York: E. P. Dutton, 1982), 68–69; and Charlotte M. Porter, "The Rise to Parnassus: Henry Fairfield Osborn and the Hall Of the Age of Man," *Museum Studies Journal* 1 (Spring 1983): 26–34. See also correspondence between Osborn and Knight, folders 8 and 9, box 12, and also box 99, Osborn Papers, AMNH, and Knight Papers. The advertisement for *The Earth Speaks to Bryan* appeared in the *New York Times Book Review,* July 12, 1925.

53. See for example, H. C. Tracy to Osborn, and Nels C. Nelson to H. C. Tracy, n.d., 1928, folder 3, box 99, Osborn Papers, AMNH.

54. Charles K. Inwood to Osborn, Dec. 19, 1924, proposing a series of paintings of caveman life, folder 1, box 99, ibid. As described, these paintings would have been far too lurid and risqué for Osborn's taste. See also the brochure "The Cave Man," in folder 4, box 99, ibid.

55. Frank L. Crow to Osborn, Mar. 9, 1932, folder 2, box 27, ibid.

56. Ernest Seeman to Osborn, July 30, 1932, folder 1, box 27, ibid.

57. Osborn, typescript statement to Harold Butcher, *New York Herald,* June 15, 1931, folder 2, box 12, ibid.

58. Moser, *Ancestral Images,* 133–141; Nicolaas Rupke, "Metonymies of Empire: Visual Representations of Prehistoric Times, 1830–90," in Renato G. Mazzolini, ed., *Non-Verbal Communication in Science prior to 1900* (Florence: Leo S. Olschki, 1993), 513–528.

59. Nelson to H. C. Tracy, no date, in response to a letter from Tracy dated Aug. 20, 1928, folder 3, box 99, Osborn Papers, AMNH.

60. "Says Our Boys Need Cave-man Training," *New York Times,* Mar. 23, 1925, 6.

61. "Bringing Cave Life to the Modern Boy," *New York Times,* Apr. 12, 1925, SM 9. See also Osborn, *Creative Education* (New York: Charles Scribner's Sons, 1927).

62. "Are We Forgetting Nature?" *New York Times,* Apr. 12, 1925, E6.

63. Osborn to Knight, July 28, 1919, folder 5, box 12, Osborn Papers, AMNH. See also Osborn, "The Cave Men Knew," *Collier's,* May 23, 1925, 23, and repetition of this theme in Henry Fairfield Osborn, *Creative Education in School, College, University and Museum* (New York: Charles Scribner's Sons, 1927).

64. Charles K. Inwood to Osborn, Dec. 29, 1924, folder 1, box 99, Osborn Papers, AMNH.

65. Field to Osborn, Dec. 28, 1915, folder 31, box 7, ibid.

66. Osborn to Field, Feb. 15, 1916, ibid.

67. Henry Fairfield Osborn, "The Influence of Habit in the Evolution of Man and the Great Apes," *Bulletin of the New York Academy of Medicine* 4 (1928): 216–249; Henry Fairfield Osborn, "The Influence of Bodily Locomotion in Separating Man from the Monkeys and Apes," *Scientific Monthly* 26 (May 1928): 387. Frémiet's sculpture is reproduced in Roselyne de Ayala and Jean-Pierre Gueno, eds., *Illustrated Letters: Artists and Writers Correspond* (New York: Harry N. Abrams, 1999), 98. See also Mark Pittenger, "Imagining Genocide in the Progressive Era: The Socialist Fiction of George Allan England," *American Studies* 35 (1994): 91–108.

68. Tyler, memorandum to Osborn, May 9, 1932, folder 20, box 39, Osborn Papers, AMNH.

69. Assistant to Osborn to Russell Spaulding, May 10, 1932, ibid.

70. Milligan, memo to Tyler, May 11, 1932, ibid.

71. Lowder to Osborn, ibid.

72. Assistant secretary to Osborn to Lowder, Sept. 24, 1932, ibid.

73. Eric Schaefer, *"Bold! Daring! Shocking! True!" A History of Exploitation Films, 1919–1959* (Durham, NC: Duke University Press, 1999), 108.

74. Gregg Mitman, *Reel Nature: America's Romance with Wildlife on Film* (Cambridge: Harvard University Press, 1999), 51.

75. Schaefer, *"Bold! Daring! Shocking! True!"* 108.

76. Briggs to Osborn, Oct. 6, 1932, folder 20, box 39, Osborn Papers, AMNH. For more on *The Blonde Captive* and other movies of this genre, see Schaefer, *"Bold! Daring! Shocking! True!"*

77. Mason to Conklin, Apr. 6, 1929, folder 10, box 15, Conklin Correspondence.

78. Reverend Lefferd M. A. Haughwont to Osborn, May 1, 1927, folder 16, box 9, Osborn Papers, AMNH.

79. Osborn, "Influence of Habit," 219.

80. "Akeley Bronze Ape to Stand in Church," *New York Times*, Apr. 9, 1924, 14.

81. Osborn to McGregor, Dec. 13, 1918, folder 8, box 12, Osborn Papers, AMNH.

82. Douglass to Osborn, Aug. 8, 1927, folder 11, box 7, ibid.

83. Reprinted in Jeffrey P. Moran, *The Scopes Trial: A Brief History with Documents* (Boston: Bedford/St. Martin's, 2002), 173; Jeffrey Moran, "Reading Race into the Scopes Trial: African American Elites, Science, and Fundamentalism," *Journal of American History* 90 (Dec. 2003): 891–911.

84. *Kansas City Times*, Jan. 10, 1930 (American Museum of Natural History neg. no. 271655). This cartoon was also reprinted in Winterton C. Curtis's reminiscence of the trial. It may have been a sign of the different levels of the hierarchy of celebrity that the image of Osborn was relatively generic rather than a recognizable caricature of Osborn himself. The truly famous cartoon subject, notably Bryan, was always an easily recognizable caricature, while the merely renowned scientist Osborn looked somewhat like a more portly version of the man himself but carried a label in order to identify him and was presented as a more or less generalized scientist—but, perhaps in a nod to Osborn's particularity, in a three-piece suit rather than in a lab coat.

85. Osborn, *Creative Education*, 47. From an address delivered at the Central High School, Philadelphia, on Nov. 22, 1926.

86. E. H. Gombrich, "The Cartoonist's Armory," *South Atlantic Quarterly* 62 (Spring 1963): 189–228.

87. W. H. Rucker, "Facts in Support of Evolution" (Itta Bena, Mississippi: W. H. Rucker, Apr. 1, 1929), 2, 7.

88. Rollin Kirby, cartoon entitled "The Little Red Schoolhouse in Tennessee," *Scientific Monthly* 21 (Sept. 1925): 224.

89. Albert A. Hopkins, "Which Races Are Best?" *Scientific American* (Feb. 1925): 77–69. The editor received a great many responses from readers and printed a selection of them in " 'Which Races Are Best?' The Views of Some of Our Correspondents on This Much-Discussed Article," *Scientific American* (July 1925): 50–51. Osborn was quoted at length, identified only as "a well-known scientist." Osborn's correspondence about this article is in folder 6, box 56, Osborn Papers, AMNH.

90. Francis P. Le Buffe, S.J., "Human Evolution and Science," pamphlet, copy in box 87, Osborn Papers, AMNH.

91. W. Maxwell Reed, *The Earth for Sam: The Story of Mountains, Rivers, Dinosaurs and Men* (New York: Harcourt, Brace and Co., 1930), 337.

92. Dutton, telegram to Osborn, Nov. 21, 1924, folder 3, box 86, Osborn Papers, AMNH.

93. Osborn, telegram to Dutton, Nov. 22, 1924, ibid.

94. Osborn to Dutton, Nov. 24, 1924, ibid.

95. John Roach Straton, "Imagination and Life," in Straton, *The Old Gospel at the Heart of the Metropolis* (New York: George H. Doran Co., 1925), 271.

96. Ibid., 273.

97. Ibid., 274.

Conclusion

1. W. E. B. Du Bois, *Dusk of Dawn: An Essay toward An Autobiography of a Race Concept* (1940; New Brunswick, NJ: Transaction Publishers, 2000), 98.

2. Reid to Osborn, Aug. 31, 1925, folder 12, box 19, Osborn Papers, AMNH.

3. Hooper to Osborn, May 23, 1933, folder 19, box 9, ibid.

4. Durant to Osborn, June 12, 1931, folder 15, box 7, ibid.; Professor J. Helder to Osborn, Jan. 12, 1928, folder 4, box 27, ibid. Professor Helder was compiling a book on the opinions of famous men on the subject of immortality. Other queries about the meaning of life and the possibilities for life after death are scattered throughout the correspondence.

5. Edwin Slosson to Edwin Grant Conklin, Nov. 21, 1921, folder 2, box 21, Conklin Correspondence.

6. "Wants Old Religion Taught in Schools. Professor Osborn Would Have Children Instructed in the Fundamental Precepts," *New York Times*, Jan. 24, 1926, sec. 2, p. 1.

7. T. D. A. Cockerell, "The Duty of Biology," *Science*, Apr. 9, 1926, 367–371.

8. Vachel Lindsay, *The Art of the Motion Picture* (1915; New York: Modern Library, 2000), 116–125.

Page numbers in bold refer to illustrations.

Osborn, William Henry, 17
Owen, Richard, 22, 139
Owen, Russell, 170–71

paleontology: and evidence for evolution, 69,
    73–78, 83, 87, 107–8, 119–20, 143; Osborn's
    approach to, 19–21, 30. *See also* orthogenesis
Paley, William, 67–71
Patten, William, 54
Pearl, Raymond, 57, 62, 64
Peay, Austin, 164
Peking Man, 117
physics, 94–97, 225
Pickens, William, 61
Piltdown, 4, 91, 118, 124–25, 129, 157, 197,
    199–200
Pithecanthropus erectus, 4, 94–95, 118, 196, 198,
    200, 202, 219
Poems of Evolution (pamphlet), 60
popularization of science: cavemen in, 205, 211,
    217, 121–22; and eugenics, 174–75; and
    human evolution, 203; illustrations in,
    28–29, 132–33, 135, 144–49, 159–60, 196,
    203; metaphors in, 93; and religion, 56,
    100–101; in response to anti-evolutionism,
    107–9; by Science Service, 55; and scientific
    method, 91, 95–96; and scientists, 15, 18, 20,
    31–32, 46–48, 225
Porter, Cole, 12
Potter, Charles Francis, 25–26, 59, 158–59, 217
Pound, Roscoe, 43
primitive, ideas of, x–xi, 3–15, 62, 66, 72,
    152–53, 207, 212–13
professionalization of science, 80, 191–92
progress, idea of: 44, 62, 67, 86, 182, 205;
    in evolution, x, 30, 83, 86–88; and fossil
    record, 71–77, 80; and human evolution, 115,
    118–19; in illustrations of evolution, 136–43,
    146–47, 150–52, 156–58, 160–61, 201; objec-
    tions to, from conservative Christians, 105,
    109, 225, 232
Progressive Era, 43–44, 196, 212
Prohibition, 1, 16, 17, 64
propaganda, concerns about, 34, 51–52, 116,
    128, 163
Protestantism, rift within, 42, 175. *See also*
    modernism, theological

psychology, 8, 9, 43–44, 72, 78
public relations, 36, 163, 228
publicity: and American Museum of Natural
    History, 21, 33–40, 127, 129, 213, 224, 229;
    and authority of science, ix, 33–40, 52; and
    Bryan, 41, 124; scientists' ambivalence toward,
    49, 52, 81, 100, 120, 175, 177, 191–93, 228–29;
    and Scopes trial, 3, 162–63, 165, 171, 173, 187
Pupin, Michael I., 101–4, 175; *From Immigrant
    to Inventor*, 54
Putnam, Frederic Ward, 7

race: and eugenics, 32, 34; and evolutionary
    theories, 30, 39, 116–18, 219, 232; and illus-
    trations, xi, 13, 108, 150–54, 224; in popular
    science, 195, 215
racism, 67, 116–18; scientific, 57, 67, 150–54
radicalism, fears of, 15, 62, 162, 176–78
radio, 18, 41, 100, 107, 166, 168
Rae-Ellis, Vivienne, 152
Ransom, John Crowe, *God Without Thunder*, 104
Rappleyea, George, 164, 176, 177
Raulston, Judge John, 166–67, 170–72, 182
Ray, John, 69–70
recapitulation, 8, 9, 71–73, 93, 212. *See also* civi-
    lization, stages of; development metaphor
reconciliation of Christianity and evolution:
    argument for, 162, 175, 180–92; books
    about, 88, 93, 98–106, 109
religion: compatibility of, with evolution, 182,
    186; and science, 120, 131, 165, 168, 171,
    173–74. *See also* reconciliation of Christianity
    and evolution
restorations, fossil, 27, 110–11, 118, 121, 123, 125,
    157, 196–201, 208–9
Ripon, Bishop of, 43
Roberts, Jon, 44
Roosevelt, Theodore, 8, 17, 213
Rutot, A., 197–98

scala naturae, 135. *See also* Chain of Being
Schickel, Richard, 10
Schulz, Adolph, 27
Science (journal), 44, 47, 49, 54, 80, 95, 102,
    129, 130, 230
Science League of America, 61–62, 178, 190.
    See also Shipley, Maynard